# 84 Advances in Polymer Science

# Electronic Applications

With Contributions by
A. F. Diaz, M. Kaneko, J. G. Kloosterboer,
H. B. Mark, Jr., J. F. Rubinson, E. J. Spiertz,
F. A. Vollenbroek, D. Wöhrle, C. P. Wong

With 108 Figures and 31 Tables

Springer-Verlag Berlin Heidelberg GmbH

Library of Congress Catalog Card Number 61-642

ISBN 978-3-662-15120-4    ISBN 978-3-540-47784-6 (eBook)
DOI 10.1007/978-3-540-47784-6
© Springer-Verlag Berlin Heidelberg 1988
Originally published by Springer-Verlag Berlin Heidelberg New York in 1988.
Softcover reprint of the hardcover 1st edition 1988

Typesetting and Offsetprinting: Th. Müntzer, GDR;

2154/3020-543210

# Editors

# Table of Contents

Network Formation by Chain Crosslinking Photopolymerization
and its Applications in Electronics
J. G. Kloosterboer . . . . . . . . . . . . . . . . . . 1

Application of Polymer in Encapsulation
of Electronic Parts
C. P. Wong . . . . . . . . . . . . . . . . . . . . . . 63

Photoresist Systems for Microlithography
F. A. Vollenbroek and E. J. Spiertz . . . . . . . . . . 85

Electrochemistry and Electrode Applications
of Electroactive/Conductive Polymers
A. F. Diaz, J. F. Rubinson, and H. B. Mark, Jr. . . . . . 113

Polymer-Coated Electrodes: New Materials
for Science and Industry
M. Kaneko and D. Wöhrle . . . . . . . . . . . . . . 141

Author Index Volumes 1–84 . . . . . . . . . . . . . . 229

Subject Index . . . . . . . . . . . . . . . . . . . . 241

# Network Formation by Chain Crosslinking Photopolymerization and its Applications in Electronics

Johan G. Kloosterboer
Philips Research Laboratories, P.O. Box 80.000, 5600 JA Eindhoven, The Netherlands

The formation of densely crosslinked networks by chain crosslinking photopolymerization is discussed in relation to selected applications in the electronics industry. All of these applications make use of the high speed and of the latitude to meet other requirements by variation of the chemical structure of the monomers. The selection comprises:
1. The coating of optical fibers;
2. The replication of optical discs;
3. The replication of aspherical lenses, used for laser read-out of these discs.
Other important processes will only be mentioned briefly for the benefit of a discussion of more fundamental results that have been obtained during the study of the selected examples. These results relate to the crosslink density, the influence of light intensity on polymer structure, the relation between shrinkage and chemical conversion, a parallel with physical aging, kinetics during vitrification, the importance of chain transfer, the build-up of peroxides in photopolymers and computer simulation of network formation by crosslinking polymerization. This serves to illustrate the continuous interaction between development of applications and fundamental research.

Abbreviations of Compounds . . . . . . . . . . . . . . . . . . . .   3

1 Introduction. Scope of this Article . . . . . . . . . . . . . . .   4

2 Chain Crosslinking Photopolymerization . . . . . . . . . . . . .   5
   2.1 Simplified Reaction Scheme and Kinetics of Photopolymerization . . .   5
   2.2 Some Special Problems Arising with Free-Radical Chain Crosslinking
       Polymerizations . . . . . . . . . . . . . . . . . . . . . . .   7
       2.2.1 Immediate Onset of the Trommsdorff Effect . . . . . . . . .   7
       2.2.2 Incomplete Conversion of Reactive Groups due to Vitrification. .   8
       2.2.3 Reactivity Ratios Change with Conversion . . . . . . . . .   9
       2.2.4 Rates of Polymerization are very Sensitive towards Chain Transfer
             to Polymer . . . . . . . . . . . . . . . . . . . . . . .  10
       2.2.5 Trapped Radicals Induce Postcopolymerization of Oxygen . . .  11
       2.2.6 No Available Analytical Theory can Fully Describe Crosslinking
             Polymerization with Extensive Ring Formation . . . . . . .  11
   2.3 Photoinitiators . . . . . . . . . . . . . . . . . . . . . . .  11
       2.3.1 Generation of Radicals by Direct Photofragmentation . . . . .  12
       2.3.2 Generation of Radicals by Hydrogen Abstraction . . . . . . .  13
       2.3.3 Photoinitiators for Cationic Polymerization . . . . . . . .  14
   2.4 Monomers and Oligomers . . . . . . . . . . . . . . . . . . .  15
   2.5 Properties of Cured Coatings. . . . . . . . . . . . . . . . .  17

Advances in Polymer Science 84
© Springer-Verlag Berlin Heidelberg 1988

**3 Optical Fiber Coatings** . . . . . . . . . . . . . . . . . . . . . . 18
  3.1 Claddings and Coatings . . . . . . . . . . . . . . . . . . . 18
  3.2 The Coating Process . . . . . . . . . . . . . . . . . . . . . 20
  3.3 Photopolymerizable Coatings . . . . . . . . . . . . . . . . 21

**4 Optical Discs** . . . . . . . . . . . . . . . . . . . . . . . . . . 22
  4.1 Optical Disc Systems . . . . . . . . . . . . . . . . . . . . 22
  4.2 Photoreplication of Discs . . . . . . . . . . . . . . . . . . 24
  4.3 Photopolymerizable Coatings . . . . . . . . . . . . . . . . 25

**5 Aspherical Lenses** . . . . . . . . . . . . . . . . . . . . . . . 27
  5.1 Laser Read-Out System . . . . . . . . . . . . . . . . . . . 27
  5.2 Photoreplication of Lenses . . . . . . . . . . . . . . . . . . 28
  5.3 Photopolymerizable Coatings . . . . . . . . . . . . . . . . 29

**6 Other Electronic Applications of Photopolymerization** . . . . . . . . . . 32
  6.1 Recording of Holograms . . . . . . . . . . . . . . . . . . . 32
  6.2 Optical Waveguides . . . . . . . . . . . . . . . . . . . . . 34
  6.3 Other Optical Components . . . . . . . . . . . . . . . . . . 34
  6.4 Wire Coatings . . . . . . . . . . . . . . . . . . . . . . . . 35
  6.5 Miscellaneous Dielectric Coatings and Adhesives . . . . . . . . 35

**7 Further Study of Network Formation with Diacrylates** . . . . . . . . . 37
  7.1 Model Monomers . . . . . . . . . . . . . . . . . . . . . . . 37
  7.2 Delayed Shrinkage. Physical and Chemical "Aging" . . . . . . . . 38
  7.3 Trapping of Radicals in a Network . . . . . . . . . . . . . . 41
    7.3.1 Structure of Trapped Acrylate Radicals . . . . . . . . . 42
    7.3.2 Fate of Trapped Acrylate Radicals. Chain Oxidation . . . . 43
    7.3.3 Inhomogeneity . . . . . . . . . . . . . . . . . . . . . 44
  7.4 Some Kinetic Features . . . . . . . . . . . . . . . . . . . . 45
    7.4.1 Oxygen assisted Chain Transfer . . . . . . . . . . . . . 45
    7.4.2 Direct Chain Transfer. Deuterium Isotope Effect on the Rate
      of Polymerization . . . . . . . . . . . . . . . . . . . 46
    7.4.3 Kinetics of a Self-Decelerating Polymerization . . . . . . 47
  7.5 Some other Studies which Relate Polymer Properties to Photocuring
    Conditions . . . . . . . . . . . . . . . . . . . . . . . . . 49

**8 Simulation of Chain Crosslinking Polymerization with a Percolation Model** . . 51

**9 Conclusions and Expectations** . . . . . . . . . . . . . . . . . . 55

**10 Acknowledgement** . . . . . . . . . . . . . . . . . . . . . . . 55

**11 References** . . . . . . . . . . . . . . . . . . . . . . . . . . . 56

**Abbreviations of Compounds**

BDDA    1,4-butanediol diacrylate
DMPA    $\alpha,\alpha$-dimethoxy-$\alpha$-phenylacetophenone
DEGDA   di-ethylene glycol diacrylate
EHA     2-ethylhexyl acrylate
HDDA    1,6-hexanediol diacrylate
HEBDM   bis(2-hydroxyethyl)bisphenol-A dimethacrylate
HMPP    2-hydroxy-2-methyl-1-phenylpropane-1-one
HCPK    1-hydroxycyclohexyl phenyl ketone
MMA     methyl methacrylate
n-PA    n-propyl acrylate
NVP     N-vinyl-2-pyrrolidone
TEGDA   tetraethylene glycol diacrylate
TMPTA   trimethylolpropane triacrylate
TPGDA   tripropylene glycol diacrylate

# 1 Introduction
## Scope of this Article

Photopolymerization is a technique which allows the rapid conversion at room temperature of liquid materials into rubbery or glassy products.

It is widely applied in several branches of the electronics industry. Applications range from optical waveguides to coatings for resistors, from lenses for laser light pens to integrated circuits.

Since 1982 more than 100 review articles have appeared on various aspects of photopolymerization. Applications as well as fundamental aspects are reviewed annually in the Specialist Periodical Reports on Photochemistry [1]. The chapters on Polymer Photochemistry in these reports contain a section on photopolymerization as well as a review of parts of the patent literature.

This review focuses on the relation between network formation and product properties.

In its most restricted sense, the term photopolymerization should be reserved for step polymerization processes in which each reaction step requires the absorption of at least one photon [2].

However, the term is also widely used in a less precise way, namely for light-induced chain reactions. Here, the light serves to generate radicals or cationic species which are capable of starting a chain process. By far most applications of photopolymerization use such chain reactions.

In these processes very efficient use is made of the radiation: the absorption of one photon may trigger the growth of a polymer chain of up to $10^5$ monomeric units. Even at moderate light intensities, chain reactions can be extremely fast.

When a step or a chain process is used for connecting linear polymer chains to form a network, the term photocrosslinking is used. The simultaneous formation of polymer chains and network is called crosslinking polymerization.

A common feature of both step and chain photopolymerization processes is the stability of the reaction mixture in the dark or e.g. in yellow light, when the system is insensitive to radiation with a wavelength of more than 500 nm. Upon switching on the actinic light, polymerization will start immediately and at a high rate.

Such an easy external control of the start of the reaction is very convenient for production processes, especially for those requiring an accurate filling of a mold.

Another common feature of step and chain processes is that the reaction may be carried out locally, by using laser irradiation or exposure via a mask. In this way, patterns can be made.

Fine patterns of polymeric material are widely used in the electronics industry. They are required in the manufacturing of integrated circuits, printed circuit boards and solder masks for the latter. However, photopolymerization finds only limited application in this area. Photocrosslinking and photochemical enhancement of the rate of dissolution are primarily used. Larger structures can also be made by screen printing. In this technique, the photopolymerizable material is deposited pattern-wise and then irradiated in an overall exposure.

The making of patterns for various applications will be discussed in detail in the chapter on photoresist systems by Vollenbroek and Spiertz [3], we will only mention some results on the making of holograms by laser-induced polymerization.

In the present chapter, we will first present a general introduction to chain crosslinking photopolymerization and next we will discuss some applications of photopolymerization which make use of the high speed and of the possibility to meet other requirements by variation of the chemical structure of the monomers.

Finally, we will present some more fundamental results on network formation, most of which were obtained during the study of the selected topics.

Our selection of applications comprises:

1. The coating of optical fibers;
2. The replication of optical discs;
3. The replication of aspherical lenses, used for laser read-out of these discs.

Other important processes such as the coating of electric wires and the making of optical waveguides and holographic devices, will only be mentioned briefly.

Likewise, topics such as photocurable adhesives and coatings for the encapsulation of electronic components, for which photopolymerization is just one of the available curing methods, will largely be left for the chapters dealing specifically with these applications.

Other applications such as the coating of sleeves for records and the printing of symbols and markings on electronic components as well as the protective coatings for compact discs will not be treated at all.

This restriction of the number of topics allows for a more general discussion on network formation with diacrylates. Results included in this section relate to the crosslink density, the influence of light intensity on polymer structure, the relation between shrinkage and chemical conversion, a parallel with physical aging, the kinetics of the self-decelerating reaction during vitrification, the importance of chain transfer, the build-up of peroxides in photopolymers and the theoretical description of network formation by chain crosslinking polymerization. Finally, we will also quote some results from a few other studies in which polymer properties have been correlated with curing conditions.

All of these points are relevant with respect to the optimization of materials properties.

So, instead of aiming at a complete list of realized and conceivable applications of photopolymerization in electronics we will attempt to illustrate the usefulness of a continuous interaction between fundamental research and the development of applications.

## 2 Chain Crosslinking Photopolymerization

### 2.1 Simplified Reaction Scheme and Kinetics of Photopolymerization

In the classical picture of radical polymerization which, apart from termination, also holds for cationic chain polymerization, the following steps are usually distinguished:

Dissociation:          $In \xrightarrow{k_d} 2 R^\bullet$

Initiation:          $R^\bullet + M \xrightarrow{k_i} RM^\bullet$

Propagation:    $RM^{\bullet} + M \xrightarrow[k_p]{} RMM^{\bullet}$

$RM_i^{\bullet} + M \xrightarrow[k_p]{} RM_{i+1}^{\bullet}$

Termination: $RM_i^{\bullet} + RM_j^{\bullet} \xrightarrow[k_{tc}]{} RM_{i+j}R$

$RM_i^{\bullet} + RM_j^{\bullet} \xrightarrow[k_{td}]{} RM_i + RM_j$

In is the initiator, M the monomer, $R^{\bullet}$ a primary radical and $RM_i^{\bullet}$ a growing radical consisting of a primary radical and i monomeric units. $RM_i$ and $RM_iR$ are 'dead' polymers. The rate constants k refer to decomposition, initiation, propagation and termination, respectively. Termination may either occur through combination (c) or through disproportionation (d). Under the usual assumptions [4] of a steady-state total free-radical concentration and of size-independent reactivities of the growing radicals, the rate equation becomes:

$$R_p = k_p[M] (R_i/2k_t)^{1/2}$$

where $R_i$ represents the rate of initiation and $k_t$ the rate constant for termination, irrespective of the mechanism. For light-induced initiation, the rate of initiation becomes:

$$R_i = 2\varphi I_a$$

in which $\varphi$ is the overall quantum efficiency for photodissociation and initiation, and $I_a$ the intensity of absorbed light. At low absorbance the latter can be expressed as:

$$I_a = 2.3\varepsilon I_0[In] d$$

where $\varepsilon$ is the decadic molar extinction coefficient at the wavelength of irradiation, $I_0$ the incident intensity and d the thickness of the sample.

The resulting rate equation for photopolymerization is:

$$R_p = k_p[M] \{2.3\varepsilon\varphi I_0[In] d/k_t\}^{1/2}$$

However, in many systems of practical use the variation of I with thickness cannot be neglected and the rate equation becomes:

$$R_p = k_p[M] \{\varphi I_0(1 - 10^{-\varepsilon[In] d})/k_t\}^{1/2}$$

In strongly absorbing systems, the rate of polymerization may vary considerably across the sample thickness; in the limit no polymerization at all will occur at the bottom of the sample. The situations which may be encountered in stirred and unstirred solutions of various absorbance levels have been discussed by Shultz [5], a simpler and less widely applicable treatment has been given by Lissi and Zanocco [6].

In the photopolymerization of thin films, advantage is often taken of using a reflecting substrate. This will level the intensity gradient across the sample considerably.

So far, we have neglected common interferences such as chain transfer to monomer, initiator, polymer, solvent or any other compound present in the system, as well as inhibition and retardation. These extensions can be found in text-books [4, 7, 8] and monographs [9, 10].

The same applies to the Trommsdorff or autoacceleration effect frequently encountered in bulk polymerization processes. We will refer to some of these while discussing a number of aspects which are typically found with or are pronounced in crosslinking systems.

## 2.2 Some Special Problems Arising with Free-Radical Chain Crosslinking Polymerizations

Network formation by chain crosslinking (co)polymerization has been thoroughly reviewed by Dušek [11]. Therefore we will only discuss some additional points which are typical of the formation of densely crosslinked networks by homopolymerization of pure divinyl compounds.

Dušek's treatment centers around the structure of the polymers and it is concluded that existing theories can only deal with a very small amount of cyclization, i.e. with systems containing only minor amounts of multifunctional monomers.

Existing theories for crosslinking polymerization cannot account for the observed inhomogeneities formed in more densely crosslinked systems [11, 12]. Since most of the applications to be reported here are related to the formation of densely cross-linked networks, and since the question of the kinetics of formation of these networks. has hardly been addressed in the literature, we will discuss a few peculiarities of these systems. At present, it is not possible to give a complete picture since our knowledge of these systems is still limited.

This holds a fortiori for cationic polymerizations of e.g. di-epoxides. Although some of the findings to be reported in this section may be of a more general character, we will, therefore, concentrate on radical chain polymerizations.

### 2.2.1 Immediate Onset of the Trommsdorff Effect

In the bulk polymerization of methyl methacrylate (MMA), the Trommsdorff effect sets in after an unperturbed period of about 15–20% conversion. This enables a separation to be made between unperturbed and autoaccelerated kinetics. Benson and North, O'Driscoll, Ito and others have attempted to describe the kinetics of autoacceleration or gelation by introducing a chain-length-dependent value of the termination rate constant [13–15].

Tulig and Tirrell extended the gelation model by including diffusion by reptation [16].

The polymerization of acrylates is much more difficult to analyse since gelation sets in near zero conversion [15–19, 162], probably due to branching and crosslinking caused by chain transfer to polymer [18, 162].

With the chain crosslinking polymerization of undiluted tetrafunctional monomers gelation starts almost immediately (Fig. 1) and at present such systems cannot be described kinetically. Since the cause of the autoacceleration is a physical one (sup-

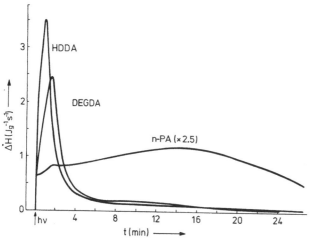

**Fig. 1.** Rate of photopolymerization of 1,6-hexanediol diacrylate (HDDA), di-ethyleneglycol diacrylate (DEGDA) and n-propyl acrylate (n-PA) vs. time. Rates are measured as exothermic heat flux. (From Ref. [24], with permission)

pression of chain termination by lack of radical mobility), the commonly used kinetic analysis for the autoaccelerated amine curing of epoxides [20, 21] cannot be applied. In the latter case, the autoacceleration has a chemical cause, namely the progressive generation of reactive hydroxyl groups which can be readily accounted for in the rate equation. This is not the case for the dependence on conversion and shrinkage of all rate constants, including those for transfer.

### 2.2.2 Incomplete Conversion of Reactive Groups Due to Vitrification

When densely crosslinked networks are prepared by low-temperature polymerization, incomplete conversion of the reactive groups is often observed [22-24].

This holds also for step reactions such as the thermal curing of epoxides with amines below $T_g$ of the fully cured polymer [21, 25].

In radical polymerization, the reduction of the mobility caused by the increasing viscosity and crosslinking first suppresses the termination process (autoacceleration) but next the mobility of free monomer is also suppressed and the rate of polymerization becomes limited by diffusion. Upon further crosslinking, the diffusion coefficient decreases and so does the rate. This may be called autodeceleration. Upon the isothermal approach to the glassy state, the rate may decrease by many orders of magnitude. The "final" extent of reaction increases with temperature and so does the crosslink density of the polymer [23, 26]. The statement, advanced in the early days of photopolymerization, that the same materials can be made at room temperature as were formerly prepared at elevated temperatures is therefore not correct [27, 28].

Stopping of polymerization due to vitrification may of course also occur with linear systems. PMMA is probably the best known example [29-36]. The rate constants for propagation and termination have been determined as a function of conversion [30] and diffusional control of the rate of polymerization at high conversion has been discussed by several authors [32-36].

Horie et al. have correlated the maximum extent of reaction with $T_g$ of the monomer/polymer system. They concluded that the reaction stops when $T_g$ equals $T_{cure}$ [32]. The influence of polymerization shrinkage on free volume and there by on the rate of polymerization was determined by Balke and Hamielec at moderate conversions [33] and more recently Panke, Stickler, Wunderlich and Hamielec studied the influence of shrinkage and free volume on the rates of propagation and termination at very high conversions [34-36]. So, with linear systems a fairly complete picture has been established.

With multifunctional monomers, however, the situation is rather different. There are several reasons for the lack of understanding of rates and maximum extents of reaction. In the case of monovinyl compounds, there is no question about the chemical structure of the polymer, but with multifunctional monomers there exists ambiguity with respect to the status of unreacted groups. These may be present either in free monomer or as pendent groups [37-39]. This ambiguity together with the poor accessibility of networks to chemical and structural analysis has caused that much less progress has been made with the study of the polymerization of multifunctional monomers.

Still another reason is that the proportionality between shrinkage and conversion, which is commonly observed with linear systems, breaks down with densely cross-linked networks (see Sect. 7.2) and finally it must be mentioned that even the chemistry may change during vitrification (Sect. 7.4.2). At this stage we will only present some qualitative considerations.

Multifunctional monomers may be reacted to high conversion by increasing the molecular mobility during the reaction in one of four ways:

(i)   elevation of the reaction temperature (provided that no degradation occurs);
(ii)  increase of the flexibility of the moieties between functional groups;
(iii) reduction of the crosslink density by increasing the length of the moieties between the functional groups;
(iv)  reduction of the crosslink density by copolymerization with bifunctional monomer;
(v)   increase of the rate of polymerization (see Sect. 7.2).

The addition of an inert solvent is usually ineffective since phase separation may occur [11, 40]. This may eventually lead to the formation of "reactive microgels" [41, 42] and macroporous structures which find application, for example, in resins for ion exchange and column packings for size exclusion chromatography [43].

## 2.2.3 Reactivity Ratios Change with Conversion

It has already been pointed out in the previous section that below complete conversion crosslink density is not related in a unique way to the extent of reaction.

The true course of the homopolymerization reaction of a multifunctional monomer is determined by the ratio of reactivities of free and pendent reactive groups. Copolymerizations are even more complicated [11] and will not be discussed here.

In Fig. 2, we have depicted the molality of crosslinks as a function of double bond conversion for a tetrafunctional monomer, 1,6-hexanediol diacrylate (HDDA) for a few special cases. The molality was chosen in order to get a plot which is independent of density changes during polymerization. In the hypothetical case of

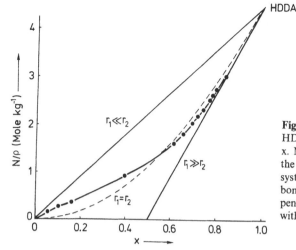

**Fig. 2.** Molality $N/\varrho$ of crosslinks in HDDA versus double bond conversion $x$. N is the molarity of crosslinks, $\varrho$ is the density of the monomer/polymer system. $r_1$ is the reactivity of double bonds in free monomer, $r_2$ that of pendent double bonds. (From Ref. [24], with permission)

$r_1 \gg r_2$ ($r_1$ refers to double bonds of the monomer, $r_2$ to pendent double bonds) the lower path will be followed, i.e. crosslinking does not start until all molecules have reacted at one side. In the other hypothetical case of $r_1 \ll r_2$, the double bonds react in pairs and the crosslink density will be proportional to conversion. All possible paths will be confined within the triangle formed by the two limiting paths. In the classical case of $r_1 = r_2$, the dashed curve will be followed. The measured curve [39] shows that initially the pendent bonds appear to exhibit an enhanced reactivity with respect to free monomer. This can be interpreted as being due to local concentration effects, arising from the inhomogeneous structure of the network [44−46].

This will be further discussed in Sect. 7.3 and 8 of this chapter.

In the final stage of the reaction a reversal of relative reactivities seems to occur. Here, overall vitrification probably sets in, affecting the mobility of pendent double bonds more strongly than those in free monomer.

So, in crosslinking homopolymerization there are physical factors such as relative mobilities and local concentrations which complicate the relation between conversion and structure. In crosslinking copolymerization, differences in chemical reactivity may cause further difficulties [11].

### 2.2.4 Rates of Polymerization are very Sensitive towards Chain Transfer to Polymer

The rapid onset of autoacceleration in the polymerization of monoacrylates has been ascribed to branching and crosslinking through hydrogen abstraction from the polymer chain [18,162]. Apart from physical effects such as localized gelation [47] no large effect on the rate of polymerization is anticipated since the newly formed tertiary alkyl radicals are roughly as reactive towards monomer as the growing chain radical itself.

With crosslinking polymerization of multifunctional acrylates a different situation may arise. Upon vitrification the propagation becomes limited by diffusion and the rate of polymerization is mainly sustained by the almost complete suppression of termination. Radicals which are hooked-up to the network can hardly diffuse but hydrogen transfer introduces a mobility of the radical sites in the polymer [48]. This will also be further discussed in Sect. 7.

### 2.2.5 Trapped Radicals Induce Post-Copolymerization of Oxygen

Oxygen is an efficient inhibitor of radical polymerization. It reacts rapidly with growing radicals which are converted into relatively unreactive peroxy radicals. Rather than participating in a propagation step, they will eventually abstract a hydrogen atom from a polymer or monomer molecule. In this step, a reactive radical is regenerated. If sufficient oxygen is present, polymerization will be suppressed by continuous oxidation.

In closed systems, an induction period is observed until virtually all of the oxygen has been consumed by reaction with radicals.

In open systems, polymerization will only occur if the generation of radicals can compete successfully with the replenishment of oxygen, that is at high rates of initiation. Some initiation systems rely on the rapid consumption of oxygen by photochemical means (see Sect. 2.3).

During vitrification, extensive trapping of radicals occurs [49-52]. Their concentration may easily reach a level of $10^{-3}$ M. Upon exposure to air, these radicals are oxidized and form hydroperoxides. It is important to note that this oxidation is a chain process [53], so that much more hydroperoxide is formed than there were radicals to start with [52]. The role of hydroperoxides in the thermal and photodegradation of various polymers, notably polypropylene and polyethylene, has been firmly established [54-56].

The presence of low concentrations of oxygen also affects the rate of polymerization through the enhancement of chain transfer to the polymer [48]. See also Sect. 7.4.

### 2.2.6 No Available Analytical Theory can Fully Describe Crosslinking Polymerization with Extensive Ring Formation

Polymeric networks can be made in one of three ways:

(i)   Step reaction of small molecules, all of which are reactive at the same time;
(ii)  Vulcanization of linear chains;
(iii) Chain reaction of activated molecules.

The first two methods can be described by the classical and elegant theory of Flory and Stockmayer [7, 57].

At present no analytical theory exists that can describe the formation of densely crosslinked networks via a chain process. However, in many applications it is desired to form such dense networks and chain processes are especially useful. Until now, the best approach seems the use of computer simulations. This will be illustrated in Sect. 8.

## 2.3 Photoinitiators

Much effort has been spent on the development of efficient generators of radicals or cationic species under UV irradiation.

Important aspects of the performance of photoinitiators are their absorption in the appropriate wavelength region (usually 350–380 nm), quantum yield for dissociation, reactivity of the radicals toward monomers, stability in the dark, solubility, absence of yellowing in the photopolymer and acceptable toxicological properties.

Since the subject of photoinitiators has been reviewed regularly and extensively [59–68], we will limit ourselves to a listing of the most important classes of photoinitiators. These will be illustrated by a few examples of commonly used compounds.

### 2.3.1 Generation of Radicals by Direct Photofragmentation

As an example of this class we will describe the benzoin ethers. These are cleaved in the following way:

Benzoin ethers have a weak absorption band with a maximum near 350 nm and a molar extinction coefficient of about 200 $M^{-1}$ $cm^{-1}$. The photochemistry of the methyl ether has been studied by Adam et al. [69, 70], the quantum yield for photodissociation at 365 nm is 0.5.

There has been some dispute in the literature whether or not the ether radical contributes to the initiation [71, 72] but recently the exclusive generation of ether radicals has been reported and it turned out that polymerization still occurred, albeit that the rate was very low, even at an unusually high concentration of the radical precursor [73].

Under conditions of relatively high radical and low monomer concentrations [72] the ether radical predominantly acts as a terminating radical.

Benzoin ethers are efficient photoinitiators. However, the pot-life of solutions in monomer is rather restricted [74], presumably due to the easy formation of hydroperoxides which undergo subsequent decomposition into initiating radicals [75].

Benzilketals are a special case. They are cleaved in a similar way but they are more efficient due to a somewhat higher extinction coefficient ($\varepsilon \sim 250$ $M^{-1}$ $cm^{-1}$) and the occurrence of a secondary cleavage:

The methyl radical is highly reactive and the secondary reaction also decreases the chance of cage recombination. The photochemistry of $\alpha,\alpha$-dimethoxy-$\alpha$-phenylacetophenone (DMPA) has been studied in the absence [76] as well as in the presence of compounds with double bonds [77, 78]. Hageman et al. investigated its photochemistry in the presence of 1,1-diphenylethylene [77], a monomer which reacts rapidly with radicals but hardly propagates, and Borer, Kirchmayer and Rist studied the formation of photoproducts in the presence of an acrylic monomer, using CIDNP-NMR [78].

Other important members of this first category are: 2-hydroxy-2-methyl-1-phenylpropan-1-one (HMPP) which is cleaved as follows [79]:

$$\text{C}_6\text{H}_5-\underset{\underset{\text{O}}{\|}}{\text{C}}-\underset{\underset{\text{CH}_3}{|}}{\overset{\overset{\text{CH}_3}{|}}{\text{C}}}-\text{OH} \xrightarrow{h\nu} \text{C}_6\text{H}_5-\underset{\underset{\text{O}}{\|}}{\text{C}}\!\cdot \;+\; \cdot\underset{\underset{\text{CH}_3}{|}}{\overset{\overset{\text{CH}_3}{|}}{\text{C}}}-\text{OH}$$

and 1-benzoyl-1-cyclohexanol which presumably reacts in a similar way [61]:

$$\text{C}_6\text{H}_5-\underset{\underset{\text{O}}{\|}}{\text{C}}-\overset{\overset{\text{OH}}{|}}{\text{C}_6\text{H}_{10}} \xrightarrow{h\nu} \text{C}_6\text{H}_5-\underset{\underset{\text{O}}{\|}}{\text{C}}\!\cdot \;+\; \cdot\overset{\overset{\text{OH}}{|}}{\text{C}_6\text{H}_{10}}$$

This compound is also called hydroxy-cyclohexyl phenyl ketone (HCPK). The mixed aliphatic/aromatic initiators are reported to cause less yellowing of the photopolymers. The efficiencies of the last three photoinitiators have been compared using laser irradiation of solutions of trimethylolpropane triacrylate in 2-propanol [80]. In air, the rates were found to be equivalent but under nitrogen the α-hydroxy-acetophenone derivatives were somewhat faster.

### 2.3.2 Generation of Radicals by Hydrogen Abstraction

The best known compound in this class is benzophenone. In its triplet excited state it readily abstracts a hydrogen atom from a suitable donor:

$$\text{C}_6\text{H}_5-\underset{\underset{\text{O}}{\|}}{\text{C}}-\text{C}_6\text{H}_5 \xrightarrow{h\nu} \left(\text{C}_6\text{H}_5-\underset{\underset{\text{O}}{\|}}{\text{C}}-\text{C}_6\text{H}_5\right)^{*}$$

$$\left(\text{C}_6\text{H}_5-\underset{\underset{\text{O}}{\|}}{\text{C}}-\text{C}_6\text{H}_5\right)^{*} + \text{RH} \longrightarrow \text{C}_6\text{H}_5-\underset{\underset{\text{OH}}{|}}{\overset{\overset{\cdot}{}}{\text{C}}}-\text{C}_6\text{H}_5 + \text{R}\!\cdot$$

The ketyl radical does not initiate polymerization in an efficient way [81] but the R· radical generally does. Frequently used hydrogen donors are tertiary amines. They have been reported to be more efficient donors than alcohols, ethers or thiols [82].

With amines, the α-hydrogen transfer may occur either directly or via the intermediate formation of an exciplex, electron transfer and finally proton transfer [64, 82, 83].

An additional advantage of the use of ketone/amine combinations is the fact that the amine radicals are not only very reactive towards oxygen but the formed peroxyradicals in turn react rapidly with α-hydrogens of the amine, thereby restarting the interrupted chain propagation.

$$R_2N-\overset{\overset{\displaystyle H}{|}}{\underset{\underset{\displaystyle H}{|}}{C}}\cdot \;+\; O_2 \quad\longrightarrow\quad R_2N-\overset{\overset{\displaystyle H}{|}}{\underset{\underset{\displaystyle H}{|}}{C}}-OO\cdot$$

$$R_2N-\overset{\overset{\displaystyle H}{|}}{\underset{\underset{\displaystyle H}{|}}{C}}-OO\cdot \;+\; CH_3NR_2 \quad\longrightarrow\quad R_2N-\overset{\overset{\displaystyle H}{|}}{\underset{\underset{\displaystyle H}{|}}{C}}-OOH \;+\; \cdot CH_2NR_2$$

In this way, oxygen is rapidly consumed in a chain process and the system is less sensitive to air inhibition. Surface tackiness of systems cured in air is also reduced. Disadvantages are the hygroscopic nature of the amines, which may facilitate corrosion at interfaces, and the rather strong yellowing which is generally observed with these systems [84]. The problem of yellowing and its suppression by the addition of radical scavengers has been discussed by Hult [85, 86]. Not unexpected, the rate of polymerization was found to decrease considerably upon the addition of scavengers. Wheathering of UV cured coatings in general has been treated by Gatechair [87].

Other members of this class are thioxanthones and 3-ketocoumarins wich absorb at longer wavelengths than aromatic ketones [88]. Thioxanthones may also act as sensitizer for other initiators [89]. However, their photochemistry seems to be rather complex. In addition to hydrogen abstraction and transfer of excitation energy, several types of cycloaddition reactions with unsaturated compounds have also been reported [90].

### 2.3.3 Photoinitiators for Cationic Polymerization

These initiators rely on the photochemical generation of strong Brönsted acids [59 -67] which are capable of initiating the chain polymerization of epoxy compounds, vinyl ethers, cyclic ethers, lactones and many other compounds.

Diaryl iodonium and triaryl sulfonium salts are among the most versatile photoinitiators. These compounds have been thoroughly investigated by Crivello [91 -95].

A diaryl iodonium salt can be depicted as $Ar_2I^+X^-$ in which $X^-$ is a counterion such as $PF_6^-$, $BF_4^-$, $AsF_6^-$, etc. Its photolysis proceeds as follows:

$$Ar_2I^+X^- \xrightarrow{\;h\nu\;} ArI^{\overset{+}{\cdot}}X^- + Ar^\bullet$$

$$ArI^{\overset{+}{\cdot}}X^- + RH \rightarrow ArI^+{-}HX^- + R\cdot$$

$$ArI^+ - HX^- \rightarrow ArI + HX$$

Together with the strong acid HX, radicals are also formed. A major advantage of these systems is that they can be polymerized in the presence of oxygen; a potential draw-back is that many of them absorb at wavelengths far below 300 nm. This makes the polymerization of, e.g., aromatic di-epoxides a rather slow process. Sensitization has proved possible but structural modification is a more useful method, especially with the sulfonium salts [95].

Iron-arene salts may also exhibit an appreciable absorption in the visible or near-UV part of the spectrum. The photochemistry of these compounds as well as their use as photoinitiators has recently been described by Lohse and Zweifel [96].

In certain applications the presence of strong acid in the photopolymer is desirable since the reaction continues after exposure.

In other applications, where corrosion of a substrate may be involved, it may be quite harmful. Under these circumstances, the use of photoinitiators recently developed by Hayase et al. may prove useful. Hayase's photoinitiators consist an aluminum complex as a catalyst and a silanol precursor. Triphenyl silanol may be generated by photolysis of (o-nitrobenzyloxy)triphenylsilane [97] or an arylsilyl peroxide [98]. Together with an aluminum complex such as for example tris(ethyl acetoacetato)-aluminum an initiating species is formed. The silanol is eventually consumed, either by self-condensation or by reaction with epoxide or secondary OH groups [99].

Therefore, these systems have the advantage that no ionic species will remain in the photopolymer but they appear to be rather slow. Rates of different cationic initiating systems for epoxies have recently been compared by Gaube [100].

## 2.4 Monomers and Oligomers

The monomers which are most widely used for photopolymerization processes are acrylates. The reason is that they polymerize fast and, by introducing chemical modifications in the ester group, materials with very different properties may be obtained without sacrifing too much of the polymerization rate. Methacrylates generally polymerize more slowly but, due to the stiffer main chain, yield harder products.

Other systems include styrene/unsaturated polyesters [101] and thiol-ene compositions [102]. These will not be discussed here.

With monoacrylates (functionality f = 2), linear chains are obtained, and upon the addition of di- and/or triacrylates (f = 4 or 6, respectively) crosslinked networks are formed. In order to avoid the presence of free monomer in the cured product, monoacrylates are sometimes omitted.

The acrylic esters of the lower mono-, di- or trialcohols or the lower ethylene or propylene glycols are liquids of low viscosity and, especially with the lower alcohols, of repellent odor. They are often used in coating formulations as reactive diluents for the much more viscous oligomers. References to their toxicological properties can be found in Refs [63] and [103].

Oligomers serve to reduce the volatility, toxicity, odor, polymerization shrinkage and to improve the properties of the cured material.

Frequently used oligomers are:

*Epoxy acrylates*, made e.g. by reacting epoxides such as DGEBA's (diglycidyl ethers of bisphenol-A) with acrylic acid (Although these compounds are no epoxides but have only been derived from epoxides, they are still generally called epoxy acrylates).

*Urethane acrylates* may be obtained by reacting hydroxyalkyl acrylates, diisocyanates and diols. A typical example of the overall reaction is:

$$2\ CH_2=\overset{H}{\underset{|}{C}}-\overset{O}{\overset{\|}{C}}-O-R-OH\ +\ 2\ O=C=N-\langle\bigcirc\rangle_{CH_3}-N=C=O\ +\ HO-R'-OH\ \longrightarrow$$

$$CH_2=\overset{H}{\underset{|}{C}}-\overset{O}{\overset{\|}{C}}-O-R-O-\overset{O}{\overset{\|}{C}}-\overset{H}{\underset{|}{N}}-\langle\bigcirc\rangle_{CH_3}-\overset{H}{\underset{|}{N}}-\overset{O}{\overset{\|}{C}}-O-R'-O-\overset{O}{\overset{\|}{C}}-\overset{H}{\underset{|}{N}}-\langle\bigcirc\rangle_{CH_3}-\overset{H}{\underset{|}{N}}-\overset{O}{\overset{\|}{C}}-O-R-O-\overset{O}{\overset{\|}{C}}-\overset{H}{\underset{|}{C}}=CH_2$$

*Polyester acrlates* can be made by reacting polyesters with acrylic acid:

$$2\ CH_2=\overset{H}{\underset{|}{C}}-\overset{O}{\overset{\|}{C}}-OH\ +\ HO\text{-}(CH_2)_6\left[O-\overset{O}{\overset{\|}{C}}-(CH_2)_4\overset{O}{\overset{\|}{C}}-O\text{-}(CH_2)_6\right]_n OH\ \longrightarrow$$

$$CH_2=\overset{H}{\underset{|}{C}}-\overset{O}{\overset{\|}{C}}-O\text{-}(CH_2)_6\left[O-\overset{O}{\overset{\|}{C}}-(CH_2)_4\overset{O}{\overset{\|}{C}}-O\text{-}(CH_2)_6\right]_n O-\overset{O}{\overset{\|}{C}}-\overset{H}{\underset{|}{C}}=CH_2\ +\ 2\ H_2O$$

*Polyether acrylates* can be made in an analogous way.

*Siloxane acrylates* may be obtained by using siloxanes instead of polyesters by reacting active hydrogens of the silane with either allyl alcohol or allyl glycidyl ether. Both reactions are catalyzed by platinum. The resulting alcohol or epoxide may be converted into the acrylate with acrylic acid.

In this way, by variation of the length and the composition of the moiety between the acrylate groups, a large number of $\alpha,\omega$ diacrylates and dimethacrylates have been obtained. By using branched oligomers, compounds of higher functionality have also been synthesized.

All systems mentioned sofar are cured by radical. chain polymerization. Other monomers such as epoxides or vinyl ethers may be polymerized by cationic chain polymerization.

*Epoxides* are widely used in thermally activated reactions but their photopolymerizations are generally slow. This has restricted their use in these processes somewhat. The photocrosslinking of epoxy resins has recently been reviewed by Lohse and Zweifel [96] and their use in dielectric coatings will be discussed in Sect. 6.5.

*Vinyl ethers* polymerize much quicker than epoxides, their rates are comparable to those of acrylates.

The photopolymerization of bis(vinylethers) has been described by Crivello et al. [104, 105] and by Dougherty et al. [106]. In these studies, it was also observed that bis(vinylethers) are excellent reactive diluents for epoxides which enhance the rate of polymerization considerably [105, 106].

All oligomers mentioned sofar have their specific advantages and limitations.

An interesting extension may become the combination of radical and cationic photopolymerization since synergistic effects may enhance the curing rates of both types [60, 107, 108].

Light-induced cationic polymerizations are often retarded by impurities such as water or basic substances [109] but will continue in the dark and in the presence of oxygen for a considerable time. Light-induced radical polymerizations, on the other hand, will start rapidly but suffer from the presence of molecular oxygen. Onium salts initiate both photopolymerization processes since they yield radicals as well as cationic species upon photolysis.

The interpenetrating networks formed in this way may have properties rather different from those expected from a linear combination of the component's properties. See also Sect. 7.5.

## 2.5 Properties of Cured Coatings

*Epoxy acrylate resins* are known for their good adhesive properties, high chemical resistance, thermal stability and non-yellowing.

*Polyester acrylate coatings* are hard, tough and solvent-resistant. Adhesion is generally poor and, therefore, they are often used in conjunction with polyurethanes. Alkali resistance is rather poor due to the sensitive ester linkage.

*Polyurethane acrylates* are sensitive to sunlight and they exhibit severe yellowing.

The introduction of polyether groups in the backbone has been reported to increase the rate of polymerization in air considerably, presumably through enhanced chain transfer of the hydrogen atoms in the $\alpha$-position to the ether oxygen to peroxyradicals, similarly to the ketone/amine systems [110]. Another approach to reduce the oxygen-sensitivity has been the advancement of gelation by adding monomers or oligomers with a high functionality [110, 111]. This increases the rate through autoacceleration.

*Polysiloxane acrylates* are very flexible materials which retain their rubbery properties even at very low temperatures. Their synthesis, curing behavior and adhesive properties has been described in Ref. [112].

Here, we have only given a few examples of some classes of UV-curable coating constituents. In practice, many more classes of compounds are in use and in most cases coatings for specific purposes are complicated mixtures.

The wide range of building blocks allows a very wide range of product properties to be realized.

The design of suitable coating compositions is sometimes considered to be an art as well as a science, many of the developments have been carried out on a purely empirical basis. This is in part due to the lack of understanding of the relation between molecular structure of the monomer, curing conditions and properties of the polymer. Hopefully, our understanding will improve the situation in that dedicated coating compositions can be made more quickly by scientific reasoning. Some attempts will be presented in following sections of this chapter.

A general description of photocurable coating compositions goes beyond the scope of this chapter. Detailed formulations for many specific applications can be found in Refs. [63, 66, 101] and [103].

Some properties of acrylic coatings which can be controlled to a certain extent are:

— mechanical (Young's modulus: 1–2000 MPa; glass transition temperature: −100 to +150 °C; tensile strength: up to 80 MPa; elongation at break: 1–500%).
— optical (refractive index: 1.42–1.60; specific absorbance; turbidity).
— interfacial (adhesion, release).

Many other properties of specific cured coating compositions can be found in Refs. [63, 100] and [103].

A new development is the use of *hybrid systems*, mentioned in the preceding paragraph. In addition to a more favorable curing behavior, shrinkage might also be considerably reduced by using these systems since epoxides generally exhibit a lower shrinkage per reactive group than acrylates.

Moreover, these hybrid systems are expected to provide even greater flexibility in the design of coatings according to prescribed specifications [106]. The difference in rate may also assist in the formation of interpenetrating networks. The formation and properties of interpenetrating networks made by the combination of photo- and thermal polymerization of acrylates and epoxides has been described by Suzuki et al. [113].

# 3 Optical Fiber Coatings

## 3.1 Claddings and Coatings

Optical fibers for telecommunication purposes generally consist of a core with a high refractive index surrounded by a cladding with a lower refractive index which confines the propagating light waves to the core (Fig. 3).

Most fibers have a core which consists of silica doped with oxides of germanium or phosphorus to raise the refractive index. The cladding usually consists of pure, undoped silica [114].

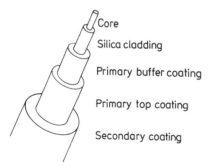

Core
Silica cladding
Primary buffer coating
Primary top coating
Secondary coating

**Fig. 3.** Schematic view of an optical fiber

Polymers such as poly(dimethylsiloxane) are also used as cladding material [115] but owing to the absorption losses upon reflection at the glass-polymer interface they are mainly used in short-distance applications.

The same restriction applies to all-polymer fibers. Modern, all-glass fibers may have losses of less than 0.5 dB $\cdot$ km$^{-1}$ at a wavelength of 1.3 μm [116, 117]. Typical diameters of the core and the cladded fiber are 50 and 125 μm, respectively, but there is a trend to switch over to so-called monomode fibers which have cores with a diameter of 5–10 μm.

All fibers have to be coated with a protective layer during the drawing process. This coating does not participate in the guiding of the lightwaves but its refractive index must be higher than that of the cladding glass. In this way cladding modes are coupled out of the core and the signal-to-noise ratio of the optical fiber system is improved.

At present most commercial fibers possess a dual coating, a soft buffer coating surrounded by a stiff external coating (Fig. 3). Typical diameters of fibers with their soft and hard coatings are 185 and 250 μm, respectively. Both can be cured in situ by means of photopolymerization.

The most important function of the coating is to protect the vulnerable surface of the glass fiber.

Since the brittle fibers are under a considerable stress which cannot relax, any surface flaw will cause a permanent stress concentration at its tip. When the local stresses exceed a certain limit, crack growth and ultimately fracture will occur.

The coating must not only protect the fiber throughout its service life but also during manufacture of the cable. Therefore, the coating is applied in line on the freshly drawn fiber. Surface deterioration occurs either through direct mechanical damage by, for example, dust particles in the air or in the liquid coating, or through interaction with moisture. Moisture will rapiddly degrade the strength of glass fibers, especially when they are under stress.

The deleterious effects of surface flaws have been discussed in detail elsewhere [118].

The second important function of the coating is to prevent losses due to microbending.

The term microbending refers to short period (about 1 mm) and small amplitude (typically a few microns) random bends which are able to couple optical modes in such a way that they leave the core [119, 120].

Microbending may be induced by lateral forces on the fiber. In order to prevent this microbending the fiber is surrounded by a soft buffer coating. Since the linear expansion coefficients of pure quartz and any organic polymeric coating differ by about two orders of magnitude, thermal fluctuations will cause axial stresses. Strong cooling may then easily induce microbending and increase transmission losses. The coating may assist in the prevention of microbending in either of two ways [120–122]:

(i) It can act as a buffer, a soft compliant enclosure which decouples the fiber mechanically from its environment.

(ii) It can act as a stiffener, a tight and hard enclosure which withstands external forces which try to cause local bending.

In addition to fulfilling the primary requirements of mechanical protection and prevention of microbending losses, the coating should be strippable for splicing and for making connections, it should be bondable into cable structures, it must possess sufficient thermal, oxidative and hydrolytic stability and the complete cable should be resistant to a wet environment, to gasoline and diesel fuel and to microbial growth.

Finally, it should not change the transmission of the cable over a wide temperature range (−60 to +80 °C) [123, 124].

## 3.2 The Coating Process

Early types of coatings included silicone oils and rubbers, polyamideimide, cellulose lacquers and blocked urethanes. Hot melts have also been proposed [114, 123]. It proved to be difficult to overcome problems in handling, stability, durability and speed of application (drawing speeds of up to 10 m · s$^{-1}$ are required).

The use of UV-curable epoxy acrylates was reported as early as in 1976 [125, 126] and at present the UV-curable coatings are widely used, either as a single coating or as a combination of a soft primary and a tough secondary buffer coating. Outstanding properties of UV-curable coatings are a high curing rate combined with good flow properties permitting a high drawing rate.

Alternatively, thermally cured silicone elastomers are often used, notably in Japan. They provide excellent microbending protection but the low rate of curing (about 1.5 m · s$^{-1}$) is an economic draw-back.

The coating and curing processes are depicted schematically in Fig. 4 [127]. More detailed information about application, thickness and adhesion control can be found in Ref. [115].

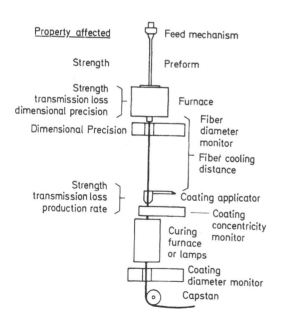

Fig. 4. Schematic view of optical fiber drawing and coating process. (From Ref. [127], with permission)

## 3.3 Photopolymerizable Coatings

Most commonly epoxy- or urethane acrylate resins are used as a basis [125-139] with additional reactive diluents to control the viscosisty of the liquid and the crosslink density of the cured coating. Work on these systems has concentrated on producing low-modulus, low-$T_g$ materials since tough secondary buffer coatings are rather easy to design.

As an illustration a few parameters of typical commercial urethane acrylate coatings are listed in Table 1 together with those of recently developed materials.

**Table 1.** Properties of urethane acrylate optical fiber coatings

|  | Commercial [129] | | New development [130] | |
|---|---|---|---|---|
| *Liquid coating* | Prim | Sec | Prim | Sec |
| $n_0$ (liquid) | 1.50 | 1.51 | 1.48 | 1.51 |
| Viscosity at 25 °C (Pa · s) | 10.3 | 9.7 | 9.1 | 8.6 |
| Viscosity at 45 °C (Pa · s) | 1.8 | 1.7 | 2.6 | 1.4 |
| *Polymerized coating* | | | | |
| Tensile strength (MPa) | 2.4 | 19 | 1.3 | 25 |
| Elongation at break (%) | 140 | 17 | 118 | 8.5 |
| Tensile modulus (MPa) | 3.0 | 345 | 1.0 | 831 |
| Dynamic modulus (MPa) | | | | |
| at 1 Hz and −60 °C | 1413 | 1585 | 631 | 1550 |
| −40 °C | 1259 | 1250 | 85 | 1380 |
| −20 °C | 891 | 930 | 22 | 1150 |
| 0 °C | 178 | 708 | 6.3 | 949 |
| +20 °C | 3.0 | 348 | 1.3 | 700 |
| +40 °C | 1.6 | 120 | 1.0 | 474 |
| +60 °C | 1.3 | 23 | 1.0 | 292 |
| +80 °C | 1.0 | 13 | 1.0 | 146 |

Refs. [131] and [132] also provide lists of optical, physical and mechanical properties.

Characterization and evaluation of performance by means of the torsional pendulum technique has been discussed by Hussain [133], Wierenga et al. have used ultramicro indentation measurements for the determination of the viscoelastic properties of coated optical fibers [134].

The main drawback of epoxy- and urethane-acrylates is the temperature of their glass transition. When used as a buffer coating the modulus increases rapidly with decreasing temperature, resulting in an enhanced sensitivity to microbending losses. Promising materials with respect to the low-temperature properties are UV-curable silicone and polybutadiene rubbers [135-137]. However, suitable silicone rubbers are still under development in order to meet the refractive index requirements. The disadvantage of polybutadiene rubbers lies in their sensitivity to oxidation. Partly hydrogenated rubbers have been reported to be much less sensitive towards oxidation [138].

Recently, polyether urethane acrylates and poly(dimethyl-co-methylphenylsiloxane) acrylates have been reported as potential fast-curing, low-modulus coatings for high strength optical fibers [139, 140].

Their structures are shown below:

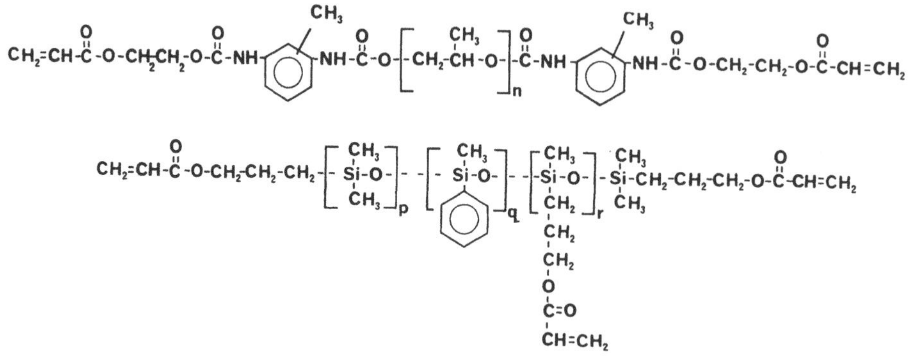

In the case of the polyether urethane acrylates, the material properties were varied by the length of the poly(propylene oxide) chain. Increasing the chain length decreases crosslink density and content of polar urethane groups and increases the volume fraction of low $T_g$, low refractive index polyether. As a result $T_g$, refractive index and E-modulus decrease. The addition of a small amount of reactive diluent ($\leq 20\%$) brings the viscosity and the curing rate to the desired level.

In the case of the poly (dimethyl-co-methylphenyl) siloxane acrylates, the material properties can be varied by changing the mole fraction of the methyl-phenylsiloxane units in the oligomer. Increase of the content of aromatic rings increases $T_g$ and the refractive index. In order to obtain a suitable viscosity and mechanical strength the molecular weight has to be relatively high. This in turn reduces the content of reactive groups to such an extent that slow curing results with $\alpha,\omega$-diacrylates. Good results are obtained when about 5 mole-% acrylate side groups are incorporated. Typical values of n are 60–160 and those of p, q and r are 160, 120 and 18, respectively, (cf. Scheme). When these coatings are provided with a hard top-coating, fibers are obtained which have a flat optical attenuation vs. temperature curve at temperatures between −60 and 80 °C. Within the refractive index requirements, the siloxane acrylates are advantageous for their low $T_g$. With respect to curing rate, polyether urethane acrylates are preferred. They cure very rapidly to complete conversion at irradiation times of less than 0.1 s [139, 140].

In a similar approach, vinyl/thiol modified poly(dimethyl-co-diphenylsiloxanes) have been proposed [141, 142].

# 4 Optical Discs

## 4.1 Optical Disc Systems

Several types of laser-readable disc systems have emerged since the first demonstration of the Laser Vision optical video disc in 1972 [143]. At present there are three main categories:

1. Laser Vision video discs, diameter 30 cm, playable on two sides and containing analog video and audio information [143–145].

2. Compact Disc (CD) audio discs, diameter 12 cm, playable on one side and containing digital information [145, 146]. A recent extension of CD is its use as a read-only memory in computer applications (CD-ROM).
3. Several types of Direct Read After Write (DRAW) discs on which data can be digitally recorded either once (write-once) or repeatedly (erasable). Initial applications are largely in professional equipment, for example as storage systems for large archieves.

Details of the systems are given in Ref. [145].

Most systems make use of a relief structure consisting of small pits, arranged in circular or spiral tracks.

The read-out principle is depicted in Fig. 5.

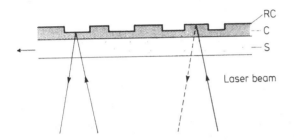

Fig. 5. Operating principle of an optical disc. (From Ref. [224], with permission)

The laser beam is focussed to a tiny spot with a diameter of about 1 μm onto the information layer of the disc. The diameter of the spot is somewhat larger than the width of the pits. Since the depth of the pits is just a quarter of the wavelength of the laser light in the disc, destructive interference will reduce the intensity of the reflected light. The rotating disc thus generates a signal of oscillating intensity. This signal is converted by a photodiode into an electrical one and next transformed into a TV, audio or data signal.

The information density which can be obtained is illustrated in the follwing example: each side of a video disc has a playing time of 30 or 45 minutes and it contains 2 or 3 * 10^{10} pits having a depth of 0.12, a width of 0.5 and a length varying between 0.5 and 2 μm. Each disc side contains about 45000 images. The information density is 1–2 orders larger than with present-day magnetic recording (see also Fig. 12).

Optical discs are manufactured by replication from a mold. Molds are made by writing the signal in a positive photoresist layer which has been deposited on a glass "master" disc [3, 147]. The writing is performed by using a laser beam. The intensity of this beam is modulated by the signal. After development of the resist layer the pattern is copied on a metallic disc by electrochemical methods. This process yields the "father" mold. Further electrochemical copying produces "mother" molds which are positive copies of the master. Another copying step then yields negative production molds.

Photopolymerization is just one of the available methods for copying discs. It has been used for several years in the large scale production of video discs. Today, injection molding is also used. In addition, photopolymerization serves for the replication of flexible molds [148] as well as for making the pre-groove structures on

DRAW discs. Pre-groove structures contain the necessary information for storage and addressing, the clock signal, etc. [149]. Low-melting tellurium alloys [150], organic dyes [151] or magneto-optical alloys [152] can be used as a recording medium for DRAW applications. A discussion of the various principles and methods for obtaining contrast can be found in Ref. [145]. Generally, the recording media are deposited as a thin layer on a substrate containing the pre-groove structure.

A critical step in the development of the photoreplication process has been the choice of a proper photopolymerizable coating. The design considerations will therefore be discussed in some detail.

Optimization generally requires a good understanding of the formation of densely crosslinked networks. This question will be addressed separately in Sect. 7.

Compact discs are replicated by compression molding or by injection molding. Hence they will not be discussed here in any detail.

## 4.2 Photoreplication of Discs

The photoreplication of Laser Vision video discs is shown in Fig. 6. In the first step a liquid photosensitive coating L is deposited on the metallic mold Mo. The slightly curved substrate S (PMMA with a thickness of 1.25 mm) is placed onto the mold with coating. The latter forms a thin layer C (20–30 μm) between substrate and mold. The coating is then irradiated for 6 s by means of an array of fluorescent lamps emitting at 350 nm. Substrate and coating are then released from the mold, turned upside down and a metallic reflective layer M is deposited on top of the

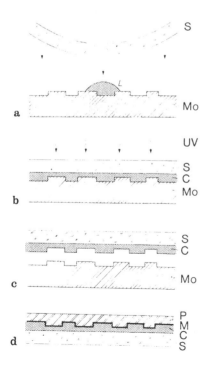

**Fig. 6 a–d.** Diagram showing four steps of the replication process: (**a**) The liquid layer L is spread over the mold Mo by deforming the substrate S to make it slightly convex; (**b**) Exposure to UV light to polymerize the liquid coating layer C; (**c**) Substrate with coating separated from mold; (**d**) Information layer coated with mirror M and protective layer P. (From Ref. [224], with permission)

organic coating. The reflective mirror in turn is covered with a protective coating P. Finally (not shown) the "P" sides of two such discs are glued together to yield a two-sided video disc [144].

## 4.3 Photopolymerizable Coatings

The coating must have a high curing rate, a low viscosity and it must wet the substrate. In the cured state, it must have high dimensional stability, durability and lack of odor, as well as easy release from the mold and good adhesion to the substrate and the reflective layer.

Coatings which meet all requirements are not easy to find, largely because some of the requirements are more or less conflicting. An example is easy release from the metallic mold and good adhesion to the metallic reflective coating. There are two ways of achieving an acceptable compromise. One way is to look for a mixture of different monomers, each providing one of the desired properties. This entails an investigation of the dependence of process and product properties on the composition. Another way is to make a careful study of the polymerization process and the relation between molecular structure and chemical behavior, and in this way to try and find a monomer that combines as many desirable properties as possible.

In the development of photopolymerizable coatings for video discs both approaches have been adopted and both were successful [24, 153].

Acrylates have been chosen as preferred monomers on account of their high rates of polymerization. DMPA proved to be the most efficient photoinitiator, its absorption spectrum strongly overlaps the emission spectrum of the fluorescent lamps [153]. The coating components were chosen with a view to ultimate properties such as behavior under metallization and adhesion to the metallic mirror.

Mapping of their functionality f versus their content of inactive, saturated hydrocarbon groups k together with hardness, adhesive and release properties has made it possible to delineate a region on the map where potentially useful monomers and mixture compositions can be found. One single monomer coating and a few mixtures have been selected for further investigation in this way (Table 2).

As an example, the influence of chemical composition on hardness and dimensional stability is shown in Fig. 7. The measurement of hardness and its influence on signal-to-noise ratio has been described [154, 155].

Hardness is related to crosslink density and the latter is determined by the extent of double bond conversion. This in turn can be determined with DSC, which also gives

**Table 2.** Properties of coatings that will give good video discs

| Coating | Composition, wt% | | Viscosity at 23 °C, mPa · s | Curing time, s | |
|---|---|---|---|---|---|
| | | | | 23 °C | 80 °C |
| 1 | 57% TPGDA | 29% NVP | 7.8 | 1.5 | 1.0 |
| | 10% TMPTA | 4% DMPA | | | |
| 2 | 61% TPGDA | 17.5% NVP | 12.3 | 1.7 | 1.0 |
| | 17.5% TMPTA | 4% DMPA | | | |
| HDDA | 96% HDDA | 4% DMPA | 6.7 | 3.0 | 1.4 |

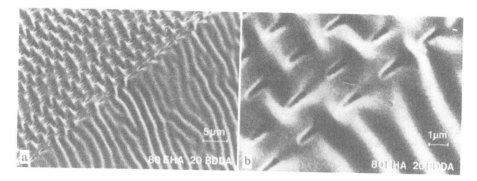

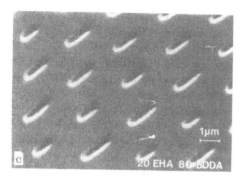

**Fig. 7a–c.** Scanning electron microscope pictures of the Laser Vision pattern after metal deposition on a coating of a mixture of 2-ethylhexyl acrylate (EHA) and 1,4-butanediol diacrylate (BDDA). Photographs (a) and (b) were obtained with the composition 80 % EHA, 20 % BDDA. After metallization, the soft coating appears as a wave pattern, due to shear deformation caused by stresses in the aluminum layer. Photograph (c) was obtained with the composition 20 % EHA, 80 % BDDA. Since the coating substrate is more densely crosslinked there is no wave pattern. (From Ref. [24], with permission)

the rate of polymerization. The experimental details have been described [26]. Unexpectedly, it came out that at room temperature the highest crosslink densities can be obtained with diacrylates [24].

After comparison of rates of polymerization (Fig. 1), purity and maximum extent of reaction of a number of diacrylates, 1,6-hexanediol diacrylate (HDDA) was selected for a single monomer coating. It has been further studied in combination with the mixtures in Table 2 with respect to curing time under process conditions, extractability of unreacted substances and separation energies, required for separation from the mold. Regarding curing rate, the mixtures polymerize faster, although at 80 °C the differences have almost disappeared. The mixtures are also less sensitive to oxygen [24]. Presumably the NVP plays a similar role as the amine in ketone-amine systems.

The free-radical polymerization of NVP is rather complex [138] but its liability to be polymerized by its own peroxide is well documented [156] as well as its strong tendency for chain transfer during polymerization [157]. However, retardation by oxygen has also been claimed [158]. Alternatively, the formation of a donor-acceptor complex has been proposed [159–161] which could yield anionic and cationic species, capable of initiation without introducing oxygen sensitivity. In view of the very restricted tendency of acrylates to polymerize by ionic mechanisms [162], this seems rather unlikely. Our observation that the rate differences between the NVP-containing mixtures and HDDA disappear when the polymerization is carried out under nitrogen also points to enhanced oxygen consumption by NVP [24].

**Table 3.** Separation of Laser Vision discs from a mold, measured at a constant pulling rate of 500 mm · min$^{-1}$.

| Coating | Total separation energy, J | Comprising: | | Separation time, s | Maximum pulling force, N |
|---------|---------|---------|---------|---------|---------|
| | | Deformation energy of the substrate, J | Separation energy of coating, J | | |
| 1 | 1.27 | 0.39 | 0.88 | 4.2 | 51 |
| 2 | 0.75 | 0.27 | 0.48 | 3.6 | 48 |
| HDDA | 0.30 | 0.08 | 0.22 | 2.4 | 31 |

A disadvantage of NVP is its incomplete conversion, yielding large amounts of extractable materials. The amount of unreacted material (monomer, photoinitiator) that can be extracted by a suitable solvent may be determined by liquid chromatography. Plotting of this quantity vs. exposure time yields an extraction curve [163]. The one-monomer system is found to contain considerably less extractable material than the mixtures [24, 153].

During release of the substrate with cured information layer from the mold, it is warped in a complicated way. One part of the total separation energy is required for the actual separation (release of adhesion), the remaining part for the (in)elastic deformation of the substrate with information layer. Table 3 shows that the release is easiest for the one-monomer system, although satisfactory results were obtained in all three cases.

With the one-monomer coating as well as with a number of mixtures, video discs of excellent quality have been made.

# 5 Aspherical Lenses

## 5.1 Laser Read-Out System

The information on an optical disc can be read-out with a light pen. One possible design is shown in Fig. 8a. The light emitted by the solid-state laser is collimated and focused on the information layer, reflected there and on its way back split into two beams which fall on a detector array. This means that when the laser spot strikes a pit, interference will reduce the amount of reflected light.

The array of four photodiodes has several functions:
(i)   to determine whether the light beam is focused on the information layer;
(ii)  to determine whether the track is followed properly;
(iii) to determine whether a pit is seen.
More details can be found in Ref. [145].

There are several possible designs for the lenses used in such a light-pen. In the case of Fig. 8a where only spherical elements are used, the requirement of only 1 μm spot size can only be fulfilled with an objective consisting of at least three elements.

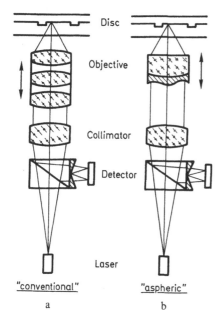

Disc

Objective

Collimator

Detector

Laser

"conventional"

a

"aspheric"

b

**Fig. 8 a and b.** Schematic drawing of Compact Disc light pens with conventional optics (**a**) and with aspherical lenses (**b**). The laser beam passes through a beam-splitting prism, is focused on the disc and the reflected beam is directed towards a set of 4 photodiodes. (From Ref. [249], with permission)

However, with the introduction of one or two aspherical surfaces, depending on the index of refraction chosen, the same optical performance can be achieved with only one element (Fig. 8b). This has the advantage of smaller size and lower weight and, if a suitable technology is used, lower price [164]. For such lenses, the asphericity, that is the deviation from the nearest sphere, needs only to be of the order of $\pm 10$ μm.

Aspherical shapes of a precision within $\pm 0.2$ μm as is required by this application can be made economically in at least three ways, which all depend on copying the aspherical surface from a mold:

(i)   Injection molding of thermoplastics. This technology has been found suitable for the production of lenses of poly(methyl methacrylate) (PMMA) with the required precision [165]. However, it is difficult to maintain proper performance over the required range of humidity and temperature.

(ii)  Hot-pressing of inorganic glasses. Objective lenses made with this technique have been described [166] but it has proved difficult to find glasses which can be pressed at acceptable temperatures and which offer sufficient environmental stability at the same time.

(iii) Replication using a UV-curable coating on a spherical glass substrate [167-170]. In this technology the thermal and environmental stability of a glass substrate is combined with the ease of replication of complicated shapes using photo-polymerization.

## 5.2 Photoreplication of Lenses

The replication process for aspherical lenses for CD players is shown schematically in Fig. 9 [168]. A small quantity of liquid coating is applied to the surface of a spherical glass substrate. This in turn is pressed into a quartz glass mold with the desired aspherical profile. The quality of the mold is decisive for the performance

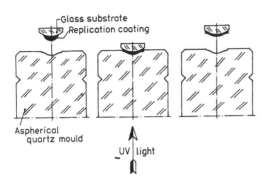

**Fig. 9.** Schematic view of aspherical lens replication. (From Ref. [168], with permission)

of its replica. High-precision molds can be produced on a computer-controlled lathe [171, 172] with an accuracy of better than 0.05 µm and a surface roughness of less than 10 nm.

On irradiation with UV light through the mold, the photopolymerization process is initiated. Then the coating is polymerized within 1 minute.

When the surfaces of substrate and mold have been subjected to proper pretreatments, adhesion may be controlled such that the cured coating can be easily released from the mold whereas it will be chemically linked to the substrate. One of the best ways to promote adhesion to the glass substrate, is surface treatment with a silane coupling agent. When silanes containing methacrylate groups are used, copolymerization of the latter with the coating will provide strong covalent bonding at the interface. Depending on the requirements additional, thermal annealing of the coating may be necessary [168].

## 5.3 Photopolymerizable Coatings

Coatings suitable for lens replication have to share a number of properties with optical disc coatings such as a high optical transparency at the wavelength of the laser light, good hardness and abrasion resistance, high chemical and dimensional stability.

Specific demands with respect to lens replication are a relatively high refractive index, minimal polymerization shrinkage and low moisture absorption.

Diacrylates have proved suitable for optical disc replication and therefore this category was also investigated for lens replication [168, 169]. A high refractive index will reduce the thickness variation across the aspherical correction layer and thereby its maximum thickness.

The refractive index of a coating may be increased by introducing strongly polarizable groups into the monomer. Examples are aromatic rings and halogen atoms, especially iodine. Since iodine atoms will strongly reduce the stability of the monomer, aromatic groups are preferred.

All monomers, but a few exceptions [173], exhibit shrinkage upon polymerization since Van der Waals bonds are converted into much shorter covalent bonds. The accompanying reduction of the entropy also favors a closer packing. When a surface with a specified curvature has to be made by replication from a mold, due allowance has to be made for shrinkage of the replication coating. This means that the shape of the mold has to be corrected for shrinkage. This can only be done when the shrinkage is highly reproducible.

Therefore, proper control of light intensity and temperature is a prerequisite for accurate replication. The smaller the shrinkage, the smaller is the shape correction of the mold. Overall shrinkage may be reduced by increasing the length of the moiety between the acrylate groups, but this occurs at the cost of a reduced cross-link density and hardness. The last disadvantage may be overcome in part by using a dimethacrylate instead of a diacrylate.

Take-up of moisture will swell the polymer and thereby change its shape. Therefore, hydrophobic coatings are preferred. Epoxyacrylates, which might be attractive at first sight, are rejected on this account.

Monomers which have proved suitable for lens replication [169, 170] are di(meth)acrylates of ethoxylated bisphenol-A in which n may vary between 0 and 3:

$$CH_2{=}C{-}C{-}\{O{-}CH_2{-}CH_2\}_n O{-}\bigcirc{-}\underset{CH_3}{\overset{CH_3}{C}}{-}\bigcirc{-}O\{CH_2{-}CH_2{-}O\}_n \underset{\underset{O}{\parallel}}{C}{-}C{=}CH_2$$

Preferred photoinitiators are HCPK and HMPP since these compounds yield completely clear coatings [63, 174] whereas DMPA often causes some yellowing. This yellowing proved to be partially reversible. In the dark, it disappears gradually but upon additional exposure it is restored.

A high light intensity not only increases the rate of polymerization but also the maximum extent of reaction which can be attained at the temperature of cure [26, 169].

Shrinkage vs. exposure time is shown in Fig. 10 [150] for a coating with $n = 1.6$. It can be seen that considerable shrinkage occurs after switching off the light. This is likely to be caused by a slow post-reaction as well as by retardation of shrinkage with respect to chemical conversion [26].

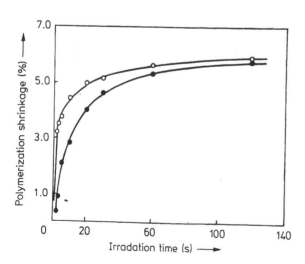

Fig. 10. Polymerization shrinkage for a coating with n = 1.6. Filled symbols: immediately after shutter closure. Open symbols: 10 min later. Intensity: $2.2\ mW \cdot cm^{-2}$. (From Ref. [168], with permission)

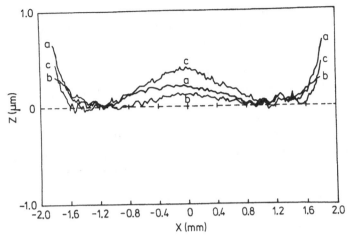

**Fig. 11.** Differential contour plot of an aspherical lens surface obtained after several irradiation times. (a): 10 s (b): 30 s (c): 120 s Intensity: 8 mW · cm$^{-2}$. Z is the deviation from the desired profile, X is the distance from the optical axis. (From Ref. [168], with permission)

The accuracy which can be obtained is illustrated in Fig. 11 [168]. The expected shrinkage for which the mold has been corrected has not been realized at all positions of the lens surface. The relation between position, exposure time and shrinkage is a complex one. Initially, shrinkage increases with exposure time (curves a and b), but later on adhesion to the mold will hamper further shrinkage more strongly at the thick central part of the coating than at its thinner sections (see Fig. 9 for the approximate shape of the coating; the Figure is not to scale, however, the layer thickness has been exaggerated for clarity).

Due to the local differences in shrinkage, more free volume will be frozen-in at the center than in the thinnest sections (Z = 0). Near the edges the volume relaxation is still thought to occur through lateral shrinkage (curves b and c).

**Table 4.** Properties of a photopolymerized lens replication coating[a])

|  | $n_D$ | $10^5 dn/dT$ | $10^5 \alpha$ | Shrink-age | Trans-mission | E'[e]) | T[f]) | $\varepsilon$[g]) | T(tan $\delta_{max}$) |
|---|---|---|---|---|---|---|---|---|---|
|  |  | (K$^{-1}$) | (K$^{-1}$) | (%) | (%)[d]) | (MPa) | (MPa) | (%) | (°C)[h]) |
| Initial | 1.569 | −16.7 | 12.1 | 6.0 | 90 | 1500 | 39.5 | 2.8 | 55 |
| Scanned[b]) | 1.567 | − 8.8 | 8.4 |  |  |  |  |  |  |
| Annealed[c]) | 1.565 | −11.2 | 8.2 |  |  | 2000 | 41.5 | 1.3 | 135 |

a) Coating: 96 w% monomer with n = 1.6 and 4 w% DMPA. Temperature 22 °C. Light intensity 2.4 mW · cm$^{-2}$
b) After 1 thermal scan to 90 °C
c) After 1 h at 120 °C
d) At 780 nm for a thickness of 100 µm. Uncorrected for reflection losses
e) Tensile modulus
f) Tensile strength
g) Elongation at break
h) Measure at 0.033 Hz

Incomplete volume relaxation provides a driving force for a slow change of properties. Hardness, measured by indentation of a diamond point with a fixed radius under a standard load, was found to increase by about 20% upon storage at room temperature for a week [168].

Similar changes have been observed with other photocured coating materials [175].

Incomplete volume relaxation also makes the coating sensitive to temperature changes and this shows up in the index of refraction and its temperature coefficient and in other properties as well (Table 4) [168,169].

Stable optical and mechanical properties may be obtained by thermal annealing.

By proper annealing, the required aspherical contour may be approached to within 0.05 µm over almost the total optical diameter. The reproducibility of the process has been tested earlier by interferometric methods [172]. A series of lenses yielded an average value of the standard deviation of the wavefront of only $0.018\lambda$.

The optical and mechanical properties of the replication coating are determined by the molecular structure of the monomer and by the conditions during polymerization. Combination of photopolymerization at room temperature with annealing at 120 °C leads to excellent physical properties over a wide temperature range (−30 to +85 °C).

# 6 Other Electronic Applications of Photopolymerization

There are many other applications of photopolymerization in (opto)electronics and their number is still increasing. A few of these will be briefly described in this section.

## 6.1 Recording of Holograms

The optical discs described in Sect. 4 rely on intensity modulation caused by destructive interference of light which is reflected at the bottom of the pits, and light which is reflected at the surface around the pits.

An alternative method of information storage uses an interferogram as the exposure source. When a photopolymerizable system is exposed to a periodic intensity profile the polymerization will preferentially occur at the maxima. This causes a localized increase of the refractive index. Refractive index image recording systems using various types of organic photochemical processes have been reviewed by Tomlinson and Chandross [176] so we will only mention a few systems in which photopolymerization is being used. Interferograms are usually generated by laser irradiation with visible light, so a common feature of the systems is that they have to be sensitized to the appropriate wavelength. Dye-sensitized photopolymerization has been reviewed by Ketley [177] with special emphasis on holographic recording. After exposure the image must be fixed. This can be done by a variety of methods. In negative photoresist systems, the unexposed area is removed, but this restricts the application to a single-layer hologram.

Volume-phase holography often makes use of the diffusion of unreacted monomer in a monomer/polymer mixture towards the exposed areas where monomer has been depleted by the reaction. Next a flood exposure is applied in order to fix the refractive index pattern [178, 179].

Alternatively, a mixture of two monomers, having different reactivities and yielding polymers with different refractive indices, may be used [180]. In a further refinement, the enhanced scattering which occurs during the generation of the pattern is avoided by forming a latent image. This has been accomplished by adsorbing the photoinitiator on the inner surface of a porous glass matrix, followed by localized photochemical destruction of the initiator. Next, the matrix is soaked with a liquid monomer solution or mixture of monomers and subjected to a flood exposure. Polymerization then proceeds at the highest rate in the previously unexposed areas and the most reactive monomer will react first, thereby generating the refractive index pattern. Several combinations of simple methacrylates and inert solvents have been reported [181].

Other workers have reported on the use of urethane (meth)acrylates for the recording of holograms. In some cases, thermal postcuring of an epoxide was used for fixation. This work has been reviewed in Ref. [182].

Volume phase holography using photopolymers has been reported to provide the highest information density in man-made systems, especially if unit volume is considered instead of unit area (Fig. 12) [183].

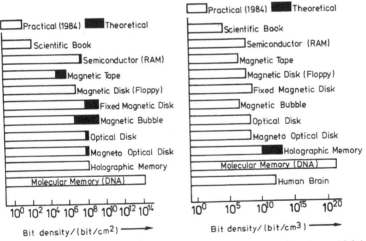

**Fig. 12.** Information density per unit area (left) and per unit volume (right) of various storage materials. (From Ref. [183], with permission)

Another interesting application of holographic recording is the faceted disc used in laser scanners for read-out of universal product codes such as the bar code on, for example grocery products [184].

The facets on a rotating disc are used to generate a 20-line scan pattern of the laser beam across the bar code label. Each facet is a separate hologram with its own combination of focal length, skew angle and elevation angle. Each facet deflects and focuses the outgoing beam, and collects and collimates a portion of the diffuse light reflected from the label.

The discs are made by photocrosslinking of gelatin/dichromate [184] but it is also possible to make them by photopolymerization [185].

## 6.2 Optical Waveguides

Optical waveguides are used in integrated optics and they serve to confine a light beam in a thin slab of material which, for example, may be connected to a glass fiber, a detector or a light source.

The coupling between a light beam and a waveguide can be achieved through a prism or through a grating. The manufacturing of holographic gratings has been outlined in the previous section.

Waveguides consist of a thin layer of a high-refractive index material surrounded by/embedded in a low-refractive material. One way of making these layers uses of photolocking of a second component in a polymeric layer, followed by removal of this component from the unexposed area [186, 187]. In this way, fiber optic sheet systems have been made as well as beam splitters, couplers, etc. [188]. In a typical example (Fig. 13) [187], a dilute solution of polycarbonate, methyl acrylate and a photoinitiator in a solvent is spin-coated on a substrate, patternwise exposed such that the methyl acrylate polymerizes in the exposed area and finally evaporation of unreacted monomer from the unexposed areas.

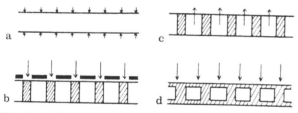

**Fig. 13a–d.** Formation of optical waveguides in a fiber-optical sheet configuration; (**a**) monomer doping in a film; (**b**) UV irradiation through a photomask; (**c**) evaporation of unreacted monomer; (**d**) lowering refractive index in the surface part. (From Ref. [187], with permission)

The unexposed areas then exhibit the high refractive index of polycarbonate whereas the exposed areas have a lower refractive index due to the presence of the poly(methyl acrylate) [186, 187]. Closed structure are obtained by polymerizing a layer of methyl acrylate on top of the system. Franke et al. designed a similar system using PMMA/styrene [189]. Here, the refractive index increased during polymerization. In both cases, the differences in refractive index were about 0.01.

## 6.3 Other Optical Components

A remarkable application of a simple optical waveguide is its use as a *chemical vapour microsensor* [190].

The waveguide is made by laser irradiation of a thin liquid layer, enclosed between two glass plates. The liquid is a divinyl oligomer containing uranyl perchlorate as a photoinitiator. After removal of the cover glass and the unexposed material, the plane polymer strip (dimensions $1 \times 2 \times 10$ mm) is coupled at its ends to optical fibers. These are connected to a HeNe laser and a photodiode, respectively. The sorption of vapors such as ammonia or toluene can be monitored by recording the transmission losses of the sensor.

The performance of UV-curable *adhesives* for the assembly of lenses, prisms, etc. has been compared with traditionally used two- or three-component epoxy adhesives by Seo [191]. Important advantages claimed are: suitable for mass production while maintaining high performance, rapid curing and elimination of mixing and storage problems.

A recent application of photopolymerization is the *protection of the surface* of optical components made of polycarbonate or acrylic resins. The surfaces of these parts are vulnerable to scratches and attack by chemicals. Coating of the surface with a thin layer (2–4 µm) of a UV-curable acrylic coating may increase the abrasion resistance and the chemical inertness considerably.

An additional advantage is that these coatings may be used on materials with low heat-distortion temperatures as well [192].

## 6.4 Wire Coatings

The coating of electric wires for magnets has been described by Feit [193]. Using a similar set-up as used for the coating of optical fibers, copper wire was coated with a mixture of a urethane monomethacrylate and a diurethane dimethacrylate. At intensities of a few mW $\cdot$ cm$^{-2}$, drawing speeds of 2 cm $\cdot$ s$^{-1}$ were obtained. Wires, coated with three layers of this material exhibited breakdown voltages of about 2 kV.

Other materials used for electron beam and ultraviolet curable wire coatings are mentioned in Ref. [124]. Many recent applications can be found in the patent literature. A few examples are given in Refs. [194–197].

## 6.5 Miscellaneous Dielectric Coatings and Adhesives

Radiation-curable coatings for *resistors and capacitors*, solder masks and conformal coatings for printed circuit boards, inks and marking coatings for electronic components as well as adhesives are briefly discussed in Ref. [124]. The characterization of the curing profile of some epoxy acrylate mixtures for printed circuit board applications has been performed using thermogravimetric analysis combined with gas chromatography / mass spectroscopy [198].

The performance of different kinds of *solder masks* (UV-cured acrylated, thermally cured epoxies and alkyd/amine systems and dry film solder masks) under humid conditions at elevated temperatures is described by Fox [199]. All systems suffer from serious reduction of insulation resistance. The soldering process partially restores the insulation resistance.

*Dual cure coatings*, based on multifunctional acrylates to which a small quantity of a polythiol has been added, have been proposed as conformal coatings for completed printed circuit boards [200]. UV exposure causes a rapid setting of the coating, thereby allowing for immediate handling of the product. At the same time, running-off during subsequent heating is prevented. The heating step is required to cure poorly or unexposed areas as well as to obtain the desired mechanical and thermal properties.

Polythiol/acrylate systems are reported to exhibit a reduced oxygen sensitivity, possibly caused by the rapid consumption of oxygen in a chain process of oxidation, hydrogen abstraction from the thiol and reinitiation [201, 202].

Dual cure coatings are also being developed for use in space and satellite applications [203]. Urethane-acrylates seem promising candidates for conformal coatings for printed circuit assemblies as well as for potting and insulating materials. The initial problem of outgassing could be tackled by performing the thermal aftercure under vacuum.

Dual functionality coatings differ from dual cure coatings in that they contain functional groups of different kinds. In the first step, partial crosslinking is accomplished through exposure to UV light, in the second step another component, for example an epoxide, is cured thermally through the activation of a latent catalyst.

Such hybrid systems can be used in the manufacturing of multilayer circuits. Here, after the photoimaging step to form the circuit tracks of the inner layers, the latter have to be bonded together with epoxy adhesive layers to form the assembly. During the final thermal curing stage, the epoxy and hardener groups from the adhesive react with those from the circuit tracks and the species become strongly bonded together via chemical bonds [64, 204].

Another hybrid system has been proposed for resist applications: the simultaneous chain polymerization of N-methylolacrylamide and the acid-catalyzed crosslinking of the resulting polymer with poly(vinyl alcohol) [205]. In dual cure plastisols, finally, a solid thermoplastic resin is dispersed in a multifunctional acrylate mixture which contains a thermal and a photoinitiator. These plastisols have been proposed for the encapsulation of integrated circuits [206].

UV-curable pressure-sensitive adhesives have been treated by Stueben and Patrylow [207] and a general discussion of adhesives applications in electronics can be found in Refs. [208] and [209].

The last class of dielectric coatings to be mentioned here is formed by the UV cationic epoxy systems. In most cases photoinitiators as invented by Crivello [95] are used to initiate the chain crosslinking polymerization of di-epoxides or mixtures thereof with monoepoxides.

The photogeneration of strong acid species to initiate cationic polymerization does not seem to be attractive from the dielectric point of view, but, when properly formulated and cured, quite good insulating and loss properties may be achieved. Zopf has described a variety of application areas, ranging from coatings to adhesives, encapsulants and dielectric barriers [210].

Rates of polymerization are considerably slower than with acrylates but this can in part be overcome by using high-intensity light sources with an electric power consumption of about 80 W per cm arc length. It is likely that the strong heating caused by these lamps contributes significantly to the curing process and to the mechanical and electrical properties of the product [100, 210]. Recently, improved electrical and corrosion protective properties have been obtained by Hayase [97]. His systems have the advantage that no ionic species will remain in the photopolymer but they appear to be rather slow. They are described in Sect. 3.3.1.

# 7 Further Study of Network Formation with Diacrylates

After having discussed a number of applications of network formation by photopolymerization, we will now turn back to the process of network formation itself. This is appropriate since the performance of the polymeric products not only depends on their chemical composition but also on the conditions during their formation. An obvious example is the extent of reaction. Further study of network formation is also justified since our knowledge of densely crosslinked networks is still limited.

During the development of the processes outlined in Sects. 3–5, we have made some observations on physico-chemical changes occurring during chain crosslinking polymerization of diacrylates. These will be reported in this section, together with results of a restricted number of other studies in which polymer properties have been correlated with the conditions during photopolymerization. Thermally induced processes, although relevant for the study of network formation in general, will not be discussed.

## 7.1 Model Monomers

Model monomers chosen for the study of network formation by photopolymerization are shown below:

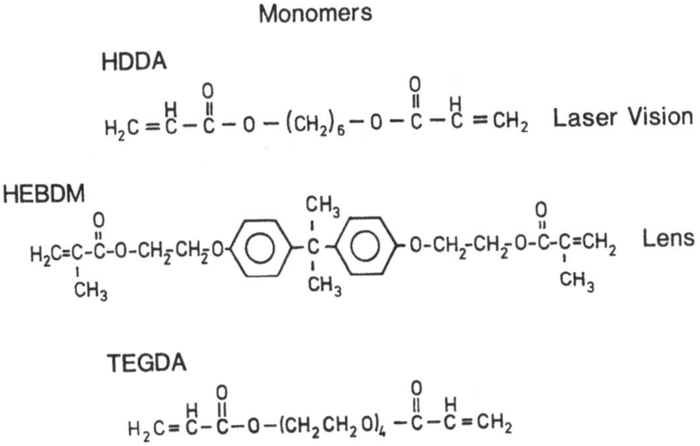

Monomers

HDDA

$$H_2C = \overset{H}{\underset{}{C}} - \overset{O}{\overset{\|}{C}} - O - (CH_2)_6 - O - \overset{O}{\overset{\|}{C}} - \overset{H}{\underset{}{C}} = CH_2 \quad \text{Laser Vision}$$

HEBDM

$$H_2C = C - \overset{O}{\overset{\|}{C}} - O - CH_2CH_2O - \underset{CH_3}{\overset{CH_3}{\bigcirc - \underset{CH_3}{\overset{}{C}} - \bigcirc}} - O - CH_2 - CH_2 - O - \overset{O}{\overset{\|}{C}} - C = CH_2 \quad \text{Lens}$$

TEGDA

$$H_2C = \overset{H}{\underset{}{C}} - \overset{O}{\overset{\|}{C}} - O - (CH_2CH_2O)_4 - \overset{O}{\overset{\|}{C}} - \overset{H}{\underset{}{C}} = CH_2$$

The first one, 1,6-hexanediol diacrylate (HDDA), is very widely used as reactive diluent in coating formulations and it can be obtained in a pure state. As we have seen, it may also be used for the replication of optical discs. The second one, bis-(hydroxyethyl) bisphenol-A dimethacrylate (HEBDM), finds application in the photoreplication of lenses.

The third one, tetraethylene glycol diacrylate (TEGDA), has been selected since it forms rather flexible networks due to the long and flexible bridge between the two acrylate groups.

The study of these compounds has yielded useful results for the optimization of coating compositions as described in Sects 4 and 5.

In addition to results obtained with these model compounds, we will describe a restricted number of studies on mixtures of epoxy acrylates or urethane acrylates with reactive diluents.

## 7.2 Delayed Shrinkage. Physical and Chemical "Aging"

Vinyl polymerizations are accompanied by a relatively large amount of volume shrinkage (MMA 23 %) [211] and therefore dilatometry has long been a standard method for monitoring conversion during polymerization [29].

In the bulk crosslinking polymerization of divinyl compounds, gelation occurs at a very low conversion of the double bonds, so most of the polymerization process proceeds in the gelled phase. This means that in order to convert the free volume generated by the chemical reaction into overall shrinkage, the whole gel has to move

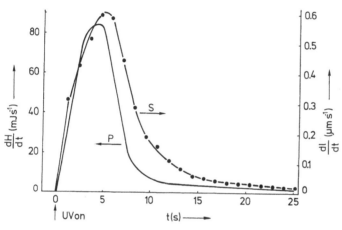

**Fig. 14.** Rate of polymerization (P) and rate of shrinkage (S) for the photopolymerization of HDDA at 20 °C. Light intensity: 0.2 mW · cm$^{-2}$. Initiator: 4 wt.-% DMPA. (From Ref. [26], with permission)

**Table 5.** Maximum extent of reaction (DSC, 20 °C, 4 wt.-% DMPA)

| Intensity: (mW · cm$^{-2}$) | 0.002 | 0.02 | 0.2 |
|---|---|---|---|
| HDDA | 0.65 | 0.72 | 0.80 |
| HEBDM[a] | 0.3[b] | 0.4[b] | 0.48 |
| TEGDA | 0.84 | 0.87 | 0.89 |

[a] In Ref. [26] somewhat higher values were quoted since the monomer used there still contained a significant amount of mono-ester

[b] Inaccurate value due to very low rate of reaction

**Table 6.** Maximum extent of reaction
(DSC, 0.2 mW · cm$^{-2}$, 4 wt. %
DMPA)

| Temperature: | 20 °C | 80 °C |
|---|---|---|
| HDDA | 0.80 | 0.89 |
| HEBDM[a]) | 0.48 | 0.72 |
| TEGDA | 0.89 | 0.97 |

[a] see note at Table 5

in a cooperative way. This process is necessarily slower than the diffusional motion of free monomer molecules. Therefore, it is not unexpected that during polymerization the shrinkage cannot keep up with the chemical conversion. This is illustrated in Fig. 14 where the rate of polymerization, obtained from isothermal calorimetry, is compared with the rate of shrinkage, measured as the decrease of the thickness of a thin layer of monomer enclosed between a thick and a very thin glass plate [26, 250]. The time lag between conversion and shrinkage generates a temporary excess of free volume which increases the mobility of unreacted $C=C$ groups.

We can take advantage of this situation: by increasing the light intensity we can increase the conversion while the mobility is still high. In this way, the ultimate conversion will increase with light intensity (Table 5) [212].

An increase of mobility may also be reached by increasing the temperature which also results in an increased ultimate conversion (Table 6) [212].

It must be noted that the extents of reaction determined by calorimetric measurements are kinetically defined. The reaction is considered to be finished when its rate falls below the limit of detection, that is when the rate of heat production has decreased by 2–3 decades with respect to the maximum rate. Below we will show that the reaction may go on for quite a long time at a very low and continuously decreasing rate.

Since the rate law for self-decelerating processes is unknown, extrapolation of conversion with time is impossible. In principle, methods like IR spectroscopy would be suitable but they often suffer from lack of accuracy, notably when small changes have to be detected near the end of the reaction.

So the physical process of shrinkage is coupled to some extent with the chemical process of polymerization but shrinkage is not a measure of conversion, as is normally observed with liquid systems.

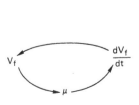

a Physical aging

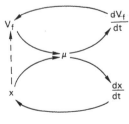

b (Photo)polymerization

**Fig. 15.** Schematic relation between segmental mobility, $\mu$, free volume, $V_f$, double bond conversion, x, and their respective rates of change. (From Ref. [39], with permission)

The mutual dependence of shrinkage and conversion and their respective rates may be compared with physical aging (Fig. 15a). Physical aging is a self-decelerating process due to the closed-loop dependence of free volume $V_f$, segment mobility $\mu$ and rate of shrinkage $dV_f/dt$ [213]. The mobility of the polymer segments is the important microscopic parameter. Physical aging is usually studied in polymers which have already been formed. However, a closed-loop dependence is expected to exist during isothermal bulk polymerization as well (Fig. 15b) [39]. During reaction, there exists an additional and similar dependence of conversion x and rate of conversion dx/dt on the mobility $\mu$ of unreacted groups, either present as pendent double bonds or as free monomer. Here too, the mobility is the important parameter.

Since chemical reaction immediately generates free volume there is also a direct relation between x and $V_f$. Chemical reaction drives the shrinkage process but a high extent of reaction reduces the rate of shrinkage. A high extent of volume shrinkage in turn reduces the rate of chemical reaction.

The higher extent of reaction obtained at the higher intensity of irradiation is reflected in the physical properties as well. For HDDA, for example, the temperature of maximum mechanical loss at 1 Hz, $T(\tan \delta_{max})$ varies by 85 K over an intensity range of 3 decades (Fig. 16) [212]. In this experiment, the times of irradiation have been obtained from the DSC experiment: irradiation was stopped when no reaction could be detected any more. This means that the sample exposed at $2 \, mW \cdot cm^{-2}$ received the highest dose. Results obtained with equal doses are shown below.

The self-retarding character of the process causes the reaction to continue at a rate below the limit of detection by the DSC, but changes still show up in the mechanical properties, notably in $T(\tan \delta_{max})$, the temperature of maximum mechanical loss. This can be seen when the experiments shown in Fig. 16 are repeated under slightly different conditions. In Fig. 17, the irradiations at the lower intensities have been continued until equal doses were applied in all cases [212].

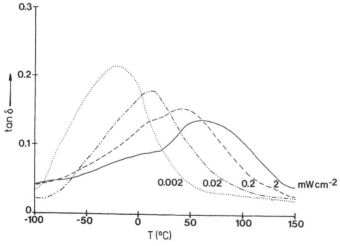

**Fig. 16.** Mechanical loss at 1 Hz as a function of temperature for samples of HDDA, photopolymerized at 20 °C using various light intensities. Initiator: 0.25 wt.-% DMPA. Required exposure times were read from DSC curves. They decrease with increasing intensity: 25, 15, 12 and 7.5 min, respectively. (From Ref. [212], with permission)

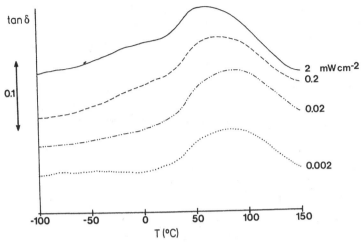

**Fig. 17.** Same experiments as in Fig. 16 but now for equal doses. Irradiation times are 7500, 750, 75 and 7.5 min, respectively. (From Ref. [212], with permission)

Obviously, the reaction has continued and now the lowest intensity even seems to be the most effective, presumably since a low intensity favors a long kinetic chain length and thereby a high extent of reaction. Similar observations have been made with TEGDA [214].

There appear to be two velocity regimes. Shortly after the rate maximum, there is a rather sudden decrease of the rate, although we have no exact criterion for separating these two regimes.

Our studies have shown that, by using high light intensities, a larger part of the reaction occurs in the "high" rate regime. This is of course important for production processes. Study of the shrinkage behavior is of special relevance for the design of molds for lens replication.

## 7.3 Trapping of Radicals in a Network

The trapping of polymeric radicals in glassy polymers is a well-known phenomenon [7]. The radicals may be produced by high-energy irradiation of glassy polymers [215], by mechanical destruction [53] or by trapping during free-radical polymerization of monomers [7]. Trapped polymeric radicals have been extensively studied by using ESR [216].

In densely crosslinked dimethacrylates, radicals with lifetimes of several days have been observed [51] and the lifetimes were found to depend on crosslink density [217]. Extremely long lifetimes at room temperature have been observed with photopolymerized di(meth)acrylates [26, 39]. In the case of photopolymerized HDDA and HEBDM, the lifetimes exceeded 5 months at room temperature and under vacuum, even in the presence of large amounts of unreacted monomer and pendent double bonds.

In poly(TEGDA) on the other hand, much shorter lifetimes were observed. Within one night, the three-line spectrum was converted into a single-line spectrum, which then in turn decayed within a few days [214]. This is presumably due to scission of the tetraoxyethylene chain since the spectrum strongly resembled that of UV-irradiated poly(oxyethylene). In that polymer, the ESR signal also decayed at 20 °C and the spectrum has been assigned to $-CH_2-CH_2-O^\cdot$ radicals, formed by chain scission [218].

The formation of such radicals would certainly increase the mobility of the radical sites as compared with acrylate groups and thereby reduce their lifetime.

Knowledge of structure and reactivity of trapped radicals is especially important for the design of thermal postcuring schedules as used with the manufacturing of lenses. This knowledge may also provide a starting point for the study of long term behavior.

### 7.3.1 Structure of Trapped Acrylate Radicals

Conflicting views about the structure of the trapped polyacrylate radicals have been presented: the three-line spectrum generally observed with polyacrylates [216, 219] is consistent with a chain-end structure (I) as well as with a mid-chain structure (II), obtained after tertiary hydrogen abstraction from the polymer chain.

$$\begin{array}{ccc} & & H \\ & & | \\ \sim\!\!\sim\!\!CH_2-C^\cdot & & \sim\!\!\sim\!\!CH_2-\overset{\cdot}{C}-CH_2\sim\!\!\sim \\ & | & & | \\ & C=O & & C=O \\ & \diagdown OR & & \diagdown OR \\ & I & & II \end{array}$$

For both it is assumed that only one of the α-protons located on a $CH_2$-group contributes to the hyperfine splitting.

Selective deuteration of the monomer at its α-positions made it possible to determine which of the structures is the correct one. Fig. 18 shows that similar spectra were obtained with normal and deuterated samples. This proves that structure II must be the correct one; in case I a two-line spectrum would be observed. Therefore, between the end of the irradiation and the time of measurement (about 5 minutes) virtually all of the propagating chain-end radicals have been terminated by combination or hydrogen abstraction.

A similar situation exists with polystyrene [220] but not with polymethacrylates [51].

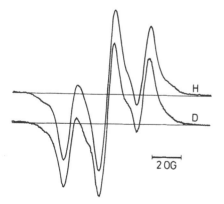

Fig. 18. ESR spectra of trapped radicals in photopolymerized HDDA. Upper curve (H): normal compound. Lower curve (D): deuterated. Intensity: $0.2\ mW \cdot cm^{-2}$. Initiator: 4 wt.-% HMPP. Exposure time: 30 min. Approximate radical concentration: $2 \times 10^{-3}$ M. (From Ref. [52], with permission)

## 7.3.2 Fate of Trapped Acrylate Radicals. Chain Oxidation

As we have seen, the trapped radicals have an appreciable stability at room temperature and under vacuum. After five months, more than 50% of the radicals were still present in the sample [52].

Heating of the samples caused a partial decay of the radical concentration which was accompanied by additional polymerization [39]. After 1 h at 80 °C, about 65% of the radicals had survived and after an additional heat treatment of 1 h at 120 °C still about 15% of the radicals could be detected [52].

The deuteration experiment shows that hydrogen abstraction does occur but the long lifetime of the radicals indicates that transfer of the radical sites by subsequent abstraction of tertiary H-atoms (which is a zero-enthalpy process) must be rather restricted. Otherwise the radicals would terminate each other rapidly. The transport of radical sites at room temperature is obviously limited to distances shorter than the average distance between trapped radicals (10–14 nm at a radical concentration of $(3 \pm 1.5) * 10^{-3}$ M).

Heating increases the local mobility of trapped radicals and of adjacent abstractable H-atoms, thereby increasing the accessible volume of a given radical site. Overlapping of accessible volumes of different radicals may lead to termination of these radicals. This picture, which is similar to that developed by Bondarev [221], explains why at higher temperatures new but lower plateaus of the radical concentration are generally observed [222] instead of second order decay kinetics with a temperature-dependent rate constant. The latter process occurs, for example, in polyester methacrylates in which the vinyl groups are separated by a long flexible bridge which allows for a high internal mobility [223].

Oxygen reacts rapidly with trapped radicals, but no peroxy radicals could be detected. Instead, the ESR signal gradually decayed to zero in about 2 h. At intermediate stages the spectrum was unchanged, only its amplitude was reduced.

Presumably, a chain oxidation process occurs, similar to the one established by

O$_2$-induced mobility of radical site

$$ROO \cdot + H\overset{|}{\underset{|}{C}}- \longrightarrow ROOH + \cdot \overset{|}{\underset{|}{C}}-$$

$$O_2 + \cdot \overset{|}{\underset{|}{C}}- \longrightarrow \cdot OO\overset{|}{\underset{|}{C}}-$$

$$-\overset{|}{\underset{|}{C}}H + \cdot OO\overset{|}{\underset{|}{C}}- \longrightarrow -\overset{|}{\underset{|}{C}} \cdot + HOO\overset{|}{\underset{|}{C}}-$$

etc.

Bresler and Kazbekov for the decay of mechanically produced radicals in PMMA [53]. If such a mechanism is operative, the sample should finally contain more hydroperoxide molecules than there were radicals to start with. The results shown in Table 7 [52] clearly show that several molecules of oxygen are required for the removal of each pair of radicals.

On the other hand, it is unlikely that oxidative transport is the only way of displacement of radical sites, since at an initial radical concentration of $3 * 10^{-3}$ M and a bulk concentration of acrylate moieties of 9 M, one out of 3000 acrylate moieties must possess a radical function.

**Table 7.** Molar concentrations of radicals and of hydroperoxides in photopolymerized HDDA

| $[R \cdot]^{a)}$ | $[ROOH]^{b)}$ | $[ROOH]$ after 1 h 175 °C$^{c)}$ |
|---|---|---|
| $3*10^{-3}$ | $2.1*10^{-2}$ | $5*10^{-3}$ |
| $3*10^{-4}$ | $1.9*10^{-2}$ | |
| blank$^{d)}$ | $1.5*10^{-4}$ | $5*10^{-4}$ |

a) Estimated accuracy $\pm 50\%$
b) Determined after irradiation and aeration; estimated accuracy $\pm 20\%$
c) Determined after irradiation, heating and aeration; estimated accuracy $\pm 20\%$
d) Blank values refer to unpolymerized monomer with initiator (0.25% HMPP); after heating the sample was polymerized

If the radicals are homogeneously distributed and if the chain oxidation occurs in a random way each termination step would require 3000 molecules of oxygen, or 1500 per radical. (More detailed calculation, which takes into account the removal of potential radical sites by the oxidation process, even yields 3000 molecules of oxygen per radical.)

This is not observed, however; so it seems likely that an additional, but spatially restricted mechanism for the transport of radical sites is available.

It might be speculated that this is hydrogen transfer, but that this process stops at sterically unfavourable positions. Oxygen may then be thought of as assisting the crossing of otherwise insurmountable barriers.

Heating may play a similar role: by increasing the temperature, a larger fraction of the barriers to radical site migration can be surpassed and hence the accessible volume of a given radical site and probability for its termination will increase.

The effect of the accumulation of hydroperoxides in the polymer certainly needs further investigation since the hydroperoxides are thermally and photochemically unstable. Upon dissociation, they may restart the chain oxidation. At present, it is not yet clear whether this has a measurable influence on the mechanical properties of the photopolymer; the oxidation process may cause chain scission as well as further crosslinking.

### 7.3.3 Inhomogeneity

DSC as well as ESR measusrements on HDDA and HEBDM samples have shown that trapping of radicals not only occurs in the end of the reaction but also in its beginning, in the presence of a considerable amount of unreacted monomer [39, 224]. Samples, polymerized in the DSC apparatus up to 15–80% double bond conversion, all showed additional polymerization (postpolymerization) when heated in the dark [26, 39].

Samples, in which the concentration of long-living polyacrylate radicals was just above the detection limit of ESR, still contained 25–30% of unreacted monomer [224]. This is interpreted as being caused by inhomogeneity. Pre-gel inhomogeneity is well-documented in the bulk copolymerization of mono and divinyl compounds but it has long been assumed that some re-homogenization occurs after gelation [11]. This appears not to be the case with HDDA, HEBDM and TEGDA up to the conversions reported in Table 6 for 20 °C.

The formation of inhomogeneous networks appears to be characteristic of chain processes. The existence of regions of widely differing mobilities has recently also been demonstrated by Meijer and Zwiers [225] for the light-induced cationic polymerization of di-epoxides.

Persistent inhomogeneity might affect the optical properties of photopolymers. However, until now this has not been observed. Presumably the size of the inhomogeneities is well below the wavelength of the light. In the examples, described in Sections 3–5 the wavelength of the light varied from 0.78 to 1.3 μm. At shorter wavelengths, scattering might become an important parameter.

## 7.4 Some Kinetic Features

Proper knowledge of the kinetics of the main process and its side reactions is required for process optimization.

As has been explained in Sect. 2.2, a complete description of the kinetics of the bulk photopolymerization of diacrylates cannot be given at the present time, however. In this section, we will describe a few typical side reactions during the bulk polymerization of diacrylates as well as a preliminary approach for the treatment of the kinetics during vitrification of the system.

### 7.4.1 Oxygen assisted Chain Transfer

In Sect. 7.3.2, it was shown that trapped radicals decay via a chain process when the sample is exposed to air.

Such a chain process also occurs when the polymerization is carried out in an atmosphere which contains oxygen. The chain reaction not only causes enhanced oxidation of the polymer, it also induces enhanced mobility of the radical sites in the network. This shows up as an enhanced termination which reduces the overall rate of polymerization.

This effect is shown in Fig. 19 [48]. It is most pronounced at the lowest light intensities where the oxidation can compete effectively with the polymerization process. At the highest intensities, the rate of initiation is so high that the oxygen cannot be replenished at a sufficient rate. The influence of oxygen on the rate of polymerization first increases with conversion. This indicates that termination by direct combination is gradually replaced by termination via oxygen induced hydrogen transfer.

This change in mechnism is caused by the strong reduction of the mobility of the macroradicals themselves.

Towards the end of the reaction, the rate reduction by oxygen diminishes again, presumably since the replenishment of oxygen is progressively suppressed by the vitrification of the sample.

The presence of oxygen also influences the apparent order of the rate with respect to light intensity. At an oxygen concentration of 2 ppm, an apparent order of 0.7 was found for HDDA at a double bond conversion of 0.3. This value was constant over 4 decades of light intensity variation. The deviation from the classical value of

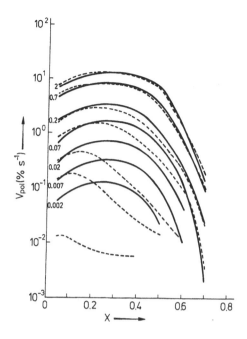

**Fig. 19.** Rate versus extent of polymerization of HDDA, recorded at various light intensities. Initiator: 0.25 wt.-% DMPA. Continuous curves: sample compartment flushed with nitrogen containing <2 ppm of oxygen. Dashed curves: with 58 ppm of oxygen. Intensities are in mW · cm⁻². (From Ref. [48], with permission)

0.5 is indicative for a competing (pseudo) first order termination process, e.g. trapping. However, at an oxygen concentration of 58 ppm the order showed a continuous decrease from 2 at a light intensity of 20 $\mu$W · cm⁻² to 0.7 at 2 mW · cm⁻². This again shows a strong influence of oxygen on the kinetics of network formation, presumably through chain transfer reactions [48].

The occurrence of chain oxidation processes during photopolymerization of multifunctional acrylates has been earlier proposed by Decker et al. [226, 227]. For liquid solutions, the apparent order with respect to light intensity was 0.5 [227], i.e. the unperturbed value, whereas for bulk systems the value of 0.85 was found [228].

Decker and Jenkins have also estimated the kinetic chain length of the peroxidation process for an epoxy acrylate. From the measured inhibition time and estimated rate of free-radical production, values of 2–8 were inferred. These did not change considerably although the intensity was varied over a large range, from 1.5 to 800 $\mu$E · cm⁻² · s⁻¹ [226].

### 7.4.2 Direct Chain Transfer. Deuterium Isotope Effect on the Rate of Photopolymerization

In the formation of linear chains, transfer to the polymer has a strong influence on the molecular weight distribution but its influence on the kinetics is rather restricted since usually a reactive radical is regenerated in the transfer step [4]. Crosslinking polymerization is different since termination by direct combination is strongly suppressed: after the (early) gelation most of the radicals will be hooked-up to the network. (This explains the large difference in polymerization rates between n-propyl acrylate and HDDA, shown in Fig. 1). Upon vitrification, the rate of propagation becomes diffusion-limited and the rate of polymerization is mainly sustained by the almost complete suppression of termination. If chain transfer occurs, this will maintain a certain mobility of the radical sites which is otherwise destroyed by their attachment to the network. Chain transfer will, therefore, assist in termination and thereby in a reduction of the rate of polymerization.

In HDDA, transfer is likely to occur through the abstraction of tertiary hydrogens from the polymer chain. The occurrence of this process during trapping has been shown in Sect. 7.3.1.

The abstraction of tertiary hydrogens may be selectively suppressed by deuterating HDDA at both of its α-positions since deuterium requires a higher activation energy for abstraction than hydrogen [229].

The importance of the transfer process is illustrated in Fig. 20 where the rates of polymerization of the deuterated and normal compounds are compared. The deuterated compound not only reacts considerably faster, but most important is that the kinetic isotope effect increases from about 1.4 in the beginning up to a factor of 5 upon overall vitrification [48].

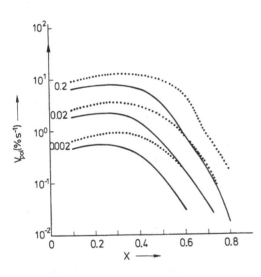

**Fig. 20.** Rate versus extent of polymerization for HDDA, recorded at various light intensities. Initiator: 4 wt.-% HMPP. <2 ppm of oxygen. Continuous curves: 1,6-hexanediol diacrylate. Dotted curves: α,α′-dideutero-1,6-hexanediol diacrylate. Intensities are in mW · cm$^{-2}$. (From Ref. [48], with permission)

This is completely in line with the findings given in the previous section, where it was concluded that termination through transfer gradually replaces termination by direct recombination.

Alternative interpretations based on secondary isotope effects on the propagation or termination rate constant would not only have to rationalize the large magnitude of the effect but also to provide an explanation for the observed increase of the isotope effect with conversion.

## 7.4.3 Kinetics of a Self-Decelerating Polymerization

The presence of long-living radicals in the polymer network will cause the reaction to continue in the dark. In HDDA, the radicals persist for several months after irradiation if the sample is kept at room temperature and under vacuum [39, 52], even

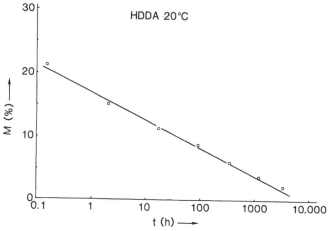

**Fig. 21.** Decay of extractable monomer [M] as a function of time. HDDA, photopolymerized in an ESR tube for 90 s at 20 °C and stored under vacuum at room temperature. Initiator: 0.25 wt.-% HMPP. Intensity: 0.2 mW · cm$^{-2}$. (From Ref. [212], with permission)

in the presence of a large amount of unreacted monomer. This was ascribed to the inhomogeneity discussed in Sect. 7.3.3.

If we allow the samples to stand before extraction and determination of free monomer, we observe virtually no decrease of the radical concentration but the concentration of free monomer [M] decays in the dark in a self-decelerating process (Fig. 21) [212].

If ordinary pseudo first-order kinetics were obeyed, log [M] vs. t should be linear; here, however, the result is just opposite; [M] vs. log t is almost linear over more than 4 decades of time. The first half-time is about 20 h and the next one 300 h. Physically, this would mean that the monomer molecules are distributed among traps and that the molecules from the shallowest traps escape and diffuse to a reactive radical site more quickly than the other ones. Further reaction will deepen the traps of the remaining molecules, so that the relaxation time of the system increases continuously.

Many relaxation phenomena in polymeric glasses can be described in an empirical way using the Williams-Watts relaxation function $\varphi(t)$ [230] in which a parameter $\beta$ accounts for the width of the relaxation spectrum:

$$\varphi(t) = \exp -(t/\tau)^\beta$$

$$0 < \beta \leqq 1$$

Stress relaxation [231], creep recovery [213], dielectric relaxation [230] and volume recovery [232] data have all been fitted with such a function. This purely empirical equation has also been connected with several kinds of models used for the description of polymeric glasses.

Models used include diffusion [233], molecular [234] and stochastic kinetics [233]. However, in view of the limited ·data and restricted knowledge of the combined but

certainly not cooperative processes of physical aging and chemical conversion we
have so far only attempted to fit our data with the empirical equation by
choosing $[M]_t/[M]_0$ as the relaxation function:

$$[M]_t/[M]_0 = \exp -(t/\tau)^\beta$$

$$\ln \ln [M]_0/[M]_t = \beta \ln t - \beta \ln \tau$$

The data of Fig. 21 yield a straight line when plotted according to the Williams-Watts
equation [212]. The slope $\beta$ is found to be 0.25 and the intercept $\tau$ has a value of 125 h.
This fit is not surprising since so many phenomena have been fitted with this empirical
equation [235].

However, zero, first or second order plots did not yield straight lines. From this
analysis it, therefore, appears that the chemical kinetics of monomer consumption
in a vitrifying system is largely determined by the physical process of decay of free
volume. A complicating factor is, however, that free volume is not only destroyed by
overall shrinkage but is also generated by the chemical reaction itself (Fig. 15b).
For comparison, it may be noted that Struik reported that $\beta$ has a value of about 1/3
for creep recovery in many different materials [213] and that similar values have been
observed for many kinds of relaxation in a large variety of vitreous materials, including metals and low-molecular weight glasses [235]. According to Hodge and Berens,
typical values for $\beta$ are in the range of 0.4 to 0.6 [232]. Our rather low value of
0.25 points to a very broad distribution of relaxation times.

The extreme width of the transitions observed in the dynamic mechanical thermal
analysis of densely crosslinked networks (Fig. 16) as compared with linear polymers
also points in this direction.

Both phenomena are in accord with an inhomogeneous structure of the network
although neither of them provides an independent proof for the existence of
inhomogeneity.

## 7.5 Some other Studies which Relate Polymer Properties to Photocuring Conditions

As has already been outlined in Sect. 7.2, the mechanical properties of a cured network
in general depend on the curing conditions, especially in cases where vitrification occurs. Other properties are likely to depend also on the extent of reaction.

Since there are many parameters which affect the curing process a multi-
technique approach is often necessary. In this section, we will report a selected
number of studies on photopolymerization processes in which the polymer properties
have been correlated with the curing conditions in several ways.

A rather comprehensive study on the relation between curing conditions and
product properties of epoxy acrylates, diluted with a reactive diluent and a vinyl-
terminated acrylonitrile/butadiene rubber has been carried out by Small et al. [236].
The systems studied served as model mixtures for dielectric coatings for printed
circuits. The curing behavior was monitored by FTIR spectroscopy, also in combination with photoacoustic spectroscopy. These methods provided information on the

initial rates of polymerization, the influence of the composition of the monomer mixture and of the initiator and oxygen concentrations on the rate as well as on the difference between the rates at the surface and in the bulk.

Thermal gravimetric analysis and solvent extraction were used to study the change in curing behavior upon changing the light intensity or the type and concentration of the crosslinking agent.

The rate of solvent absorption was used to estimate the crosslink density.

It was concluded that at high light intensity the systems are much less sensitive to oxygen, viscosity and initiator screening effects than at low intensities.

However, no account was taken of the possibility of heating of the samples at the high intensities used (up to 75 mW · cm$^{-2}$). Other measurements have shown that, due to the high rate and exothermicity of the reaction, temperature rises exceeding 100 °C may already occur at intensities as low as 2 mW · cm$^{-2}$ [48]. This will of course reduce the viscosity and affect the rate of polymerization to a considerable extent.

Systems which contained about 10% of rubber did not show phase separation. A parallel study of epoxides which were photopolymerized cationically in the presence of various amounts of epoxy-terminated rubbers, showed a range of morphologies from a homogeneous (no domains larger than 0.1 µm observed) to a very inhomogeneous structure. The rubber particle size increased with rubber content due to a reduction in rate of polymerization, thereby facilitating phase separation during photopolymerization. Post-baking also increased the domain sizes, more strongly than increase of irradiation time. Thermomechanical curves, measured with a torsional pendulum, were also reported for the epoxides [237].

Levy and Massey studied the effect of composition of mixtures on the mechanical properties of the cured polymers [238]. Their mixtures contained a urethane diacrylate, 2-ethoxy-ethoxy ethyl acrylate (EEEA) and/or N-vinyl pyrrolidone (NVP). The EEEA was found to react not only with radicals but also with the photoinitiator (DMPA). It also formed a significant amount of homopolymer, thereby reducing the modulus, tensile strength and elongation at break. These effects were ascribed to the low rate of polymerization of EEEA as compared with the diacrylate.

Priola et al. have studied the mechanical properties of mixtures of epoxy acrylates, copolymerized with various reactive diluents [239–241]. Rates and conversion were measured with infrared spectroscopy and the influence of variation of molecular structure and functionality on flexibility, T (tan $\delta_{max}$), elastic modulus, abrasion resistance, hardness, adhesion, impact resistance, resistance to chemicals and wheathering behavior was studied. The influence of chain transfer on network structure was also investigated albeit this effect could not be separated from the plasticizing effect of the transfer agent itself since large amounts (5–15%) were added.

The use of different light intensities and gas atmospheres with samples prepared for infrared and mechanical measurements, respectively, makes it difficult to correlate mechanical properties to conversion in a precise way. Here too, no control of temperature was applied during photopolymerization. This probably explains the very high values of T (tan $\delta_{max}$) of up to 146 °C obtained after polymerizations started at room temperature [240].

Rheological measurements during UV curing of epoxy acrylates have been reported by Otsubo et al. who measured dynamic viscosities using an oscillating plate rheometer [242].

The influence of the order of two different curing steps, UV induced radical polymerization of dimethacrylates and thermally induced homopolymerization of di-epoxides, on the mechanical properties of networks has been studied by Suzuki et al. [113]. Three kinds of interpenetrating networks were obtained:

(i)   normal IPN's; photopolymerization followed by thermal polymerization;
(ii)  inverse IPN's; thermal polymerization of a mixture in which the dimethacrylate had been substituted by a non-reactive di-isobutyrate, extraction of the latter from this pre-swollen network, swelling of the epoxy network with the dimethacrylate and photopolymerization;
(iii) simultaneous IPN's, simultaneous thermal polymerization of both oligomers.

In all cases, the modulus showed a minimum when plotted vs. mixture composition which was most pronounced for the normal IPN. With the normal IPN, the plot resembled a superposition of the linear plots obtained with the pre-swollen networks. The hypothesis was advanced that the first network is topologically close to a pre-swollen network and that the second one may have a constrained structure but is at least partially similar to a pre-swollen network.

This hypothesis explains the minimum but not the large quantitative difference between normal IPN's on one hand and simultaneous and inverse IPN's on the other hand.

It may well be that the temperature of formation of the first network is also of much importance, the normal IPN's showed considerably higher $T(\tan \delta_{max})$ values than the inverse and simultaneous IPN's. There was also some indication for inhomogeneity of the normal IPN's, not for the other systems.

Although the precise causes are not yet clear it is beyond doubt that the properties of these networks depend strongly on the order of preparation.

# 8 Simulation of Chain Crosslinking Polymerization with a Percolation Model

The classical model for network formation has been proposed by Flory [7] and Stockmayer [57]. This combinatorial model provides an adequate description for the formation of networks via step reactions and for crosslinking of linear chains. Basic assumptions are:

(i)  equal and independent reactivity of chemical groups of the same type,
(ii) no ring formation.

This model has been extended by Gordon et al. [58] by taking into account unequal reactivity and substitution effects within monomer units and by a mean field treatment of very restricted cyclization. However, all these models are limited to systems which can be described by Markovian statistics without any long range correlations affecting the apparent reactivity of a functional group.

For the chain crosslinking (co)polymerization these assumptions are not fulfilled. Here, the strong cyclization or intramolecular crosslinking causes formation of inhomogeneities consisting of densely crosslinked regions surrounded by less densely crosslinked regions. In the vicinity of the reactive end of a growing chain, the concentration of pendent groups is much higher than average. This manifests itself as an apparently enhanced chemical reactivity of the pendent group with respect to the free monomer. Therefore, the apparent reactivity ratio of pendent groups with respect to the groups in the free monomer changes with conversion. These complications cannot be accounted for adequately by any presently existing analytical model.

There exist several attempts to describe the chain crosslinking (co)polymerization using an analytical mean field theory. Thus, Gordon and Malcolm [58a)] treated chain crosslinking copolymerization of a monovinyl and a divinyl monomer with equal and independent reactivity of vinyl groups and for the ring-free case only.

Dušek and Ilavský have presented a mean-field model which can describe the amount of cyclization reasonably well when a cyclization parameter is properly adjusted [243a)]. However, this model requires a very high chain flexibility [11)] and it does not treat shielding of pendent groups and trapping of radicals.

A related version of this model has been tested on the initial stage of the co-polymerization of styrene with ethylene glycol dimethacrylate by Dušek and Spěváček [243b)]. The copolymerization could be well described. However, it was also concluded that, with the exception of very low contents of the divinyl component, the usual assumptions allowing a theoretical treatment (tree-like models and spanning tree approximation) are not fulfilled. So far, testing has been limited to the pre-gel region.

Simulation models can deal with strong cyclization, high divinyl contents and conversions beyond gelation since local concentrations are easily introduced [243)].

The formation of densely crosslinked networks obtained by chain polymerization has been modeled by Boots [44-46)], by adaptation of programs developed by Manne-ville and De Sèze [243)] and by Herrmann et al. [244, 245)].

In this percolation model, the monomers are considered to be points on a cubic lattice. Then an arbitrary site is changed into a free-radical. Every unit of time, a neigbor is chosen. If this neighbor has not fully reacted, a chemical bond is formed and the radical function is transferred to this neighboring site. Thus, the radical performs a random walk on the lattice, thereby connecting a series of monomer units by chemical bonds. In the case of polymerization of tetrafunctional monomers such as divinyl compounds, each site may be visited at most twice. The walk continues until the radical is trapped between fully reacted sites. Then a new radical is generated at an arbitrarily chosen free monomer site, etc. This procedure simulates the lowest possible initiation rate and termination only occurs through trapping. Termination through combination can be introduced by increasing the radical concentration.

Step reactions can be modeled in a similar way by randomly forming bonds between neighboring sites which have not yet fully reacted.

The results can be visualized by making snapshots of simulations on a two-dimensional lattice.

Figure 22a shows such a snapshot obtained at 25 % bond conversion of a divinyl monomer. The formation of inhomogeneous structures is obvious, as is the presence of trapped radicals. Upon reduction of the maximum kinetic chain length by introducing an upper limit for the latter, the inhomogeneity becomes smaller and smaller (Fig. 22b and c).

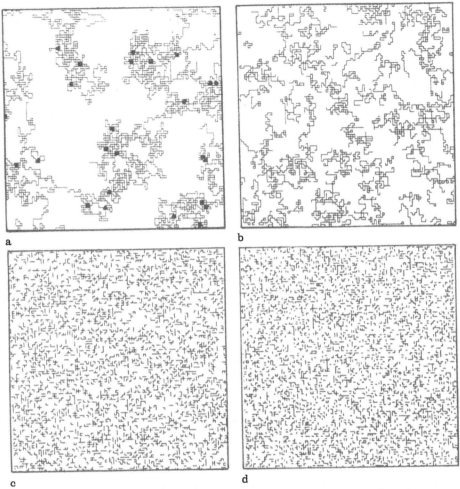

a

b

c                                                d

**Fig. 22a–d.** Snapshots of polymerization of tetrafunctional monomer in two dimensions on a $100 \times 100$ lattice at a bond conversion of 25%. Bonds between units are indicated, units themselves are not. (**a**) is obtained by a chain reaction of one 'living' radical at a time; when a radical is trapped (indicated by crossed squares), a new radical starts. The living radical is indicated by X. To show the decrease of inhomogeneity on decreasing the kinetic chain length, the maximum kinetic chain length is limited to 20 steps in (**b**) and to 2 steps in (**c**), respectively, (here radicals are not indicated). (**c**) is as homogeneous as snapshot (**d**) for a step polymerization of a tetrafunctional monomer at a reactive group conversion of 25%. (From Ref. [46], with permission)

The snapshot of Fig. 22c, for chain lengths of at most two steps, is as homogeneous to the eye as the snapshot of a step reaction of a tetrafunctional monomer shown in Fig. 22d.

The trapping of radicals is also predicted by the simulation in three dimensions.

The fraction of fully reacted sites in the polymer is plotted versus monomer conversion in Fig. 23, together with experimental results [44] and values predicted on the assumption of equal reactivity (mean field assumption). Contrary to the latter, the percolation model nicely predicts the sudden increase in crosslink density at low con-

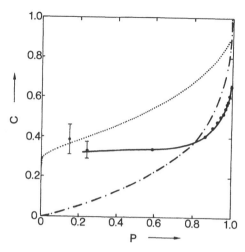

**Fig. 23.** The fraction of fully reacted monomer units in the polymer (C) as a function of monomer conversion P. Experimental points on continuous curve are for HDDA [44]. Dashed-dotted line results from mean-field assumption (equal reactivity of pendent and free double bonds, cf Ref. [38]). Dotted line results from the percolation model for the polymerization of a pure divinyl compound in three dimensions

versions. At high conversions, it underestimates the shielding of pendent double bonds. This discrepancy is tentatively explained on the basis of the mobility of polymer units over distances of the order of the reaction distance. Reorientations and small displacements, required for the reaction to occur, are not included in the model, which treats the monomer units as points. This is a good approximation at low conversion where local motions are extremely rapid but does not apply to pendent bonds in a vitrifying system. Under these conditions, free monomer molecules are probably more reactive than pendent bonds. Although there certainly exist limitations caused by the tremendous simplification of the polymerization process, the percolation model correctly shows the formation of inhomogeneities and their dependence on initiation rate, the trapping of radicals and the initially enhanced reactivity of the pendent double bonds, albeit only in a qualitative way.

A logical extension of the percolation model is the representation of tetra-functional monomer units on the lattice as randomly distributed pairs of connected bifunctional sites. These may or may not be neighbors. This will allow the inclusion of some aspects of the molecular structure of the monomer in the model, namely the length and flexibility of the moiety between the reactive groups.

There is a difference between the present percolation model and the one originally proposed by Herrmann, Stauffer and Landau [244, 245]. The latter authors treated the initiation as a discontinuous, stepwise process: all of the reactive sites were generated right at the beginning of the process. Boots used continuous initiation [44–46].

Recently it has been pointed out that the former method produces a peculiar effect, namely a strong oscillation of the cluster-size distribution as a function of conversion. This is caused by the fact that initially all clusters are of the same size. Merging then causes abrupt changes which are absent when continuous initiation is used [245a].

# 9 Conclusions and Expectations

We have shown that photopolymerization is a versatile technique which can be used for many different applications in electronics.

Applications range from optical fiber coatings to waveguides and from dielectric coatings to optical discs.

Main reasons for the steady increase of applications are the easy external control of the process combined with an almost infinite pot life, the high speed and the wide variation in properties that can be obtained by modification of the chemical structure of the monomer without sacrificing the high rate.

In most of the applications densely crosslinked networks are desired. These are easy to make but difficult to characterize. The investigation of the relation between monomer structure, polymerization conditions and product properties will demand a continuing effort. At present we are still far from being able to make accurate predictions, although considerable progress has been made.

Specific areas which are likely to become important are:

(i)   the use of oriented monomers; polymerization in an oriented state may lead to products with anisotropic properties. Experiments with monomers oriented in an elongational flow during extrusion have yielded filaments with highly anisotropic linear thermal expansion coefficients:
$2 \cdot 10^{-5} \text{ K}^{-1}$ in the parallel direction, $8 \cdot 10^{-5} \text{ K}^{-1}$ in the perpendicular direction and an isotropic value of $7 \cdot 10^{-5} \text{ K}^{-1}$ [246]. Modulus and strength also increased.

(ii)  the use of liquid crystalline monomers; polymerization and optionally crosslinking in a liquid crystalline monomer will extend the magnitude and the temperature region of orientation [247].

(iii) the use of dual cure and/or hybrid systems (cf. Sect. 6.5). Localized, laser-induced curing of an adhesive may be of considerable help with the automated mounting of small parts in complicated production processes.

(iv)  the use of cationic curing of epoxides by high-intensity light sources such that the photopolymerization will occur at elevated temperature (cf. Sect. 6.5). This procedure improves the thermal properties of the product. The use of vinyl ethers may also expand the range of fast curing monomers.

(v)   the use of cationic photoinitiators which do not leave considerable amounts of strong acid in the polymer. This modification will improve the dielectric properties and reduce corrosion (cf. Sect. 6.5).

(vi)  the use of low-shrinkage monomers [248]. The application of these systems will facilitate the replication of complicated structures such as screens and large aspherical lenses for projection TV as well as high-precision mounting of components in automated production processes.

# 10 Acknowledgement

The helpful discussions with H. m. J. Boots and many other colleagues and the experimental support from (chronologically): H. J. L. Bressers, H. P. M. van Genuchten, L. A. J. Swerts-Willekens, G. M. M. van de Hei and G. F. C. M. Lijten are gratefully acknowledged.

# 11 References

1. Photochemistry (1970–1986). A Specialist Periodical Report. Vols. *1–16*, Bryce-Smith D (Senior reporter) The Royal Soc. of Chem., London
2. Phillips, D., in Ref. 1, *4*, 870 (1973)
3. Vollenbroek FA, Spiertz EJ (1987) Adv. Polym. Sci. *84*
4. Odian G (1981) Principles of Polymerization. 2nd Ed. Wiley-Interscience, New York
5. Shultz AR (1984) J. Polym. Sci., Polym. Phys. Ed. *22*:1753
6. Lissi EA, Zanocco A (1983) J. Polym. Sci. Polym. Chem. Ed. *21*:2197
7. Flory PJ (1953) Principles of Polymer Chemistry. Cornell Univ. Press, Ithaca, New York
8. Hiemenz P (1984) Polymer Chemistry. The Basic Concepts. Marcel Dekker, New York
9. North AM (1966) The Kinetics of Free Radical Polymerization. Pergamon, Oxford
10. Eastmond GC (1976) in: Bamford CH, Tipper CFH (eds) Comprehensive Chemical Kinetics. Vol. 14A, Elsevier, Amsterdam
11. Dušek K (1982) in: Haward RN (ed) Developments in Polymerisation-3. Ch. 4. Appl. Science, London
12. Dušek K, Galina H, Mikeš J. (1980) Polym. Bull. *3*:19
13. Benson SW, North AM (1959) J. Amer. Chem. Soc. *81*:1339; (1962) *84*:935
14. O'Driscoll KF (1981) Pure Appl. Chem. *53*:617
15. Tulig TJ, Tirrell (1981) Macromolecules *14*:1501
16. Ito K (1980) Polymer J. *12*:499
17. Breitenbach JW, Fally A (1951) Mona n. Chem. *82*:1118
18. Burnett GM, Duncan GL (1961) Makromol. Chem. *51*:177
19. Mangaraj D, Patra SK (1967) ibid. *104*:125
20. Horie K, Hiura H, Savada M, Mita I, Kambe H (1970) J. Polym. Sci., Polym. Chem. Ed. *8*:1357
21. Sourour S, Kamal MR (1976) Thermochim. Acta *14*:41
22. Loshaek S, Fox TG (1953) J. Amer. Chem. Soc. *75*:3544
23. Moore JE (1978) in: Pappas SP (ed) UV Curing: Science and Technology. Chap. 5, Technol. Market. Publ., Stamford (Conn.)
24. Kloosterboer JG, Lippits GJM (1984) J. Rad. Cur. *11*(1):10
25. Enns JB, Gillham JK (1983) J. Appl. Polym. Sci. *28*:2567
26. Kloosterboer JG, Van de Hei GMM, Gossink RG, Dortant GCM (1984) Polym. Commun. *25*:322
27. Schick JP (1978) Plastica *31*(1):4
28. Schroeter SH (1973) in: Gardon JL, Prane JW (eds) Nonpolluting Coatings and Coating Processes. Plenum Press, New York, p. 109
29. Schulz GV, Harborth G (1947) Makromol. Chem. *1*:106
30. Hayden P, Melville HW (1960) J. Polym. Sci. *43*:201 (See ref. 10, p. 77 for a numerical correction)
31. Burnett GM, Duncan GL (1961) Makromol. Chem. *51*:177
32. Horie K, Mita I, Kambe H (1968) J. Polym. Sci., Part *A1*, *6*: 2663
33. Balke ST, Hamielec AE (1973) J. Appl. Polym. Sci. *17*:905
34. Panke D, Stickler M, Wunderlich W (1983) Makromol. Chem. *184*:175
35. Stickler M (1983) ibid. *184*:2563
36. Stickler M, Panke D, Hamielec AE (1984) J. Polym. Sci. Polym. Chem. Ed. *22*:2243
37. Aso C (1959) J. Polym. Sci. *39*:475
38. Malinský J, Klaban J, Dušek K (1971) J. Macromol. Sci.-Chem. *A5*:1071
39. Kloosterboer JG, Van de Hei GMM, Lijten GFCM (1986) in: Kleintjens LA, Lemstra PJ (eds) Integration of Fundamental Polymer Science and Technology. Elsevier Appl. Sci., London, p. 198
40. Dušek K, Prins W (1969) Adv. Polym. Sci. *6*:1
41. Funke W, Beer W, Seitz U (1975) Progr. Colloid Polym. Sci. *57*:48
42. Funke W (1983) Plast. Rubb. Proc. Appl. *3*:243
43. Wheaton RM, Lefevre LJ (1981) in: Kirk-Othmer Encyclopedia of Chemical Technology. 3rd. Ed. *13*:678, Wiley, New York
44. Kloosterboer JG, Van de Hei GMM, Boots HM (1984) Polym. Commun. *25*:354

45. Boots HMJ, Kloosterboer JG, Van de Hei GMM, Pandey RB (1985) Brit. Polym. J. *17*:219
46. Boots HMJ (1986) in: Kleintjens LJ, Lemstra PJ (eds) Integration of Fundamental Polymer Science and Technology. Elsevier Appl. Sci., London, p. 204
47. Kloosterboer JG, Bressers HJL (1980) Polym. Bull. *2*:205
48. Kloosterboer JG, Lijten GFCM (1987) Polym. Commun. *28*:2
49. Ref. 7, p 128
50. Atherton NM, Melville H, Whiffen DH (1959) J. Polym. Sci. *34*:199
51. Zimbrick J, Hoecker F, Kevan L (1968) J. Phys. Chem. *72*:3277
52. Kloosterboer JG, Lijten GFCM, Greidanus FJAM (1986) Polym. Commun. *27*:268
53. Bresler SE, Kazbekov EN (1964) Adv. Polym. Sci. *3*: 688
54. Carlsson DJ, Wiles DM (1976) J. Macromol. Sci.-Rev. Macromol. Chem. *C14*:65
55. Garton A, Carlsson DJ, Wiles DM (1980) in: Allen NS (ed) Developments in Polymer Photochemistry. Vol. 1, Chap. 4. Elsevier Appl. Sci., London
56. Billingham NC, Calvert PD (1983) in: The Degradation and Stabilisation of Polyolefins. Chapters 1 and 5. Applied Science, London
57. Stockmayer WH (1943) J. Chem. Phys. *11*:45
58. Gordon M, Scantlebury GRJ (1967) J. Chem. Soc. *B* 1967:1
58a. Gordon M, Malcolm GN (1966) Proc. Roy. Soc. London *A295*:29
59. Pappas SP, McGinniss VD (1978) in: Pappas SP (ed) UV Curing: Science and Technology. Chap. 1, Technol. Market. Publ., Stamford (Conn.)
60. Pappas SP (1985) ibid., Vol. 2, Chap. 1
61. Vesley GF (1986) J. Rad. Cur. *13*(1):4
62. Hageman HJ (1983) Progr. Org. Coat. *13*:123
63. Roffey CG (1982) Photopolymerization of Surface Coatings. Wiley, Chichester
64. Green GE, Stark BP, Zahir SA (1981–82) J. Macromol. Sci.-Rev. Macromol. Chem. *C21*:188
65. Mishra MK (1982–83) ibid. *C22*:409
66. Senich GA, Florin RE (1984) ibid. *C24*:240
67. Green PN (1985) Polym. Paint Colour J., *175*: 246
68. Baumann H, Timpe HJ (1983) Z. Chem. *23*:197
69. Adam S, Güsten H, Steenken S, Schulte-Frohlinde D (1974) Liebigs Ann. 1831
70. Adam S, Güsten H, Schulte-Frohlinde D (1974) Tetrahedron *30*:4249
71. Carlblom LH, Pappas SP (1977) J. Polym. Sci. Polym. Chem. Ed. *15*:1381
72. Hageman HJ, Van der Maeden FPB, Janssen CGM (1977) Makromol. Chem. *180*:2531
73. Pappas SP, Asmus RA (1982) J. Polym. Sci. Polym. Chem. Ed. *20*:2643
74. Heine HG, Trancker HJ (1975) Progr. Org. Coat. *3*:115
75. Ogata Y, Sawaki Y (1976) J. Org. Chem. *41*:373
76. Sandner MR, Osborn CL (1974) Tetrahedron Lett. *5*:415
77. Groenenboom CJ, Hageman HJ, Overeem T, Weber AJM (1982) Makromol. Chem. *183*:291
78. Borer A, Kirchmayr R, Rist G (1978) Helv. Chim. Act. *61*:305
79. Eichler J, Herz CP, Naito I, Schnabel W (1980) J. Photochem. *12*:225
80. Decker C, Fizet M (1980) Makromol. Chem. Rapid Commun. *1*:637
81. Kuhlmann R, Schnabel W (1976) Polymer *17*:419
82. Cohen SG, Parola A, Parsons GH (1973) Chem. Rev. *73*:141
83. Sandner MR, Osborn CL, Trecker DJ (1972) J. Polym. Sci. Part A-1, Polym. Chem. *10*:3173
84. Osborn CL (1976) J. Rad. Cur. *3*(3):2
85. Hult A, Yuan YY, Rånby B (1984) Polym. Degrad. Stab. *8*:241
86. Hult A, Rånby B (1984) ibid. *9*:1
87. Gatechair LR (1985) in: Pappas SP (ed) UV Curing: Science and Technology. Vol. 2, Technol. Market. Publ., Stamford (Conn.), p 283
88. Davis MJ, Doherty J, Godfrey AA, Green PN, Young JRA, Parrish MA (1978) J. Oil Col. Chem. *61*:256
89. McGinniss VD (1979) Photogr. Sci. Eng. *23*:124
90. Kamphuis J (1985) Diss., Univ. of Utrecht, The Netherlands
91. Crivello JV, Lam JHW (1976) J. Polym. Sci. Polym. Symp. *56*:1
92. Crivello JV, Lam JHW (1977) Macromolecules, *10*:1307

93. Crivello JV (1978) in: Pappas SP (ed) UV Curing: Science and Technology. Chap. 2. Technol. Market. Publ., Stamford (Conn.)
94. Crivello JV (1981) in: Allen NS (ed) Developments in Polymer Photochemistry-2. Appl. Science, Essex, p 1
95. Crivello JV (1984) Ann. Rev. Mat. Sci., 13:173
96. Lohse F, Zweifel H (1986) Adv. Polym. Sci. 78:61
97. Hayase S, Onishi Y, Suzuki S, Wada M (1985) Nippon Kagaku Kaishi 1985 (3): 328
98. Hayase S, Onishi Y, Suzuki S, Wada M (1986) Macromolecules 19:968
99. Hayase S, Suzuki S, Wada M, Inoue Y, Mitui H (1984) J. Appl. Polym. Sci. 29:269
100. Gaube HG (1986) Proceedings "Radcure '86", Assoc. Finish. Processes of SME., Dearborn (Mich.), p 15–26
101. Gruber GW (1978) in: Pappas SP (ed) UV Curing: Science and Technology. Chapters 6 and 7. Technol. Market. Publ., Stamford (Conn.)
102. Morgan CR, Ketley AD (1980) J. Rad. Cur. 7(2):10
103. Tu RS (1985) in: Pappas SP (ed) UV Curing: Science and Technology. Vol. 2, Chap. 5. Technol. Market. Publ., Stamford (Conn.)
104. Crivello JV, Conlon DA (1983) J. Polym. Sci. Polym. Chem. Ed. 21:1785
105. Crivello JV, Lee JL, Conlon DA (1983) J. Rad. Cur. 10(1):6
106. Dougherty JA, Vara FJ, Anderson LR (1986) Proceedings "Radcure '86", Assoc. Finish. Processes of SME, Dearborn (Mich.), p 15–1
107. Ketley AD, Tsao J (1979) J. Rad. Cur. 6(2):22
108. Crivello JV, Lam JHW (1979) J. Polym. Sci. Polym. Lett. Ed. 17:759
109. Watt WR (1979) in: Bauer RS (ed) Epoxy Resin Chemistry. ACS Symposium Series 114, p 17
110. Martin B (1985) in: Pappas SP (ed) UV Curing: Science and Technology. Vol. 2, Chap. 4. Technol. Market. Publ., Stamford (Conn.)
111. Oraby W, Wash WK (1979) J. Appl. Polym. Sci. 23:3227
112. Eckberg RP (1984) Proceedings "Radcure '84", Assoc. Finish. Processes SME., Dearborn (Mich.), p 2–1
113. Suzuki Y, Fujimoto T, Tsunoda S, Shibayama K (1980) J. Macromol. Sci.-Phys. B17:787
114. Blyler LL, Aloisio CJ (1985) ACS Symp. Ser. 285:907
115. Blyler LL, Eichenbaum BR, Schonhorn H (1979) in: Chynoweth AG, Miller SE (eds) Optical Fiber Telecommunications. Chap. 10. Academic Press, New York
116. Murata H, Inagaki N (1981) IEEE J. Quantum Electron., QE-17:835
117. Nagel SR, MacChesney JB, Walker KL (1982) ibid. QE-18:459
118. Kalis D, Key PL, Kurkjian CR, Tariyal BK, Wang TT (1979) in: Chynoweth AG, Miller SE (eds) Optical Fiber Telecommunications. Chap. 12. Academic Press, New York
119. Marcuse D, Gloge D, Marcatili EAJ (1979) in: Chynoweth AG, Miller SE (eds) Optical Fiber Telecommunications. Chap. 3. Academic Press, New York
120. Gloge D (1975) Bell Syst. Tech. J. 54:245
121. Gloge D, Gardner WB (1979) in: Chynoweth AG, Miller SE (eds) Optical Fiber Telecommunications. Chap. 6. Academic Press, New York
122. Yoshizawa N, Yabuta T, Kojima N, Negishi Y (1981) Appl. Opt. 20:3146
123. Lawson K, Cutler OR (1982) J. Rad. Cur. 9(2):4
124. Mabrey DW, Surber S (1984) ibid. 11(1):2
125. Schonhorn H, Kurkjian CR, Jaeger RE, Vazirani HN, Albarino RV, DiMarcello FV (1976) Appl. Phys. L. 29:712
126. Vazirani HN, Schonhorn H, Wang TT (1977) J. Rad. Cur. 4(4):18
127. Blyler LL, DiMarcello FV, Hart AC, Huff RG (1986) Polym. Mat. Sci. Eng. 55:536
128. Paek UC, Schroeder CM (1981) Appl. Opt. 20: 1230
129. DeSoto Inc (1979) Des Plaines (Ill.). Product Bull. 10
130. Broer DJ, Philips Res. Labs, personal communication
131. Schlef CL, Narasimham PL, Oh SM (1982) J. Rad. Cur. 9(2):11
132. Sinka JV, LieBerman RA (1983) ibid. 10(4):18
133. Hussain A (1985) ibid. 12(3):18
134. Wierenga PE, Broer DJ, Van der Linden JHM (1985) Appl. Opt. 24:960

135. Aulich HA, Douklias N, Rogler W (1983) in: Melchior H, Solberger A. (eds) ECOC83-9th Eur. Conf. Opt. Commun. Elsevier, Amsterdam, p 377
136. Kimura T, Sakaguchi S (1984) Electron. Lett. 20(8):317
137. Kimura S, Yamakawa S (1984) ibid. 20:202
138. Ohno R, Kikuchi T, Matsumura Y (1985) Int. Wire and Cable Symp. Proceed., p 76
139. Broer DJ, Mol GN (1986) J. Lightwave Techn. LT-4:938
140. Broer DJ, Mol GN (1986) in: Kleintjens LA, Lemstra PJ (eds) Integration of Fundamental Polymer Science and Technology. Elsevier Appl. Science Publ., London, p 669
141. Lutz MA, Lee C, Clark JN (1986) Polym. Mat. Sci. Eng. 55:740 (Short abstract only)
142. Dennis WE (1986) Laser Focus / Electro-Optics July, p 90
143. Compaan K, Kramer P (1973) Philips Tech. Rev. 33:178
144. Haverkorn van Rijsewijk HC, Legierse PEJ, Thomas GE (1982) ibid. 40:287; (1984) J. Rad. Cur. 11(1):2
145. Bouwhuis G (ed) (1985) Principles of Optical Disc Systems. Adam Hilger Bristol
146. Carasso MG, Peek JBH Sinjou JP (1982) Philips Tech. Rev. 40:151
147. Legierse PEJ, Schmitz JHA, Van Hoek MAF, Van Wijngaarden S (1984) Plat. Surf. Finish. 71(12):20
148. Kerfeld DJ (1981) (3M Comp.), WO81/02236
149. Bulthuis K, Carasso MG, Heemskerk MPJ, Kivits PJ, Kleuters PJ, Zalm P (1979) IEEE Spectrum 16(8):26
150. Vriens L, Jacobs B (1984) Philips Tech. Rev. 41:313
151. Gravestijn DJ, Van der Veen J (1984) ibid. 41: 325
152. Hartmann M, Jacobs BAJ, Braat JJM (1985) ibid. 42:37
153. Kloosterboer JG, Lippits GJM, Meinders HC (1982) ibid. 40:298
154. Van den Broek AJM, Legierse PEJ (1984) Photogr. Sci. Eng. 28:128
155. Legierse PEJ, Van den Broek AJM (1986) J. Rad. Cur. 13(3):4
156. Cech F (1957) Diss. Univ. Wien
157. Breitenbach JW (1957) J. Polym. Sci. 23:949
158. Kaplan RH, Rodriquez F (1975) Appl. Polym. Symp. 26:181
159. Lorenz DH, Azorlosa JL, Tu RS (1977) Radiat. Phys. Chem. 9:843
160. Tu RS (1983) J. Rad. Cur. 10(1):17
161. Dowbenko R, Friedlander C, Gruber G, Prucnal P, Wismer M (1983) Progr. Org. Coat. 11:71
162. Vollmert B (1973) Polymer Chemistry. Springer, Berlin, pp 45, 68
163. Kloosterboer JG, Van Genuchten HPM, Van de Hei GMM, Lippits GJM, Melis GP (1983) Org. Coat. Appl. Polym. Sci. Proc. 48:445
164. Braat J Ref. 145, p 7
165. Kiriki T, Izumiya N, Sakurai K, Klima T (1984) Proc. CLEO Meeting, Paper WB3, Anaheim
166. Maschmeyer RO, Andrysick CA, Geyer TW, Meissner HE, Parker CJ, Sanford LM (1983) Appl. Opt. 22:2410
167. Braat JJM, Smid A, Wijnakker MMB (1985) ibid. 24:1853
168. Zwiers RJM, Dortant GCM (1985) ibid. 24:4483
169. Zwiers RJM, Dortant GCM (1986) in: Kleintjens LA, Lemstra PJ (eds) Integration of Fundamental Polymer Science and Technology. Elsevier Appl. Science Publ., London, p 673
170. Fantone SD (1983) Appl. Opt. 22:764
171. Gijsbers TG (1980) Philips Tech. Rev. 39:229
172. Visser D, Gijsbers TG, Jorna RAM (1985) Appl. Opt. 24:1848
173. Bailey WJ, Endo T (1976) J. Polym. Sci. Polym. Chem. Ed. 14:1735
174. Gatechair LE, Wostratzky D (1983) J. Rad. Cur. 10(3):4
175. Ohngemach J, Neisius KH, Eichler J, Herz CP (1980) Merck Kontakte 3/80:15
176. Tomlinson WJ, Chandross EA (1980) Adv. Photochem. 12:201
177. Ketley AD (1982) J. Rad. Cur. 9(3):35
178. Colburn WS, Haines KA (1971) Appl. Opt. 10:1636
179. Ingwall RT, Fielding HL (1985) Opt. Eng. 24:809
180. Tomlinson WJ, Chandross EA (1976) Appl. Opt. 15:534
181. Chandross EA, Tomlinson WJ, Aumiller GD (1978) ibid. 17:566

182. Lipatov YuS, Grishchenko VK, Gudzera SG (1985) Vistn. Akad. Nauk. Ukr. RSR *1985/2*:45
183. Lechner MD (1985) Springer Ser. Sol. State Sci. *63*:301
184. Dickson LD, Sincerbox GT, Wolfheimer AD (1982) IBM J. Res. Dev. *26*:228
185. Walker P, private communication
186. Kurokawa T, Oikawa S (1977) Appl. Opt. *16*:1033
187. Kurokawa T, Takato N, Oikawa S, Okada T (1978) ibid. *17*:646
188. Kurokawa T (1982) ibid. *21*:1940
189. Franke H, Festl HG, Krätzig E (1984) Colloid Polym. Sci. *262*:213
190. Giuliani JF, Kim KH, Butler JE (1986) Appl. Phys. Lett. *48*:1311
191. Seo N (1984) Proceedings "Radcure '84", Assoc. Finish. Processes of SME., Dearborn (Mich.), p 2–27
192. Lupton EC, Simmonds DFC, Longo R (1985) Plast. Eng. *41*:59
193. Feit ED (1973) in: Platzer NAJ (ed) Polymerization Reactions and New Polymers. ACS Adv. Chem. Ser. *129*, Washington, D.C., p 269
194. Yamazaki M, Shingyochi K, Takahata N, Hitachi Cable Ltd. Jap. Pat. No. 85143518
195. Matshushita Electric Works Ltd. Jap. Pat. No. 84 74112
196. Furukawa Electric. Co. Ltd. Jap. Pat. No. 83163107
197. Sumitomo Electric Ind. Ltd. Jap. Pat. No. 83163108
198. La Perriere DM, Ors JA, Wight FR (1982) Proceed. 7th Int. Conf. Therm. Anal. Vol. *2*:1373
199. Fox NS (1984) in: Davidson T (ed) Polymers in Electronics. ACS Symposium Ser. *242*, p 367
200. Morgan CR, Kyle DR, Bush RW (1984) in: Davidson T (ed) Polymers in Electronics. ACS Symposium Ser. *242*, p. 373
201. Morgan CR, Ketley AD (1980) J. Rad. Cur. *7*(2):10
202. Morgan CR, Kyle DR (1983) ibid. *10*(4):4
203. Anderson EA, Rawls RM (1984) Proceedings "Radcure '84", Assoc. Finish. Processes SME., Dearborn (Mich.), p 11–30
204. Irving E, Stark BP (1983) Brit. Polym. J. *15*:24
205. Crivello JV (1984) in: Davidson T (ed) Polymers in Electronics. ACS Symposium Ser. *242*, p 1
206. Morgan CR (1983) J. Rad. Cur. *10*(4):8
207. Stueben KC, Patrylow MF (1982) ibid. *9*(2):16, 20 and 24
208. Stadtmüller R (1984) Adhäsion *1984*(6):22
209. Lee L-H (ed) Adhesive Chemistry. Developments and Trends. Plenum Press, New York, 1984
210. Zopf R (1982) Rad. Cur. *9*(4):10
211. Rauch-Puntigam H, Völker T (1967) Acryl- und Methacrylverbindungen. Springer, Berlin, p 205
212. Kloosterboer JG, Lijten GFCM (1987) in: Kramer O (ed) Biological and Synthetic Networks. Elsevier Appl. Science, London
213. Struik LCE (1978) Physical Aging in Amorphous Polymers and Other Materials. Elsevier, New York   .
214. Kloosterboer JG, Lijten GFCM (1987) Polymer *28*:114g
215. Geuskens G, David C (1973) Makromol. Chem. *165*:273
216. Rånby B, Rabek JF (1977) ESR Spectroscopy in Polymer Research. Springer, Berlin
217. Szöcs F, Lazár M (1969) Eur. Polym. J. (Suppl.) p 337
218. Frunze NK, Berlin AA (1969) Vysokomol. Soed. *A11*:1444; (1969) Polym. Sci. USSR *A11*: 1639
219. Liang RH, Tsay FD, Gupta A (1983) Polym. Mat. Sci. Eng. *49*:143
220. Yokota K, Hirabayashi T, Takahashi K (1980) Polym. J. *12*:177
221. Bondarev BV (1985) Chem. Phys. *97*:73
222. Yoshida H, Kodaira T, Tsuji K (1964) Bull. Chem. Soc. Japan *37*:1531
223. Korolev GV, Smirnov BR, Volkhovitinov AB (1962) Polym. Sci. USSR *4*:506
224. Kloosterboer JG, Lippits GJM (1986) J. Imaging Sci. *30*:177
225. Meijer EW, De Leeuw DM, Greidanus FJAM, Zwiers RJM (1985) Polym. Commun. *26*:45
226. Decker C, Jenkins AD (1985) Macromolecules *18*:1241
227. Fizet M, Decker C, Faure J (1985) Eur. Polym. J. *21*:427

228. Decker C, Bendaikha T (1984) Eur. Polym. J. *20*:753
229. Bell RP (1973) The Proton in Chemistry. 2nd. ed., Chap. 11. Chapman and Hall, London
230. Williams G, Watts DC (1970) Trans. Far. Soc. *66*:80
231. Knott WF, Hopkins IL, Tobolsky AV (1970) Macromolecules *4*:750
232. Hodge IM, Berens AR (1982) ibid. *15*:762
233. Simha R, Curro JG, Robertson RE (1984) Polym. Eng. Sci. *24*:1071
234. Chow TS (1984) ibid. *24*:1079
235. Struik LCE (1980) Europhys. Conf. Abstr. *A4*:135
236. Small RD, Ors JA, Royce BSH (1984) in: Davidson T (ed) Polymers in Electronics. ACS Symposium Ser. *242*, p 325
237. Ors JA, Enns JB (1984) in: Davidson T (ed) Polymers in Electronics. ACS Symposium Ser. *242*, p 345
238. Levy N, Massey PE (1981) Polym. Eng. Sci. *21*:406
239. Priola A, Renzi F (1985) J. Mat. Sci. *20*:2889
240. Priola A, Renzi F, Cesca S (1983) J. Coat. Technol. *55*:63
241. Giuliani G, Priola A (1982) *23*:761
242. Otsubo Y, Amari T, Watanabe K (1984) J. Appl. Polym. Sci. *29*:4071
243. Manneville P, de Sèze L (1981) in: della Dora I, Demongeot J, Lacolle B (eds) Numerical Methods in the Study of Critical Phenomena. Springer, Berlin
243a. Dušek K, Ilavský M (1976) J. Polym. Sci., Polym. Symp. Ed. *53*, 57 and *53*, 75
243b. Dušek K, Spěváček J (1980) Polymer *21*: 750
244. Herrmann HJ, Stauffer D, Landau DP (1983) J. Phys. *A16*:1221
245. Stauffer D, Coniglio A, Adam M (1982) Adv. Polym. Sci. *44*:103
245a. Chhabra A, Matthews-Morgan D, Landau DP, Herrmann HJ (1986) Phys. Rev. *B34*:4796
246. Broer DJ, Mol GN (1986) Polym. Mat. Sci. Eng. *55*:540
247. Broer DJ, Finkelman H, Kondo K, Makromol. Chem., in press
248. Cohen MS, Bluestein C, Dunkel M (1984) Proceedings "Radcure '84", Assoc. Finish. Processes SME., Dearborn (Mich.) p 11–1
249. Gossink RG (1986) Angew. Makromol. Chem. 145/146:365
250. Kloosterboer JG, Lüten GFCM (1988) in: Dickie RA, Bauer RS, Labana S (eds) Chemistry, Properties, and Applications of Crosslinking Systems, ACS Symp. Ser., Washington DC.

Editor: K. Dušek
Received March 12, 1987

# Application of Polymer in Encapsulation of Electronic Parts

C. P. Wong

AT & T Engineering Research Center P.O. Box 900, Princeton, NJ 08540, U.S.A.

The rapid development of integrated circuit (IC) technology from small-scale integration (SSI) to very large-scale integration (VLSI) has had great technological and economic impact on the electronic industry. The exponential growth of the number of components per IC chip, and the exponential decrease of device dimensions and the steady increase in IC chip size have imposed stringent requirements, not only on the IC physical design and fabrication, but also on the IC encapsulants. This chapter reviews the VLSI technology trends and their economic impact on the electronic industry. The purpose of encapsulation, encapsulation techniques and a general overview of the application of inorganic and organic polymer materials as electronic device encapsulants will also be addressed.

1 Introduction . . . . . . . . . . . . . . . . . . . . . . . . . . . . . . . . . . 64

2 Purpose of Encapsulation . . . . . . . . . . . . . . . . . . . . . . . . . . 65
  2.1 Moisture . . . . . . . . . . . . . . . . . . . . . . . . . . . . . . . . . 65
  2.2 Mobile Ion Contaminants . . . . . . . . . . . . . . . . . . . . . . . 66
  2.3 Ultraviolet — Visible Light . . . . . . . . . . . . . . . . . . . . . . 66
  2.4 Hostile Environments . . . . . . . . . . . . . . . . . . . . . . . . . 66

3 Encapsulation Techniques . . . . . . . . . . . . . . . . . . . . . . . . . . 67
  3.1 On-Chip Encapsulation Technique . . . . . . . . . . . . . . . . . . 68
    3.1.1 Thermal Deposition . . . . . . . . . . . . . . . . . . . . . . . 68
    3.1.2 Hot-Wall, Reduced-Pressure Reactor Process . . . . . . . . 68
    3.1.3 Continuous, Atmospheric Pressure Reactor . . . . . . . . . 69
  3.2 Plasma Depositions . . . . . . . . . . . . . . . . . . . . . . . . . . 70
    3.2.1 Parallel-Plate Plasma Assisted CVD . . . . . . . . . . . . . 70
    3.2.2 Hot-Wall Plasma Assisted CVD . . . . . . . . . . . . . . . . 71
    3.2.3 Radiation Stimulated Deposition . . . . . . . . . . . . . . . 71
  3.3 Chip Packaging Encapsulation Techniques . . . . . . . . . . . . . 72
    3.3.1 Cavity-Filling Processes . . . . . . . . . . . . . . . . . . . . 72
    3.3.2 Saturation and Coating Processes . . . . . . . . . . . . . . 73

4 Device Encapsulants . . . . . . . . . . . . . . . . . . . . . . . . . . . . . 76
  4.1 Inorganic Encapsulants . . . . . . . . . . . . . . . . . . . . . . . . 76
  4.2 Organic Encapsulants . . . . . . . . . . . . . . . . . . . . . . . . . 76
    4.2.1 Silicones . . . . . . . . . . . . . . . . . . . . . . . . . . . . . 76
    4.2.2 Epoxies . . . . . . . . . . . . . . . . . . . . . . . . . . . . . . 79
    4.2.3 Polyimides . . . . . . . . . . . . . . . . . . . . . . . . . . . . 80

5 Conclusions . . . . . . . . . . . . . . . . . . . . . . . . . . . . . . . . . . 82

6 References . . . . . . . . . . . . . . . . . . . . . . . . . . . . . . . . . . . 82

Advances in Polymer Science 84
© Springer-Verlag Berlin Heidelberg 1988

# 1 Introduction

The rapid development of integrated circuit (IC) technology from small-scale integration (SSI) to very large-scale integration (VLSI) has had great technological and economic impact on the electronic industry throughout the world. The exponential growth
of the number of components per chip [1], the exponential decrease of device dimensions [2] (Fig. 1), and the steady increase in IC chip size (Fig. 2) have imposed stringent
requirements, not only on the IC physical design and fabrication, but also on the
IC encapsulants. In addition to the technological growth in the IC area, the economic
growth has become apparent. In the United States alone, the growth of electronic

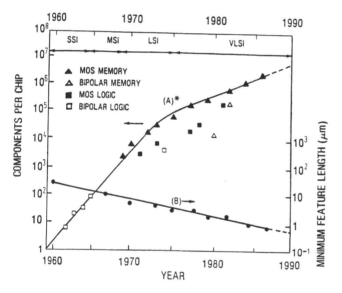

**Fig. 1.** Integrated circuit technological trends [1]

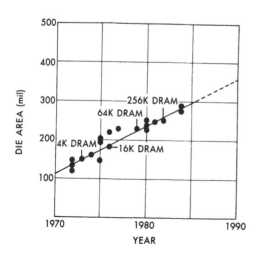

**Fig. 2.** Integrated circuit dimension trends

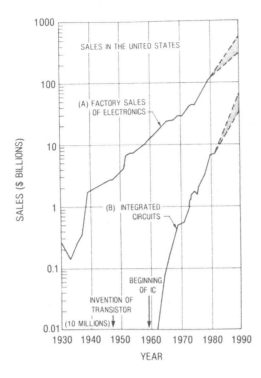

**Fig. 3.** Economics of integrated circuit sales [4]

industry has increased by a factor of 10 since the early 1960s. Figure 3 shows the electronic sales reached 114 billion dollars in 1981 and are projected to continue at a rate of 10–15% per year. Integrated circuit sales are expected to reach 50 billion dollars by 1990 [3, 4]. The increase of integration in VLSI technology has resulted in the miniaturization of the device size which has reduced the propagation delay due to higher density packaging and interconnection. This new VLSI technology operates at a faster speed, consumes less power, and subsequently dissipates less heat during operation. As a consequence, VLSI technology has increased the reliability and reduced the cost per function of the devices which has had profound impact on the electronic industry.

## 2 Purpose of Encapsulation

The purpose of encapsulation is to protect the electronic IC devices from moisture, mobile ion contaminants, ultraviolet-visible and alpha particle radiation, and hostile environmental conditions.

### 2.1 Moisture

Moisture is one of the major sources of corrosion of the IC device. Electro-oxidation and metal migration are associated with the presence of moisture. The diffusion rate of moisture also depends on the encapsulant material [5]. Figure 4 shows the permea-

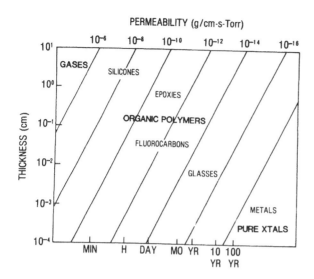

Fig. 4. Diffusion of moisture in various materials [5]

bility of various materials. In general, the moisture diffusion rate is proportional to the material thickness and the diffusion time.

## 2.2 Mobile Ion Contaminants

The new mobile ion, such as sodium or potassium, tends to migrate to the p-n junction of the IC device where it picks up an electron and deposits as the correspondent metal on the p-n junction which destroys the device [6]. Chloride ions, even in trace amounts (in ppm level), could cause the dissolution of aluminum metallization of complementary metal-oxide semiconductor (CMOS) devices [7]. Unfortunately, CMOS is likely to be the trend of the VLSI technology and sodium chloride is a common contaminant. The protection of these devices from the effects of these mobile ions is apparent.

## 2.3 Ultraviolet — Visible Light

There is an increasing amount of light sensitive opto-devices that need UV-VIS protection. Alpha particle radiation is caused by a very low level of uranium and cosmic radiation present as background radiation in the device package and the atmosphere, respectively, and which could generate a temporary "soft error" in operating dynamic random access memory (DRAM), such as; the 64 K, 256 K, 1 and 4 Megabits DRAM devices. This type of alpha particle radiation has become a major concern, especially in high density memory devices.

## 2.4 Hostile Environments

Hostile environments, such as, extreme temperatures, high and low, high relative humidity, and operating bias operating environments of the device are part of the real life operation. It is critical for the device to survive these operation-life cycles. Besides, encapsulants must also enhance the mechanical and physical properties of the IC devices.

In addition to the above functions, the encapsulant must be an ultrapure material, with superior electrical and physical properties, and ease of application and repair in production and service. With the proper choice of encapsulant, the encapsulation could enhance the fragile IC device, improve its mechanical and physical properties, and its manufacturing yields, and prolong the reliability of the IC device which is the ultimate goal of the encapsulation [8].

# 3 Encapsulation Techniques

Prior to the discussion of encapsulation techniques, the IC assembly sequence will be reviewed. The wire bonded IC device will be used as an example of a common IC package. This IC packaging example will provide a better understanding of the packaging sequence and encapsulation process. Let us assume that the IC devices have gone through the completed fabrication process and are still in a wafer form. These devices

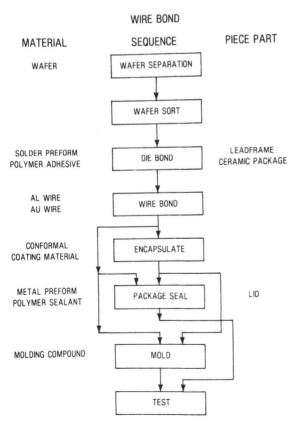

**Fig. 5.** Flow chart of IC packaging [2]

are then tested for functions and defects, and separated or sorted by dicing. The functional devices are then die bonded to the package or lead frame with a solder preform (tin-lead solder eutectic) or polymer adhesive (typically silver-filled epoxy, or metal-filled silicone or polyimide die attach adhesives.) The IC device is further wire bonded with a thin (1.0 to 1.25 mil diameter) gold or aluminum wire to the package or to the lead-frame for outside chip interconnection. The wire-bonded chip is now ready for encapsulation with the encapsulant and package lid-sealing with metal preform or polymer sealant in the package, or for some lead frame package, an epoxy-type molding compound is used to post-mold the package. At this stage, the package is ready for testing. A flow chart is shown in Fig. 5 for this packaging sequence. Since the package step is close to the end of the IC device manufacturing process, the major concerns are the ease of the encapsulation process, high yield and repairability, especially in terms of the high cost customer designed VLSI devices.

Encapsulation techniques consist of two major levels: 1) on-chip encapsulation (deposition of passivation dielective layers right after IC chips are fabricated), prior to chip testing, dicing and sorting steps and (2) chip packaging encapsulation, where devices are being fabricated and going through final packaging process.

## 3.1 On-Chip Encapsulation Technique

Silicon dioxide, silicon nitride, silicon-oxy-nitrite and polyimides are commonly used in device passivation. These passivation layers are known to have excellent moisture and mobile ion barriers of the devices. As for the sodium ion barrier, silicon dioxide is inferior to silicon nitride. However, the use of phosphorous (a few weight percent, usually less than 6%) doped silicon dioxide has greatly improved its mobile ion barrier property. A thin layer (in 1–2 um thickness) of one of these dielectric materials is deposited uniformly on the finished device, except at the bond pad areas for bonding. Most of these inorganics are deposited in one of the three major processes [9, 10, 11, 12].

3.1.1 Thermal Deposition

Chemical vapor deposition (CVD) by thermal process is one of the widely used methods in preparing silicon dioxide, polysilicon, silicon nitride or silicon-oxy-nitride passivation layers [13, 14, 15, 16, 17, 18, 19, 20, 21, 22, 23, 24]. This thermal process is further divided into two methods.

3.1.2 Hot-Wall, Reduced-Pressure Reactor Process

For this process, the reactive gases (see Table 1) are passed from one end of the reactor and pumped out through the quartz tube reactor chamber (Fig. 6). The fabricated devices which are in wafer form (from 50–200 wafers per run) are vertically stacked in the reactor chamber. Since the deposition rate is a function of both reactive gas concentration and temperature, there is a reactive gas concentration gradient in the reactor chamber — being rich at the beginning of the reactor and poor at the end of the reactor. Therefore, the oxide or nitrite tends to deposit faster at the beginning of chamber and progressively less as they move down from the reaction chamber. A non-uniform thickness deposit could result. To resolve this non-uniform deposition problem, a

Typical reactions for depositing dielectrics and polysilicon
**Table 1.** Reactive gases [9]

| Reactants | Deposition temperature (°C) | Product |
|---|---|---|
| $SiH_4 + CO_2 + H_2$ | 850–950 | Silicon dioxide |
| $SiCl_2H_2 + N_2O$ | 850–900 | |
| $SiH_4 + N_2O$ | 750–850 | |
| $SiH_3 + NO$ | 650–750 | |
| $Si(OC_2H_5)_4$ | 650–750 | |
| $SiH_4 + O_2$ | 400–450 | |
| $SiH_4 + NH_3$ | 700–900 | Silicon nitride |
| $SiCl_2H_2 + NH_3$ | 650–750 | |
| $SiH_4 + NH_3$ | 200–350 | Plasma silicon nitride |
| $SiH_4 + N_2$ | 200–350 | |
| $SiH_4 + N_2O$ | 200–350 | Plasma silicon dioxide |
| $SiH_4$ | 600–650 | Polysilicon |

three-zone heating furnace with increasing furnace temperature range between 300°
to 900 °C is used to compensate for the difference in this type of deposition. The pressure of the reactant chamber is maintained at $\sim 0.25$–2.0 torr and the gas flow is between 100–1000 std. cc/min. The advantages of this process are high quality uniform films, large loading and batch-process in production. However, the disadvantages are the toxic reactive gases used in this process.

### 3.1.3 Continuous, Atmospheric Pressure Reactor

An alternative CVD process employs a continuous throughput, conveyor belt process (Fig. 7). The reactive gas is purged from the center of the reactor and flows uniformly through the entire conveyor heated wafers at atmospheric pressure. Silicon dioxide and silicon nitride are usually formed in this process. The advantage of this process is the high throuput, good quality and uniform films. However, the disadvantage is a large consumption of reactive gas and the formation of particulates that require frequent chamber cleaning.

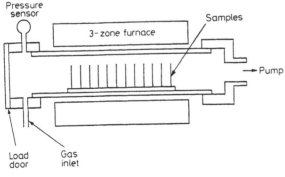

**Fig. 6.** Chemical vapor deposition: hot-wall, reduced-pressure reactor process [9]

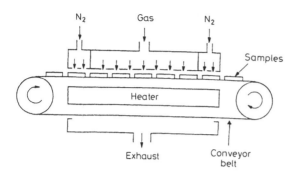

Fig. 7. Chemical vapor deposition: continuous, atmospheric pressure reactor process [9]

## 3.2 Plasma Depositions

Thermal processes, previously discussed, in general provide high quality and good uniform films. However, a higher deposition temperature is required. For CMOS technology, the aluminum metallization tends to have intermetallic diffusion problems at high temperatures and results in the formation of "Hillocks" which are mainly due to the high temperature deposition generated stress-release phenomena. The low temperature ($\sim 400\,^\circ$C) plasma assisted CVD becomes an attractive alternative. There are generally two plasma CVD processes [25, 26, 27, 28, 29, 30, 31, 32, 33].

### 3.2.1 Parallel-Plate Plasma Assisted CVD

The parallel plate Plasma assisted CVD is typically a cylinder glass reactor with parallel aluminum plates acting as electrodes on the top and bottom (Fig. 8). The lower plate is the grounded electrode which wafers are horizontally placed on the heated electrode ($\sim 100$–$400\,^\circ$C). A radio frequency voltage is applied on the top electrode to generate a glow discharge between the two plates. Reactive gases (see Table 1) flow through the discharged area and deposit on the wafer. Silicon nitride and silicon oxide are usually deposited by this process. The advantage of this process is the lower temperature deposition. However, the disadvantages are: (1) low through-put capacitor, (2) manual process for each wafer, and (3) particles and particulates generated in the reactor may damage the reactor.

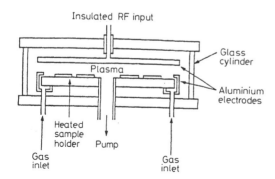

Fig. 8. Plasma assisted chemical vapor deposition: parallel-plate process [9]

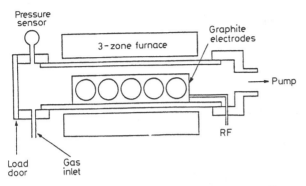

**Fig. 9.** Plasma assisted chemical vapor deposition: hot-wall process [9]

## 3.2.2 Hot-Wall Plasma Assisted CVD

This process eliminates most of the parallel-plate processes as mentioned before (Fig. 9). The process usually takes place in a three-zone heated quartz reactor with wafers vertically placed parallel to the reactive gas flow. The reactive gases flow from one end to the other with the similar setup as the hot-wall, reduced pressure CVD process. The radio frequency electrode which supports the wafers consists of aluminum or graphite slabs. Alternate electrodes are located on top of the furnace for plasma discharge. The major advantage of this process is low temperature deposition. However, the disadvantages are that particulates are generated from the furnace and the process requires the manual loading and unloading of wafers.

## 3.2.3 Radiation Stimulated Deposition

Radiation stimulated deposition is a more recently developed process by J. Peters of Hughes Aircraft Company [34, 35]. The process uses mercury lamps with ultraviolet (UV) radiation to stimulate the mercury catalyst, silane and ammonia (or hydrazine) reaction for the deposition of the silicon nitride passivation layer. Photo-deposition does not rely on thermal energy to initiate the deposition. The UV photon energy is first absorbed by the reactants which lead to the dissociation of their chemical bondings. In addition, Ehrlich and coworkers [36] of MIT Lincoln Laboratories and Boyer and coworkers of Colorado State University [37] developed the low temperature laser deposition process. The laser provides a high intensity deep UV source with a wavelength of 200 nm for the photodissociation of the reactive gases (metal hydrides) which is then capable of depositing metal without masking. The high cost of UV laser and the purity of film prevents its commercial application. However, the mercury sensitized radiation stimulated process which uses only low cost UV lamps has become quite attractive. Recently, Tylan Company has licensed and commercialized this process. Figure 10 shows the reaction diagram where nitrogen carrier gas was used to purge the reaction chamber. Silane, ammonia, or hydrazine (gases) and mercury catalyst were premixed in a manifold chamber and introduced to an UV activated chamber. Wafers were loaded horizontally on the heated tray ($\sim 100$ °C) and silicon nitrite was deposited on the wafer. The quality of the film is inferior to the thermal growth films and trace amounts of mercury contaminants (incorporated in the deposit-

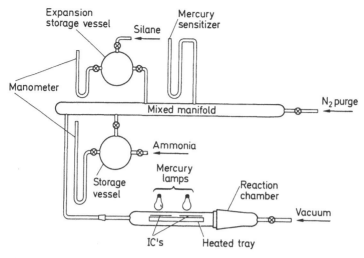

**Fig. 10.** Radiation stimulated deposition process

ed film) may cause some reliability concern in certain VLSI processes [38]. However, this film has potential as passivation layer for mercury-cadnuim-tellurate (Hg-Cd-Te) temperature sensor of the II–VI compounds.

## 3.3 Chip Packaging Encapsulation Techniques

Chip packaging encapsulation techniques consist mainly of Cavity-Filling and Saturation and Coating [39].

### 3.3.1 Cavity-Filling Processes

Potting, casting and molding are common processes for cavity-filling. Potting is the simplest. It involves filling the electronic component within a container with a liquid resin and then curing the material as an integrate part of the component. Polymeric resins (such as epoxies, silicones, polyurethanes, etc.) are usually used as potting materials. Containers (such as metal can or rugged polymeric casing) enhance the effectiveness of the encapsulant. In the fast growing automation manufacturing process, rugged machine insertable components, such as surface mounted chip carriers, dual-in-line (DIP), single-in-line (SIP) packages and discrete components are highly desirable for automation processes.

Casting is similar to potting, except the outer casing is removed after the polymer cavity-filling process is completed and cured. No heat or pressure is applied in the process.

Molding has become an increasingly important process in modern device encapsulation. It involves injecting a polymeric resin (one of the molding compounds) into a mold and then curing. The process involves the following steps: (i) The molding compound is preheated until it melts and the resin flows through runners, gates, and finally fills up the cavities. (ii) The resin is then cured (hardened) and released from the mold

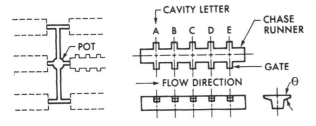

**Fig. 11.** Plastic modeling process

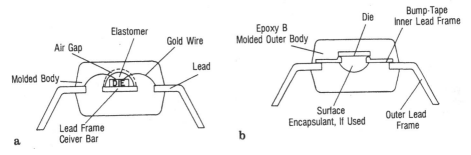

**Fig. 12.** New and old molded device parts

to predetermined shapes. The exact control of the mold pressure, viscosity of the molten molding compound, the delicate balance of runners, gates and cavity designs are very critical in optimizing the increasing molding plastic IC. Finite element analyses of the plastic molding process are becoming an integrate part in solving this process. Pressure, injection and conformal moldings are some of current molding process. Figures 11 and 12 illustrate the typical conformal molding process and the old and new molded components.

### 3.3.2 Saturation and Coating Processes

Impregnation, dipping, conformal and surface coatings are common saturation coatings. Impregnation coating is performed by the saturation of a low viscosity resin into the component which includes a thin film coated on the component surface as well. This process is usually used with cavity filling or conformal coating process. Dip coating is performed by dipping the component into an encapsulant resin. The component is then withdrawn, dried then cured. Coating thickness is usually a function of resin viscosity, withdrawal rate and temperature of resin.

Conformal coating is the most common technique used in IC device encapsulation. Spin coating and flow coating are commonly used in the electronic process. A suitable rheological property of the encapsulant is critical in obtaining a good flow coating package, especially in hydrid IC encapsulation, where the encapsulant tends to run-over from the substrate and wick the leads of the hybrid devices [40].

**Table 2.** Typical properties of electronic encapsulants

| Polymer-system | Typical cured properties of basic encapsulants [39,48] | | | | | | | | |
| | Electrical | | | | | | Thermal | | |
| | Dielectric strength | Specific resistivity | Water absorption | Dielectric constant | Loss tangent | Relative arc track resistance[c] | Heat distortion temperature | Safe use temperature | Linear expansion |
| | (volts/mil) | (ohm-cm) | (%)[b] | $(10^{10}$ cps) | $(10^{10}$ cps) | | (°F) | (°F) | x $10^5$/°F |
| **Thermoplastics:** | | | | | | | | | |
| Asphalt and tars[k] | 300 | $10^{10}$ | .08 | 3.5 | 0.04 | 5 | 130 | 160 | 8. |
| Fluorocarbon | 450 | $10^{18}$ | .00 | 2.1 | .0003 | 1 | 250 | 500 | 5.5 |
| Polyethylene[k] | 500 | $10^{16}$ | <.01 | 2.3 | .0005 | 3 | — | 240 | 9.5 |
| Polystyrene[k] | 550 | $10^{18}$ | .04 | 2.5 | .0003 | 3 | 180 | 185 | 4. |
| Polyvinyl chloride[k] | 400 | $10^{15}$ | .15 | 2.8 | .006 | 3 | 150 | 210 | 3. |
| Wax[k] | 400 | $10^{17}$ | .02 | 2.6 | .001 | 3 | 80 | 135 | 11. |
| Silicone-polyimide[a] | 1500–2800 | $10^{15-17}$ | <1 | 3.0 | 0.007 | 2 | 300–460 | 750 | 30–47(C) |
| Parylene[d] | 500–7000 | $10^{13-16}$ | 0.03 | 2.8 | 0.01–.003 | 2 | 535–760 | 250 | 3.5–6.9 |
| **Thermosets:** | | | | | | | | | |
| Alkyd[k] | 350 | $10^{14}$ | .4 | 3.8 | .025 | 2 | 220 | 250 | 4. |
| Allyl ester[k] | 400 | $10^{14}$ | .7 | — | — | 3 | >190 | 210 | 4. |
| Butadiene-styrene[k] | 600 | $10^{16}$ | .03 | 2.4 | .006 | 3 | 260 | 475 | 5. |
| Epoxide[k] | 450 | $10^{14}$ | .20 | 2.9 | .018 | 2 | 400 | 450 | 3. |
| Phenol-aldehyde[k] | 350 | $10^{12}$ | .3 | 4.7 | .04 | 4 | 175 | 175 | 4. |
| Polyester[k] | 350 | $10^{13}$ | .4 | 3.5 | .05 | 3 | 190 | 325 | 6. |
| Silicone[k] | 600 | $10^{15}$ | .03 | 2.8 | .002 | 2 | 100 | 500 | 7. |
| Polyimides[e] | 3400 | $10^{16}$ | — | 3.6 | .002 | 2 | >600 | <800 | 2–7 |
| Silicone-epoxy[f] | 246–338 | $10^{15}$ | 0.1 | 3.6 | 0.004 | 2 | — | <390 | 3–6 |
| **Elastomers:** | | | | | | | | | |
| Buna-S rubber | 500 | $10^{14}$ | — | 2.5 | .01 | 4 | — | 250 | 6. |
| Chloro rubber | 400 | $10^{12}$ | — | 2.7 | .05 | 3 | — | — | 9. |
| Natural rubber | 500 | $10^{16}$ | — | 2.1 | .03 | 4 | — | 150 | 4. |
| Silicone rubber[k] | 600 | $10^{13}$ | — | 3.0 | .05 | 2 | >450 | 500 | — |
| Thioplast[k] | 150 | $10^{11}$ | — | 14. | .15 | 4 | — | 250 | 10. |
| Urethane[k] | 350 | $10^{11}$ | .4 | 3.5 | .04 | 4 | >150 | 200 | 10. |
| **Inorganics:** | | | | | | | | | |
| $SiO_2$ | 5000 | $>10^{16}$ | | 3.5–4 | | 1 | 1400 | 1400 | .3–.5 |
| $Si_3N_4$ | 5000 | $10^{12}$ | | 7–10 | | 1 | 1400 | 700–1400 | 2.5–3 |

[a] = M & T Chemical Co.
[b] In 24 hours, 1/8 inch thick.
[c] 1 best, 5 poorest.
[d] = Union carbide
[e] = DuPont
[f] = Dow corning

Physical[h]

| Ultimate tensile strength (psi) | Ultimate elongation (%) | Hardness | Relative adhesion[e] | Remarks |
|---|---|---|---|---|
| 600 | 5. | SD 60 | 4 | Lowest cost. |
| 3000 | 200. | SD 60 | none | Good solvent resistance |
| 4000 | 1000. | SD 65 | 5 | Flexible. |
| 7000 | 1.5 | M 80 | 4 | Rigid. |
| 3000 | 100. | SD 80 | 3 | Coating material |
| 300 | 5. | SD 30 | 4 | Melt, pour, and chill. |
| 2000 | 200 | — | 1 | Good solvent resistance, high temp. |
| 10,000 | 200 | — | 4 | Conformal coating Excellent solvent resistance |
| 8000 | — | SD 90 | 2 | |
| 5500 | — | M 70 | 3 | |
| 4000 | 4. | SD 80 | 4 | |
| 10,000 | <1. | M 90 | 1 | Excellent solvent resistance |
| 7000 | 1.5 | M 126 | 2 | |
| 8000 | <5. | M 100 | 3 | |
| 2500 | 8. | M 60 | 4 | Excellent THB performance |
| 14,000–20,000 | 10–80 | — | 3 | Good solvent resistance, high temp. |
| 8000 | — | SD 60 | 3 | Good for molding |
| 300 | 400. | SA 50 | 2 | Flexible |
| 2500 | 500. | SA 70 | 3 | Flexible |
| 3000 | 700. | SA 50 | 2 | Flexible |
| 650 | 100. | SA 60 | 4 | Flexible |
| 300 | 400. | SA 40 | 2 | Poor electrical properties high temperature. |
| 5000 | 400. | SA 60 | 1 | Poor electrical properties high temperature. |
| 14,000–56,000 | 0. | | 4 | Excellent passivation properties |
| 14,000–140,000 | 0. | | 4 | Excellent passivation property |

[g] M = Rockwell M.
SA = Shore Durometer A.
SD = Shore Durometer D.
[h] At 70–90° F.
[k] Unfilled.

# 4  Device Encapsulants

Device encapsulants are divided in two groups of materials.

## 4.1  Inorganic Encapsulants

Silicon dioxide and silicon nitride which have previously been described on Sect. 3.1.1 to 3.1.3 are commonly used inorganic encapsulants. (See Sect. 3.1.1 to 3.1.3).

## 4.2  Organic Encapsulants

Organic polymeric encapsulants are divided into three categories: 1) thermosetting materials, 2) thermoplastic polymers, and 3) elastomers. Thermosetting materials are crosslinking polymers which cannot be reversed to monomers after curing. Silicones, polyimides, epoxies, silicone-modified polyimides, polyesters, butadiene-styrenes, alkyd resins, allyl esters and silicone-epoxies are examples. Thermoplastic polymers when subjected to heat will flow and solidify upon cooling without cross-linking. These thermoplastic processes are reversible and become a suitable engineering plastic material. Polyvinyl chloride, polystyrene, polyethylene, fluorocarbon polymers, asphalt, acrylics, tars, parylene and recently developed preimidized silicone-modified polyimides are examples of thermoplastic polymers. Elastomers are materials that have high elongation properties, have crosslinking in their systems and belong to thermosetting polymers. Silicone rubbers, silicone gels, natural rubbers, and polyurethanes are examples (see Table 2). However, for IC technology applications, only a few of the above which have ultrapure properties, such as silicones, epoxies, polyurethanes and polyimides have been shown to be acceptable IC encapsulants.

### 4.2.1  Silicones

RTV silicone elastomer, an organosiloxane, is one of the most effective encapsulants used for temperature cycling and moisture protection of IC devices. Since World War II, silicones (organosiloxane polymers) have been used in a variety of applications where properties of high thermal stability, hydrophobicity, and low dielectric constant are necessary, e.g., as encapsulants or conformal coatings for integrated circuits. In 1969 it was demonstrated that room temperature vulcanized (RTV) silicones exhibited excellent performance as moisture protection barriers for IC devices and a number of different RTV silicone have been adapted for use in the electronics industry [41, 42, 43, 44, 45, 46, 47, 48, 49, 50, 51, 52, 53, 54, 55, 56, 57, 58].

The primary commercial process for preparing silicone polymers is the Rochow process, wherein a stream of alkyl or aryl monohalide, usually the chloride, is passed through a heated bed of pure silicon alloyed with copper metal as catalyst. The exact mechanism of this Rochow process is not well characterized. It presumably proceeds through an organocopper intermediate. The main product of the reaction is a diorganodihalosilane; however, some byproduct monoorganotrihalo, -triorganomonohalo-, tetraorgano-, and tetrahalosilanes are observed. The product is purified by distillation, then catalytically hydrolyzed to disilanols, which are unstable and combine

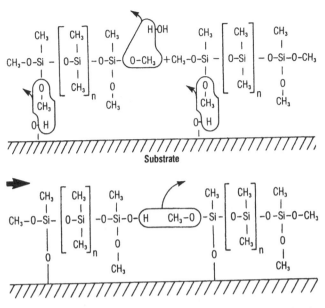

**Fig. 13.** Room temperature vulcanized (RTV) silicone cure mechanism

to form a mixture of cyclic siloxane oligomers (primarily trimer and tetramer) and linear hydroxyl end-blocked (HEB) siloxane polymers. The cyclic oligomers can be ring-opened to form linear polymers. The molecular weight of the HEB siloxanes is controlled by the reaction conditions. Endblocking can be changed by a variety of reactions. The HEB siloxanes are fluids with visocosities of approximately $10^{-6}$ m$^2$/s (a few cS) or resins and gums with much higher viscosity (1 m$^2$/s). In these cases, they are not suitable as coatings and must be cross-linked or vulcanized by free radical-initiated cure, platinum addition cure or condensation cure process.

The RTV condensation cure system consists of four types: carboxylate cure, alkoxide cure, oxime cure, and amine cure. For electronic applications, only the alkoxide cure system is preferred because the byproduct alcohol generated during cure is non-corrosive.

Ionic materials, whether from the device surface, encapsulation materials or the environment, affect the electrical reliability of encapsulated IC devices. For this reason, silicone is subjected to intense purification. The concentration of Na$^+$, K$^+$, and Cl$^-$ mobile ions is less than a few ppm, and alpha particle emission is less than 0.001 alpha/cm$^2$ hr. Thus it offers excellent alpha particle shielding for eliminating soft error in dynamic random access memory devices, such as; 64 K, 256 K, and megabit chips [59]. The drawback of RTV silicone as an IC encapsulant is its poor solvent resistance and weak mechanical properties [60]. Highly fluorinated alkyl substituted siloxane has also showed improvement in solvent resistance. However, a recently developed silicone material with high crosslinking and high filler loading system seems to improve the solvent resistance of the silicone encapsulant [61, 62].

Heat curable silicone (either elastomer or gel) has become an attractive device encapsulant. This heat curable silicone gel tends to have a slightly higher thermal

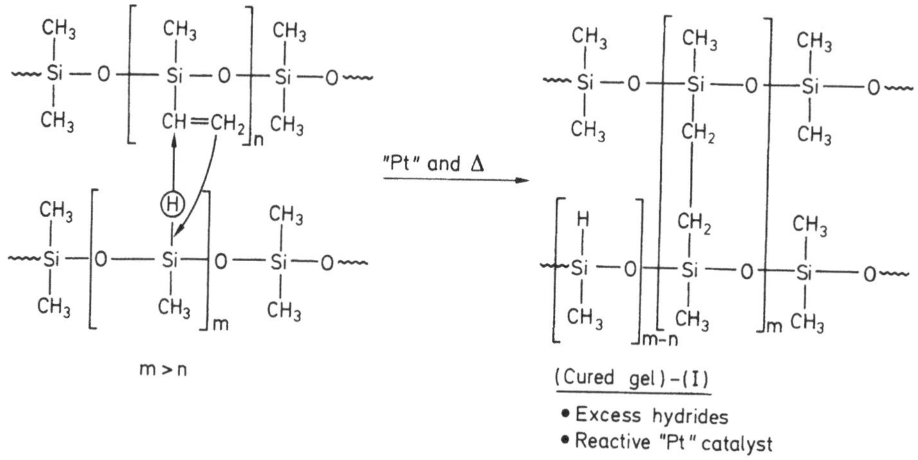

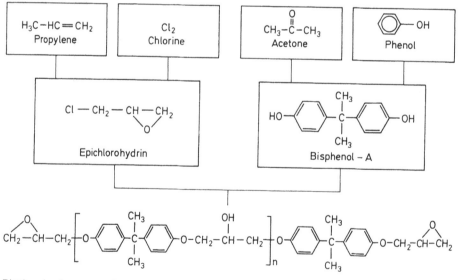

**Fig. 14.** Heat curable silicone cure mechanism

property than the conventional RTV silicone. With its excellent jelly-like (very low in modulus) intrinsic solftness, silicone gel becomes a very attractive encapsulant in wire bonded IC devices. The two part heat curable system which consists of the vinyl and hydride reactive functional groups, and the platinum catalyst additional cure system provides a fast cure system without any byproduct. (See Figure 14 for cure mechanism). This solventless type of heat curable silicone gel will have increased usage in electronic applications [62].

**Fig. 15.** Commercial epoxy preparation scheme

## 4.2.2 Epoxies

Epoxies were first prepared in early 1930 [63]. They have become one of the most favored polymeric material used for electronics. Their unique chemical and physical properties such as excellent chemical and corrosion resistances, super electrical and physical properties, excellent adhesion, insulation, low shrinkage properties, and reasonable material cost have made epoxy resins very attractive in electronic applications [63, 64, 65, 66]. The commercial preparation of epoxies are based on bisphenol A, which upon reaction with epichlorohydrin produces bisglycidyl ethers. (See Fig. 15). The repetitive group, n, varies from zero (liquid) to approximately 30 (hard solid). The reactants ratio (bisphenol A versus epichlorohydin) determines the final viscosity

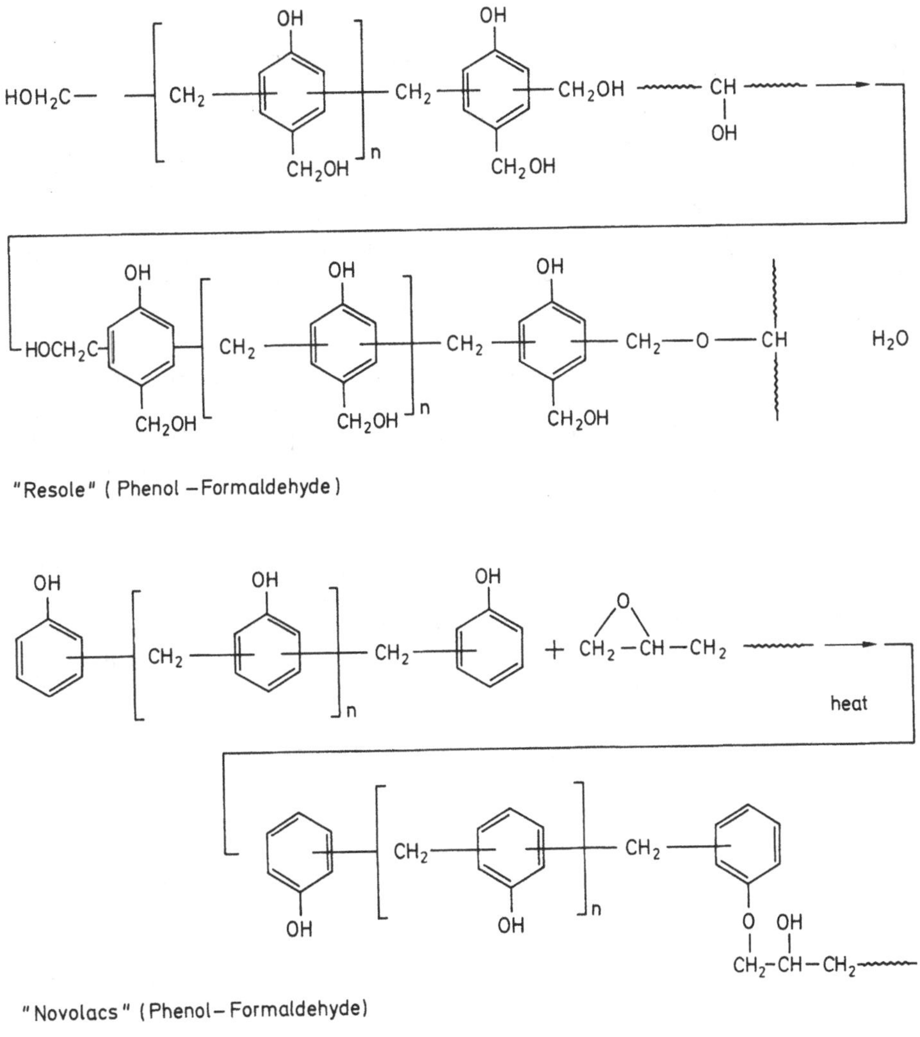

"Resole" ( Phenol – Formaldehyde )

"Novolacs" (Phenol–Formaldehyde)

Fig. 16. Novolac epoxy chemical structure

of the epoxies. In addition to the bisphenol A resins, the Novolac resins (see Fig. 16) with multifunctional groups which lead to higher cross-link density and better thermal and chemical resistance have gained increasing acceptance in electronic applications. Typical epoxy curing agents are amines, anhydrides, dicyandiamides, melamine/formaldehydes urea/formaldehydes, phenol/formaldehydes, and catalytic curing agents, such as anhydrides and amines are two of the most frequently used curing agents.

Selecting the proper curing agents is dependent on application techniques, curing conditions, pot-life required and the desired physical properties. Besides affecting viscosity and reactivities of the epoxy formulations, curing agents determine the degree of cross-linking and the formation of chemical bond in the cured epoxy system. The reactivity of some anhydrides with epoxies is slow, therefore an accelerator, usually a tertiary amine is used to assist the cure. "Novolacs" and "Resole" are two major commonly used phenol-formaldehyde epoxies. Novolacs is a acid-catalyzed phenol-formaldehyde, epoxidized polymer. The phenolic groups in the polymer are linked by a methylene bridge which provides highly cross-linked systems, high temperature and excellent chemical resistance polymer. "Resole" is a base-catalyzed phenol-formaldehyde epoxy polymer. In most phenolic resins commonly used with epoxies, the phenolic group is converted into ether to give improved basic resistance. Phenolic resins are cured through the secondary hydroxyl group on the epoxy backbone. Highly temperature curing is required in this system and it provides excellent chemical-resistant and solvent properties epoxies. Recently developed high purity epoxies have become a very attractive encapsulant for electronics.

These new type of resins contain minimum amount of chloride and other mobile ions, such as sodium and potassium, and have become widely used in device encapsulation and molding compounds. The incorporation of fused silica as filler in the epoxy system has drastically reduced the thermal coefficient of expansion of these materials which make them more comparable with the IC die attached substrate materials. The incorporation of a small amount of the elastomeric material (such as, silicone elastomer) to the rigid epoxy has drastically reduced the modulus of the material and reduced the thermal stress of the epoxy material [67]. This new type of low stress epoxy encapsulant has some potential application in molding large IC devices. The epoxy "glob-top" type material is increasingly becoming more acceptable in "chip-on-board" type encapsulation of die-and-wire type devices. When the epoxy material is properly formulated and applied, this "glob-top" type epoxy encapsulant is quite reliable in performance. The continuous advancements in epoxy material development will have a greater impact in device packaging and material utilization.

## 4.2.3 Polyimides

Polyimide is one of the fastest growing materials in polymers for electronic applications. During the past couple of decades, there has been a tremendous interest in this material for electronic applications. The superior thermal (up to 500 °C) mechanical and electrical properties of polyimide have made its use possible in many high performance applications, from aerospace to microelectronics. In addition, polyimides show very low electrical leakage in surface or bulk and form an excellent interlayer dielectric insulators and excellent step coverage in multilayer IC structures. They

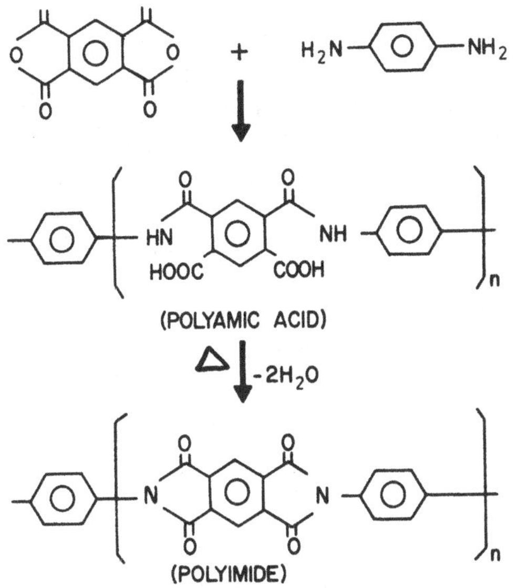

Fig. 17. Polyimide cure mechanism

have excellent solvent resistance and ease of application. They could be easily either spun-on or flow-coated and imaged by conventional photolithography and etch process.

Most polyimides are aromatic diamine and dianhydride compositions. Polyamic acids are precursors of the polyimides. Thermal cyclization of polyamic acid is a simple curing mechanism for this material (See Fig. 17). Siemens of Germany developed the first photodefinable polyimide material. Photo-degradation of the diazo compound is believed to be the photodefinable polyimide reaction mechanism [68]. However, Ciba Geigy has recently announced a new type of photodefinable polyimide which does not include a diazo material [69]. Both of these photodefinable materials are negative resisty type polyimides. A positive resisty type polyimide which reduces the processing step in IC fabrication is not yet commercially available. Hitachi has recently announced an ultra-low thermal coefficient expansion (TCE) polyimide which has some potential in reducing the thermal stress of the silicon chip and the polyimide encapsulant. The rod-like, rigid structure of the polyimide backbone structure is the key in preparing a low TCE polyimide [70]. By simply blending a high and a low TCE polyimide, one will be able to achieve a desirable TCE encapsulant which could match the TCE on the substrate, and reduce the thermal stress problem in encapsulated device temperature cycling testing. However, the affinity for moisture absorption, a high temperature cure and high cost of the polyimide are the only drawbacks that prevent its use in general electronic application. Preimidized polyimides which cure by evaporation of dissolved solvent may reduce the drawback of high temperature cure of the material. Advances in polyimide synthesis have reduced the material moisture absorption and improved the adhesion of the material. This new type of polyimide development will have significant implications in device packaging.

# 5 Conclusions

The rapid development of the IC technology has created a critical need for the advanced polymeric materials as a device interlayer dielectric, passivation layer or encapsulant. Recent advances in high performance materials, such as improved silicone elastomers, silicone gels, epoxies and polyimides have provided some relief in the VLSI technological application [71]. However, the demands for low dielectric constant, high breakdown voltage strength, high sheet resistance and less dielectric change with humidity, high performance polymeric materials will persist. Their application in on-chip interconnections, wafer-scale integration architecture structure with very fast interconnecting network will become apparent. It is a challenge that the collaborative efforts between chemists, material scientists and device engineers will face in the near future.

# 6 References

1. Moore, G.: "VLSI, What Does the Future Hold", Electron, Aust., 42, 14 (1980)
2. "VLSI Technology", (Sze, S. M., Ed.), McGraw-Hill, New York (1983)
3. Electronic Market Data Book 1982, Electronic Industries Association, Washington, D. C. (1982)
4. "World Markets Forecast for 1982", Electronics, 55, No. 1, 121 (1982)
5. Traeger, R. K., in: Proc. 25th Electronic Components Conferences, 361 (1976)
6. Adams, A. C., in: VLSI Technology, John Wiley and Sons, New York (1967)
7. Michael, K. W., Antonen, R. G., in: The Proceedings of the International Society for Hybrids and Microelectronics Conferences, Anaheim, California, USA 1978 and references therein
8. Wong, C. P., in: Polymers in Electronics, 2. Ed. Encyclopedia Polym. Sci. and Eng., Vol. 5, 638, John Wiley and Sons, New York (1986)
9. Adams, A. C., in: "VLSI Technology", (Sze, S. M., Ed.), McGraw-Hill, New York (1983)
10. Kern, W., Ban, V. S., in: (Vossen, J. L., Ker, W., Eds.), "Thin Film Processes", Academic, New York, 257 (1978)
11. Kern, W., Schnable, G. L.: IEEE Trans. Electron Devices, Vol. ED-26, 647 (1979)
12. Douglas, E. C.: Solid State Technol., 22, 61 (1979)
13. Vossen, J. L., Kern, W.: Phys. Today, 33, 26 (1980)
14. Hammond, M. L.: Solid State Technol., 23, 104 (1980)
15. Rosler, R. S.: Solid State Technol., 20, 63 (1977)
16. Brown, W. A., Kaminus, T. I.: Solid State Technol., 22, 51 (1979)
17. Chemical Vapor Deposition — Sixth International Conference, (Gieske, R. J., McMullen, J. J., Donaghey, L. F., Rai-Choudhury, P., Tauber, R. N., Eds.), Electrochemical Society, Princeton, (1977)
18. Hitchman, M. L., in: (Sedgwick, T. O., Lydtin, H., Eds.), Chemical Vapor Deposition — Seventh International Conference, Electrochemical Society, Princeton, New Jersey (1979)
19. Brtant, W. A.: Thin Solid Films, 60, 19 (1979)
20. Van Den Brekel, C. H. J., Bollen, L. J. M.: J. Cryst, Growth, 54, 310 (1981)
21. Kamins, T. I.: J. Electrochem. Soc., 127, 686 (1980)
22. Mandurah, M. M., Saraswat, K. C., Kamins, T. I.: J. Electrochem. Soc., 126, 1019 (1979)
23. Logar, R. E., Wauk, M. T., Rosler, R. S., in: (Donaghey, L. F., Rai-Choudhury, P., and Tauber, R. N., Eds.), Chemical Vapor Deposition — Sixth International Conference, Electrochemical Society, Princeton, New Jersey (1977)
24. Huppertz, H., Engl, W. L.: IEEE Trans. Electron Devices, ED-26, 658 (1979)
25. Adams, A. C., Capio, C. D.: J. Electrochem. Soc., 126, 1042 (1979)
26. Watanabe, K., Tanigake, T., Wakayama, S.: J. Electrochem. Soc., 128, 2630 (1981)
27. Maeda, M., Nakamura, H.: J. Appl. Phys., 52, 6651 (1981)
28. Tobin, P. J., Price, J. B., Campbell, L. M.: J. Electrochem. Soc., 127, 222 (1980)

29. Adams, A. C., Capio, C. D.: J. Electrochem. Soc., *128*, 423 (1981)
30. Pliskin, W. A.: J. Vac. Sci. Technol., *14*, 1064 (1977)
31. Adams, A. C., Schinke, D. P., Capio, C. D.: J. Electrochem. Soc., *126*, 1539 (1979)
32. Nagasima, N.: J. Appl. Phys., *43*, 3378 (1972)
33. Peercy, P. S., Stein, H. J., Doyle, B. L., Picraux, S. T.: J. Electron. Mat., *8*, 111 (1979)
34. Hall, T. C., Peters, J. W.: Insulation/Circuits, Jan. 1981
35. Peters, J. W.: U. S. Pat. *4*, 371, 587 (1981)
36. Elrlich, D. J., Osgood, R. M., Jr., Deutsch, T. F.: IEEE J. Quantum Electronics, 1233 (1980)
37. Boyer, P. K., Roche, G. A., Collins, G. J.: Electrochem. Soc. Extended Abstr. 82-1, 102 (1982)
38. Okake, H.: Photochemistry of Small Molecules, John Wiley and Sons, New York (1978)
39. Volk, M. C., Lefforge, J. W., Stetson, R.: Electrical Encapsulations, Reinhold Publishing Co., New York (1962)
40. Wong, C. P., Rose, D. M.: IEEE Trans. Comp. Hybrids Manufact. Technol., Vol. *CHMT-6(4)*, P. 485 (1983) and references therein
41. White, M. L.: Proc. IEEE, *57*, 1610 (1969)
42. Mancke, R. G.: IEEE Trans. Comp. Hybrids Manufact. Technol., Vol. *4*, 482 (1981)
43. Jaffe, D., Soos, N.: Proc. Electronic Components Conf., 213 (1978)
44. Wong, C. P.: Int. J. Hybrids and Microelectronics, *4* (2), 315 (1981)
45. Wong, C. P., Maurer, D. E.: "Improved RTV Silicone for IC Encapsulants", National Bureau of Standards, Special Publication 400-72, Semiconductor Moisture Measurement Technology, 275 (1982)
46. Wong, C. P., in: Polymer Materials for Electronics Applications, ACS Symp. Ser. No. *184*, 171 (1982)
47. Wong, C. P.: ACS, Organ. Coat. and Appl. Polymer Sci. Proc., Vol. *48*, 602 (1983)
48. Wong, C. P., Rose, D. M.: 33rd Electronic Components Conference Proceedings, 505 (1983)
49. Wong, C. P.: "Integrated Circuit Devices Encapsulants", An Intensive Short Course in "Polymers in Electronics", University Extension, University of California at Berkeley, 1–30, August (1983)
50. Wong, C. P., in: Polymers for Electronic Applications P1–35, Materials Research Laboratory and Center for Continuing Education, State University of New York, New Paltz, November (1983)
51. Wong, C. P., in: Polymers in Electronics, ACS Symp. Ser. No. *242*, 285–304 (1984)
52. Wong, C. P.: Polymer Sci. and Engin. Proc., ACS, Vol. *55*, 803 (1986)
53. Wong, C. P.: U. S. Pat. *4*, 278, 784 (July 14, 1981)
54. Wong, C. P.: ibid. *4*, 318, 939 (March 9, 1982)
55. Wong, C. P.: ibid. *4*, 330, 637 (May 18, 1982)
56. Wong, C. P.: ibid. *4*, 396, 796 (Aug. 2, 1983)
57. Wong, C. P.: ibid. *4*, 508, 758 (Apr. 2, 1985)
58. Wong, C. P.: ibid. *4*, 552, 818 (Nov. 12, 1985)
59. Riley, J. E.: J. Radioanal. Chem., *72*, 89 (1982) and references therein
60. Noll, W.: Chemistry and Technology of Silicones, Academic Press, New York (1968)
61. Wong, C. P.: U. S. Pat. *4*, 564, 562 (Jan. 14, 1986)
62. Wong, C. P.: ibid. *4*, 592, 959 (June 3, 1986)
63. Lee, H., Neville, K.: Handbook of Epoxy Resins, McGraw-Hill, New York (1967)
64 May, C. A., Tanaka, Y.: Epoxy Resins, Marcel Dekker, New York (1973)
65. Encyclopedia of Polymer Sci. Techn., Vol. *6*, p. 209, John Wiley and Sons, New York (1967)
66. Advances in Electronic Circuit Packaging, Vol. *1–4*, Plenum Press, New York (1960–1963)
67. Ito, S., Uhara, Y., Tabata, H., Suzuki, H.: 36th Electronic Components Conf. Proc., p. 360 (1986)
68. Rubner, R.: Siemens Forsch. Entwickl.-Ber. *5* Springer. Heidelberg (1976)
69. Pfeifer, J., Rhode, O.. Proc. Second Intern. Conf. on Polyimides, p. 130, Ellenville, New York (1985)
70. Numata, S., Fujisaki, K., Makino, D., Kinjo, N.: Proc. Second Intern. Conf. on Polyimides, p. 492, Ellenville, New York (1985)
71. Wong, C. P., 5th VLSI Packaging Workshop, p. 45, November 17–18, 1986, Paris

Editors: G. Henrici-Olivé and S. Olivé
Received February 16, 1987

# Photoresist Systems for Microlithography

Frans A. Vollenbroek and Elly J. Spiertz
Philips Research Laboratories, P.O. Box 80.000
5600 JA Eindhoven, The Netherlands

Many different photoresist systems are used in the manufacture of various industrial products such as integrated circuits, compact discs, cathode ray tubes and printed circuit boards. The resist systems are utilized to deliver the relief images, which are needed in one or more steps of the production processes. In this review, the technological applications and chemical design of the various photoresist systems are discussed. In the technology of integrated circuits and optical recording, the emphasis is mainly on resolution, as very small details are required. During the last decade, this has given rise to a lot of research in this field. For cathode ray tubes and printed circuit boards, the required dimensions are less critical. Here the emphasis is on specific properties which the resist should possess, e.g. good chemical resistance and good adhesion to the substrate.

The relative ease with which relief images can now be produced has led to the introduction of microlithography in many other applications. These include the production of liquid crystal displays, liquid crystal television, and solid state cameras. These applications are reviewed only briefly. Finally the impact of polymers on photoresists is discussed more explicitly.

1 Introduction . . . . . . . . . . . . . . . . . . . . . . . . . . . . . 87

2 Chemistry of Photoresist Systems . . . . . . . . . . . . . . . . 88
 2.1 Negative Photoresists . . . . . . . . . . . . . . . . . . . . . . . 88
 2.2 Positive Photoresists . . . . . . . . . . . . . . . . . . . . . . . . 89

3 Integrated Circuits (ICs) . . . . . . . . . . . . . . . . . . . . . . 90
 3.1 Techniques to Improve Resolution . . . . . . . . . . . . . . . 92
  3.1.1 Contrast Enhancement Layer (CEL) . . . . . . . . . . . . 92
  3.1.2 Built On Mask (BOM) . . . . . . . . . . . . . . . . . . . 93
  3.1.3 Profile Modification Technique (Promote) . . . . . . . . . 94
  3.1.4 Image Reversal (ImRe) . . . . . . . . . . . . . . . . . . . 95
  3.1.5 Portable Conformable Mask (PCM) . . . . . . . . . . . . 96
  3.1.6 Trilayer Systems Involving Plasma Etching . . . . . . . . . 97
  3.1.7 Bilayer Systems Involving Plasma Etching . . . . . . . . . 98
  3.1.8 Dry Developable Resists . . . . . . . . . . . . . . . . . . 98
 3.2 Pattern Stabilization . . . . . . . . . . . . . . . . . . . . . . . 99
  3.2.1 Deep UV Hardening . . . . . . . . . . . . . . . . . . . . 99
  3.2.2 Plasma Hardening . . . . . . . . . . . . . . . . . . . . . 100
  3.2.3 Chemical Hardening . . . . . . . . . . . . . . . . . . . . 100
  3.2.4 Molding . . . . . . . . . . . . . . . . . . . . . . . . . . . 100
 3.3 Erosion in Plasma Etch Systems . . . . . . . . . . . . . . . . 100
 3.4 Lift-off Metallization . . . . . . . . . . . . . . . . . . . . . . . 101

Advances in Polymer Science 84
© Springer-Verlag Berlin Heidelberg 1988

**4 Discs for Optical Recording** . . . . . . . . . . . . . . . . . . . . . . . 102

**5 Printed Circuits** . . . . . . . . . . . . . . . . . . . . . . . . . . . . . 102
5.1 Wet Photoresists . . . . . . . . . . . . . . . . . . . . . . . . . . . 102
5.2 Dry-Film Photoresists . . . . . . . . . . . . . . . . . . . . . . . . 103
5.3 Screen Printing . . . . . . . . . . . . . . . . . . . . . . . . . . . 103

**6 Cathode Ray Tubes** . . . . . . . . . . . . . . . . . . . . . . . . . . . 104

**7 Miscellaneous Products** . . . . . . . . . . . . . . . . . . . . . . . . . 105
7.1 Liquid Crystal Displays . . . . . . . . . . . . . . . . . . . . . . . 105
7.2 Liquid Crystal Television . . . . . . . . . . . . . . . . . . . . . . 105
7.3 Colour Filters for Image Sensor Applications . . . . . . . . . . . . 105

**8 Polymers in Photoresists** . . . . . . . . . . . . . . . . . . . . . . . . 106
8.1 Negative Photoresists . . . . . . . . . . . . . . . . . . . . . . . . 106
8.2 Positive Photoresists . . . . . . . . . . . . . . . . . . . . . . . . . 107
8.3 Miscellaneous . . . . . . . . . . . . . . . . . . . . . . . . . . . . 107
8.3.1 Photoresists with High Thermal Flow Stability . . . . . . . . 107
8.3.2 Photoresists that are Resistant to Oxygen Plasma . . . . . . 108
8.3.3 Highly Sensitive Photoresists . . . . . . . . . . . . . . . . . 108

**9 Summary and Outlook** . . . . . . . . . . . . . . . . . . . . . . . . . . 108

**10 References** . . . . . . . . . . . . . . . . . . . . . . . . . . . . . . . . 109

# 1 Introduction

The manufacture of such very different products as integrated circuits (IC), compact discs (CD), cathode ray tubes (CRT) and printed circuit boards (PCB) requires a relief image in a polymer layer on top of a substrate. These images are generally obtained by photolithographic techniques which involve imagewise exposure to near UV light and subsequent development of a photosensitive material (photoresist). High energetic radiation such as deep UV, electron beam and X-ray is also utilized in lithographic techniques, but mostly for specific applications. The resists which are used in this field are different from the near UV photoresists and are not included in this review. Near UV photoresists consist of one or more polymers, one or more light-sensitive compounds and one or more solvents. Generally the photoresist is applied on a substrate by dispensing the solution in the centre of the substrate, followed by spinning. After drying of the photoresist layer a chemical reaction is induced locally by patternwise exposure to UV light. The result is a difference in dissolution rate (in a developer) between exposed and non-exposed areas, which gives rise to a relief image after a certain development time. A resist which shows a decreased dissolution rate upon irradiation is called a negative resist, while an increased dissolution rate refers to a positive resist.

Both positive photoresist systems and negative photoresist systems can be obtained using different light-sensitive compounds. Positive resists make use of a dissolution inhibitor which is destroyed upon exposure to UV light, whereas negative resists make use of crosslinking agents.

For applications, where resolution is not a major requirement, mostly negative photoresists are utilized. For example in the production of cathode ray tubes (CRT) the red, green and blue fluorescing powders are deposited in periodic structures using water-soluble polymers like poly(vinyl alcohol) in combination with a crosslinking agent like ammonium dichromate. These negative-working systems are developable in water. Although swelling occurs during development, the required resolution of approximately 100 µm can easily be achieved. Another negative working system, which has been widely used in printed circuit board fabrication, is the poly(isoprene)-bisazide system. More recently printed circuit board fabrication moved from wet photoresists to laminated dry-film systems because of easier handling. The resolution of these systems is also limited by swelling during development.

For applications, where a high resolution is required, e.g. IC technology and optical recording, resist systems based on novolak (a cresol-formaldehyde resin) or poly(vinyl phenol) are utilized. These resists are developed in aqueous base, which causes no swelling of the polymer and therefore no loss of pattern definition.

In this review, the above-mentioned resist systems will be discussed in terms of the chemistry involved in their photochemical reaction and in their applications in the various production processes.

Finally, the influence of polymers and their properties on the performance of photoresists is discussed.

# 2 Chemistry of Photoresist Systems

In the production of e.g. IC's, CTR's, PCB's and CD's, photoresists are used which are sensitive to UV light in the wavelength region between 300 and 450 nm. In this region, the resists can be manipulated in filtered daylight (yellow) without problems. Using light in this wavelength-region has the additional advantage that high optical resolution can be obtained. In IC lithography, where resolution is the major requirement, there is a growing interest in photoresists that are sensitive to even shorter wavelengths than 300 nm. As already mentioned, these resists are not included in this review.

The chemistry involved in negative and positive photoresist systems is rather different. Therefore, they will be discussed separately in the next sections.

## 2.1 Negative Photoresists

Photoresists of this type have the longest history. Apart from the solvents, these resists consist of a polymer and a crosslinking agent. Examples are poly(vinyl alcohol)-ammonium dichromate dissolved in water, poly(vinyl cinnamate) dissolved in cellosolve acetate and poly(isoprene)-bisazide dissolved in xylene. These resists are (or have been) applied, respectively, in CRT, PCB and IC technologies. The chemistry of these resist systems was reviewed earlier in the monograph by DeForest[1]. The negative action of these resists is due to photochemically induced crosslinking. In poly(vinyl alcohol)-ammonium dichromate, chromium (VI) is photochemically reduced to chromium (III), which complexes with hydroxyl groups of the polymer[2]. Poly(vinyl cinnamate) is crosslinked via a (2+2) cyclo-addition of different cinnamoyl groups[1]. Finally, in poly(isoprene)-bisazide the azides are photochemically converted to nitrenes, giving up a nitrogen molecule. These nitrenes are very reactive and insert, for example, in a carbon-hydrogen bond[1]. In all cases, a three-dimensional network is the result. Although these systems show a large difference in dissolution rate between exposed and non-exposed areas, the resolution is limited because the crosslinked areas exhibit serious swelling in the developer. The reason for this phenomenon is that the crosslinked polymer still has a strong affinity to the developer. As a matter of fact, the developer is the same solvent as that used to make up the photoresist.

Recently, some non-swelling negative resists have been developed, which are based on poly(vinyl phenol) or novolak (a cresol-formaldehyde resin) and a crosslinking agent[3,4]. These resists are made up with organic solvents while development is done in aqueous base. No swelling occurs during development because the polymer is hydrophobic in nature and dissolution occurs only after salt formation via the phenolic OH group (see also Sect. 8). Some resists based on this principle are known at this moment. One system contains poly(vinyl phenol) as the polymer and various bisazides as crosslinking agents[3]. Another system makes use of photochemical formation of an acid, which induces acid-catalyzed crosslinking during a post-exposure bake[4]. Because the mechanism of the crosslinking involves a catalytic reaction, a very high sensitivity can be reached.

Apart from the wet photoresists discussed above, photoresists can also be applied on a substrate using dry-film systems. Such films may consist of a polymeric binder, e.g. a mixture of unsaturated diallyl phthalate prepolymers, a multifunctional acrylic monomer, e.g. pentaerythritol triacrylate, and a photoinitiator, e.g. phenantroquinone. Upon exposure to UV light, a three dimensional network results due to free radical polymerization [1]. Development is carried out in an organic solvent e.g. 1,1,1-trichloroethane.

## 2.2 Positive Photoresists

Positive photoresists are generally based on aqueous base developable polymers (non-swelling) [1, 5].

The main components of most commercially available positive photoresists are novolak as a binder and naphthoquinone-diazide as a light-sensitive component. This light-sensitive compound is not base soluble and acts as a dissolution inhibitor for the novolak, which results in a very low dissolution rate of unexposed resist in aqueous base developer. Upon exposure a reaction is induced to yield indene carboxylic acid [6] via a ketene:

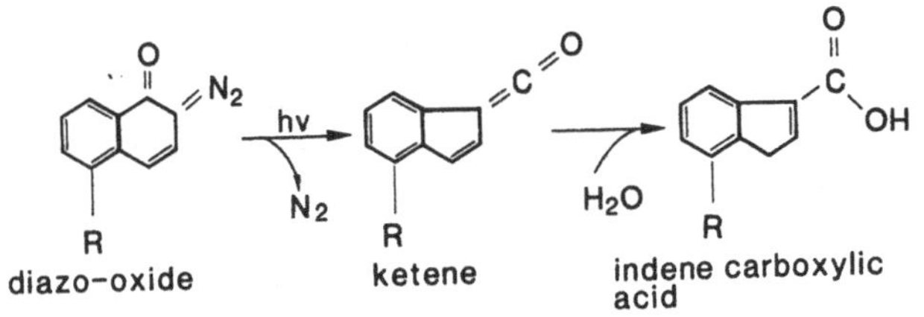

*Scheme 1*

The R-substituent is usually a sulfonic ester of trihydroxybenzophenone of which one or more hydroxyl groups are esterified.

The dissolution rate of the resist is now approximately 1000 times enhanced as compared to unexposed resist because of the absence of inhibitor and because of microporosity which resulted from the nitrogen extrusion [7].

It has been shown that in the absence of water in the resist layer another reaction route occurs. In this case, the intermediate ketene reacts with the phenolic OH group to form an ester linkage [6]:

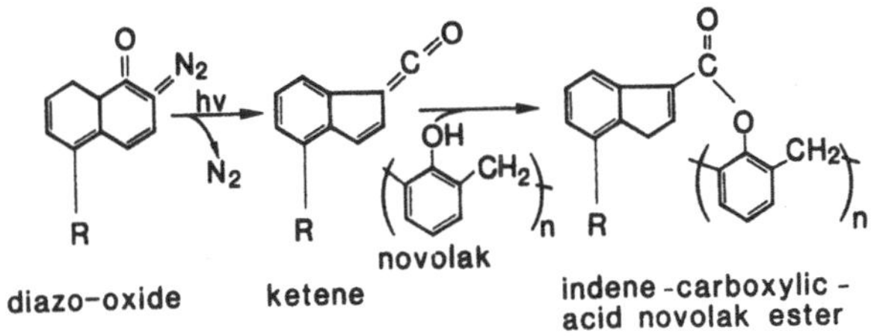

*Scheme 2*

One of the techniques used to improve resolution is based on this reaction (see Sect. 3.1.3).

Another interesting feature of the diazo-oxide chemistry is the possibility to decarboxylate the indene carboxylic acid by thermal activation in the presence of a catalyst (e.g. ammonia) [8]:

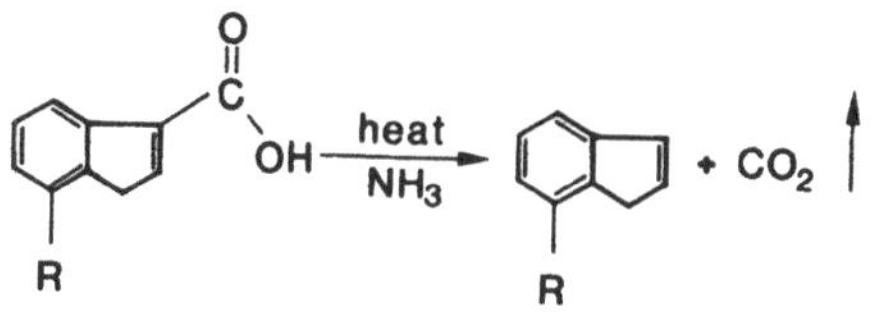

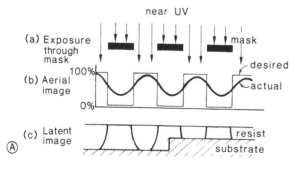

*Scheme 3*

This reaction forms the basis for another technique used to improve the resolution (see Sect. 3.1.4).


# 3 Integrated Circuits (ICs)

The manufacturer of IC's makes use of a lithographic technique involving the projection of a mask pattern on a 1–2 μm thick photoresist, which is coated on a silicon wafer [9]. For the patterning of structures down to 1 μm, full wafer exposure is widely used in machines with mirror optics (1:1 imaging of the mask on the resist). For submicron lithography, step-and-repeat machines are necessary. In these machines, a small area of a wafer covered with photoresist is irradiated imagewise using a 5× or 10× reduction lens and then an adjacent area of the wafer is moved to the optical system. The photoresists most commonly used in this technology are the positive photoresists, based on novolak and diazo-oxides.

These resists are very successful in this field because they are aqueous-base-developable and do not swell during development. However, in the submicron region even these resists are no longer adequate (in their conventional mode) to

**Fig. 1A.** Conventional imaging in a positive photoresist. Imagewise irradiation through a mask (**a**), gives an aerial image (**b**) and results in a latent image (**c**). Upon development, patterns are obtained with sloping profiles (see Fig. 1 B)

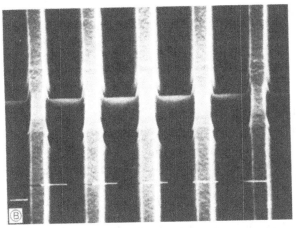

**Fig. 1 B.** Scanning electron microscope (SEM) picture of 1.2 µm lines and spaces in a positive photoresist over a substrate with a SiO$_2$ step, nominal resist thickness 1.3 µm. The width at the top of the lines is different from that at the bottom, due to the limited contrast of the aerial image. This also gives rise to variation of the linewidth at the bottom over the step on the substrate

unravel the information in the image, which originates from the mask (aerial image). Due to diffraction, the intensity distribution of the aerial image is such that the masked areas of the resist also receive a considerable amount of energy (see Fig. 1A). This gives rise to a latent image in which the indene carboxylic acid concentration does not change abruptly at the edge of the masked area. Furthermore, because of absorption of UV light by the photoresist layer, the concentration of indene carboxylic acid decreases from the top to the bottom. Upon development, a sloping profile of the resist lines is the result. The resolution, which can be achieved, is strongly correlated to the angle of this slope. Thus, if the lateral dimensions of the resist lines approach the resist thickness, the resolution is seriously hampered. Apart from this an additional detrimental phenomenon is observed on wafers with topography. As the thickness of the resist layer varies on these substrates, sloping profiles give rise to linewidth variation over steps. This is illustrated in Fig. 1 B.

A solution to these problems can be found in shortening the wavelength ( <300 nm) of the UV light used for projection printing. It is even possible to use X-rays to project the mask pattern onto the silicon wafer [9], but these solutions are not mature yet and require much more research and investment on projection tools and new deep UV and X-ray resists.

In the last decade, several techniques have been developed in order to improve the resolution that can be obtained with optical lithography. This review describes a number of these techniques. Some techniques make use of a photosensitive layer on top of the photoresist, which results in enhancement of the contrast of the aerial image (Sects. 3.1.1 and 3.1.2). Other techniques use the possibility of slope control during development (Sects. 3.1.3 and 3.1.4). Also discussed are some techniques involving anisotropic plasma etching.

After the resist pattern has been delineated, it has to withstand processes like etching (wet or plasma) or ion implantation. A high thermal flow stability and low corrosion rate in plasma etch systems are, therefore, important properties. Several methods have been developed to improve these properties and are discussed in Sects. 3.2 and 3.3.

## 3.1 Techniques to Improve Resolution

### 3.1.1 Contrast Enhancement Layer (CEL)

This layer is applied on top of the prebaked photoresist and consists of a photobleachable dye and a polymer. Upon imagewise exposure to UV light, the non-masked areas bleach relatively fast, whereas the masked areas bleach only marginally (see Fig. 2). The result is an "in situ" mask on top of the photoresist which enhances the contrast of the aerial image. After imagewise exposure, the CEL layer is stripped and the photoresist developed.

The principle of contrast enhancement by photobleachable layers has been known in photography [10] for quite a long time. New formulations were developed for application in lithography and used for the production of cathode ray tubes [11]. More recently, a family of arylnitrones have been investigated and described as effective contrast-enhancing materials (CEM) for use in IC-lithography [12a, b]. The photochemical reaction which occurs is a cyclization reaction:

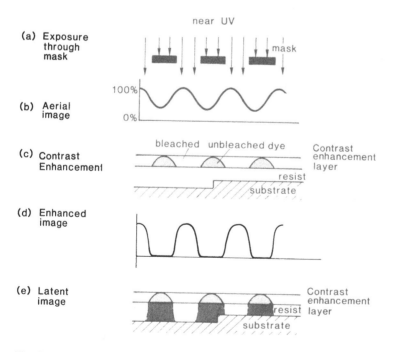

**Fig. 2a–e.** Imaging by using contrast enhancement layer. Irradiation through a mask (**a**) gives aerial image (**b**), which contrast is improved by a photobleachable dye on top of the photoresist (**c**), thus giving an enhanced image (**d**) and hence a better latent image (**e**). Upon stripping the lop layer and development an improved pattern is obtained

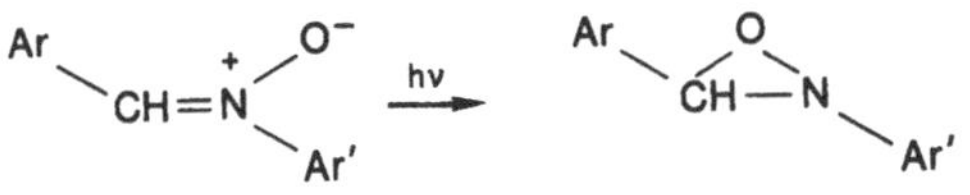

*Scheme 4*

The absorption maximum of these components (unbleached) should correlate with the resist sensitivity. The desired sensitivity can be reached by a proper choice of the substituents. To be able to spincoat this material on top of a photoresist, it is dissolved together with a polymer in an organic solvent that does not attack the photoresist. A suitable combination is toluene as the solvent and styrene copolymerized with allyl alcohol as the polymer. As this polymer is not soluble in aqueous base, it should be stripped very carefully before development of the photoresist.

Intermixing at the interface of the CEL layer and the photoresist can give problems with stripping and, therefore, a barrier layer in between is sometimes applied. This layer is a approximately 1000 Å thick poly(vinyl alcohol) layer which is spincoated from a solution in water [12a]. As this polymer exhibits no intermixing with the photoresist, an alternative CEL system was based on this polymer mixed with a water-soluble diazonium salt [13].

Although the CEL technique gives roughly a twofold improvement in resolution, a disadvantage is the longer exposure time needed ($2-3 \times$) [12a]. This is an inherent property of the CEL method, since the CEL layer should not bleach faster than the photoresist layer.

### 3.1.2 Built On Mask (BOM) [14]

This technique also uses a photosensitive layer on top of a conventional positive photoresist. Upon imagewise exposure to UV light, the absorption maximum of the exposed areas is shifted from e.g. 365 nm to 436 nm (see Fig. 3). Subsequently, the resulting image is fixed by heat treatment, thus producing a permanent mask on top of the photoresist (Built On Mask). This gives the opportunity to transfer the image in the top layer to the photoresist by means of a blanket exposure. If this exposure is done with UV light of 365 nm, a positive image of the mask is obtained in the photoresist after stripping and development, whereas a blanket exposure to 435 nm light results in a negative image (see Fig. 3). The BOM procedure has the advantage over CEL that in principle the imagewise exposure can be done in a much shorter time. Another advantage is that both positive and negative images can be obtained, which enables the IC manufacturer to choose the mask tone with the least transparent areas. Using these masks, the effect of dust particles is minimized.

The chemistry which can be used to complement the BOM technique may be one known from the various copying methods. One of these include an imine which shifts the absorption maximum upon protonation [14]. The proton is produced by photo-decomposition of an activator, e.g. $CBr_4$. In this case, fixation can easily be obtained by evaporation of $CBr_4$. The binder and the solvent used to spincoat a layer on top of the photoresist are the same as used for the CEL technique.

Although it was demonstrated that the BOM concept works with the described materials, there are some shortcomings in the present chemistry. The sensitivity

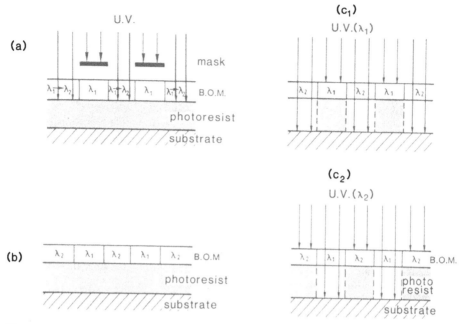

**Fig. 3a and b.** Outline of BOM: **a** imagewise exposure to UV. Absorption maximum shifts from $\lambda_1$ to $\lambda_2$ in irradiated areas; **b** desensitization of top layer (fixation); $C_1$, blanket exposure to $\lambda_1$ gives positive image; $C_2$, blanket exposure to $\lambda_2$ gives negative image

is relatively low, which results in some degree of bleaching of the photoresist during imagewise exposure. Furthermore, the shelf life of the system is not sufficient.

### 3.1.3 Profile Modification Technique (Promote) [15]

This technique does not use a supporting layer but accepts the aerial image as such. However, after imagewise irradition through the mask the latent image in the photoresist is manipulated. This is done by placing the latent image in a dry atmosphere (in vacuum and/or at elevated temperature) and subsequently by giving of flood exposure to deep UV light (see Fig. 4). The diazo-oxide in the masked areas of the imagewise exposure now converts to an ester (see Sect. 2.2) because of the absence of water. This conversion is limited to the upper part of the layer because deep UV light is strongly absorbed by the novolak resin. Subsequently, the latent image is returned to normal atmosphere and flood exposed to near UV. This results in the conversion of the residual photoactive compound to indene carboxylic acid (see Fig. 4). Upon development, a positive image of the mask is obtained, but the profile of the resist lines changes from a positive slope, via vertical, to a negative slope. This is due to the low dissolution rate of the esterified part of the layer as compared to the dissolution rate of the underlying material, which contains mainly indene carboxylic acid. If the development is stopped at the moment the profiles are vertical, high resolution images can be obtained. The method is also applicable to produce resist patterns with negative slopes. These patterns are mainly used for the lift off metallization technique (see Sect. 3.4).

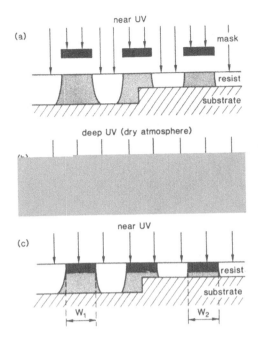

**Fig. 4a–c.** Profile Modification Technique. After imagewise exposure (**a**), a top region of the resist is converted into insoluble material by irradiation with deep UV in a water-free environment (**b**), followed by blanket exposure to near UV in normal atmosphere (**c**). During development, the slope of the resist wall changes from positive to negative. The linewidth is determined by the topwidth of the esterified area

### 3.1.4 Image Reversal (ImRe) [8, 16]

ImRe makes use of another possibility to manipulate the latent image in the photoresist. After imagewise exposure, the indene carboxylic acid in the exposed areas is converted into indene by a base-catalyzed decarboxylation reaction (see Sect. 2.2). Then a flood exposure to near UV is applied to convert all residual diazo-oxide into indene carboxylic acid (see Fig. 5). Because of the decarboxylation reaction in the imagewise exposed areas, these areas now have a low dissolution rate as compared to the masked areas, which contain mainly indene carboxylic acid. Thus, upon

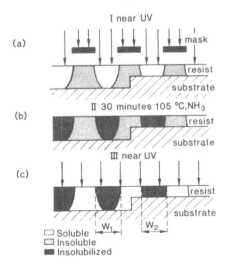

**Fig. 5a–c.** Image Reversal. After imagewise exposure (**a**) the irradiated areas are insolubilized by decarboxylation of the indene carboxylic acid, followed by blanket exposure to near UV (**c**). During development, the slope of the resist wall changes from positive to negative. The linewidth is determined by the top width of the insolubilized area

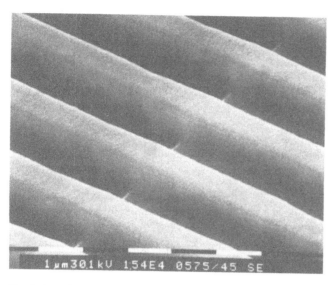

**Fig. 6.** Scanning electron microscope (SEM) picture of 0.8 μm lines and spaces in a photoresist used in an image reversal process. The development was stopped at the moment the profiles of the resist walls were vertical

development a negative image of the mask is obtained, but as in the Promote technique (see Sect. 3.1.3), the profile of the resist pattern changes from a positive slope, via vertical, to overhanging. In this case, this effect naturally arises from the distribution of indene carboxylic acid in the latent image after the imagewise exposure. As this concentration decreases from the top to the bottom (because of absorption), the insolubilization is more effective in the upper part of the decarboxylated area than in the lower part. This causes an increase in dissolution rate from the top to the bottom after the blanket exposure has been applied. If the development is stopped at the moment the profiles are vertical, high-resolution images can be obtained. In Fig. 6, a resist pattern is shown which demonstrates the submicron capability of the ImRe technique. As indicated, Promote (positive image) as well as ImRe (negative image) offer the opportunity to control the slope of resist lines and, therefore, the choice between both can be based on mask requirements.

### 3.1.5 Portable Comformable Mask (PCM) [17]

PCM is a bilayer system with a deep UV sensitive photoresist as the bottom layer and a conventional photoresist on top. First the top resist is imagewise irradiated through a mask and developed. Subsequently, the resulting relief image in the top resist is used as a contact mask in a blanket exposure to deep UV, which converts the bottom resist in the exposed areas into developable material (see Fig. 7). This resist is for example poly(methyl methacrylate) (PMMA), which is fragmented during deep UV exposure. The top resist has very good masking properties for deep UV light as the novolak absorbs this light very strongly. A

(a)

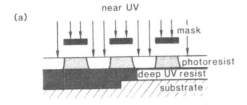

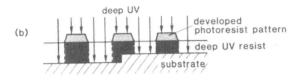

(b)

**Fig. 7a and b.** Portable Conformable mask. A substrate with step is covered (planarized) with a deep UV-sensitive resist (e.g. PMMA) and a top layer which consists of a normal positive photoresist. After imagewise irradiation through a mask (**a**), the top layer is developed (**b**) and blanket exposed to deep UV. This results in a sharp image (see text)

sloping profile of the top resist patterns thus hardly influences the dimensions of the bottom resist pattern.

The advantage of the PCM system is that the topography on the substrate can be planarized with the deep UV resist, providing the opportunity to use a relatively thin photoresist on top with a very uniform thickness. Both characteristics are beneficial for archieving high resolution in the top resist. As the image transfer from the top resist pattern to the bottom resist is done by contact printing with short wavelength light, the pattern in the bottom resist exhibits a very steep profile after development.

Although it has been shown that high resolution can be reached with PCM, its feasibility in a process is still problematic. As in CEL, intermixing occurs at the interface of the bottom and top layers, but in the case of PCM this phenomenon is more severe. It not only hampers the development of the bottom resist, but it also acts as a grey filter during deep UV exposure in the areas which should be transparent to deep UV. Several improvements have been described recently. It was shown that the composition of the top resist has a considerable influence on the degree of intermixing [18a]. On the other hand, using poly(dimethylglutarimide) as the bottom resist also reduces intermixing. This reduction contributes to the high solvent resistance of this polymer [18b].

## 3.1.6 Trilayer Systems Involving Plasma Etching [19,20]

This technique includes a planarizing bottom layer, e.g. hard baked novolak (1–2 μm thick), an intermediate inorganic layer, e.g. $SiO_2$ (0.1 μm thick) and a conventional positive photoresist (0.5 μm thick) as the top layer. Only the photoresist is delineated by means of optical lithography, i.e. the intermediate layer and the bottom layer are patterned by plasma etching. Similarly in the PCM technique, the advantage is that the top resist is relatively thin and has a very uniform thickness. The pattern in the top resist, which may have sloping profiles, is first transferred into the thin inorganic layer by plasma etching in a gas which contains fluorine [19].

Subsequently, the bottom layer is etched by Reactive Ion Etching (RIE) in an oxygen plasma. During this process, the pattern in the inorganic layer protects the underlying material against the etchant. This results in vertical profiles in the bottom layer, because of the anisotropic nature of the RIE technique.

The method is very powerful and submicron structures down to 0.5 μm have been shown. However the process is complex and thus expensive, while the yield is relatively low. A lot of effort is currently being put into reducing the complexity, the aim being to develop a bilayer system which includes all the advantages of the trilayer system. This system is discussed in the next section.

### 3.1.7 Bilayer Systems Involving Plasma Etching

As in the trilayer technique, a bottom layer (e.g. hard baked novolak) planarizes the topography on a substrate. The second layer is a photoresist, which can be delineated conventionally by wet development, but contains an element that can be converted to a nonvolatile oxide during RIE-plasma etching in an oxygen gas. Thus, the top resist takes over the function of the intermediate layer which is used in the trilayer technique.

The concept was first demonstrated using poly(methylvinylsiloxane) as the top resist [21]. Meanwhile, many specially designed Si-containing resists have been demonstrated for application in the bilayer system [22], but only a few of them are suited for near UV lithography. These resists will be discussed in some more detail in Sect. 8.3.2.

Apart from using resists in which silicon is incorporated in the resin, it is also possible to introduce silicon in the pattern of the top resist after wet-development. This can be done by gas-phase silylation of functional groups in the resist (e.g. phenolic OH).

This approach may be easy to implement in a process as conventional novolak containing photoresists can be utilized.

### 3.1.8 Dry Developable Resists

If a resist shows a difference in plasma etch rate between irradiated and non-irradiated areas, plasma etching can be utilized to develop a resist pattern. As this approach does not involve any wet development it is called "dry development".

The first attempts to work out this concept include the use of a mixture of poly-(2,3-dichloro-1-propyl acrylate), N-vinylcarbazole monomer and phenanthroquinone sensitizer [23]. Upon imagewise exposure of a layer of this material to 290–350 nm light the monomer reacts with the polymer in the exposed region. Subsequent heating in vacuum causes evaporation of the unreacted monomer in the masked areas. During development in an oxygen plasma the unexposed regions etch faster than the exposed regions, but the difference is to small for a reliable lithographic process.

A higher contrast material was described being a mixture of poly(methyl isopropenyl ketone) and aromatic bisazides [23 a]. Exposure to UV light results in the formation of nitrenes, which insert into C—H bonds of the polymer. After exposure, a difference in etch rate in an oxygen plasma is observed but this difference increases considerably upon baking. The reported etch rates after baking are 100 nm/min for exposed material and 600 nm/min for unexposed material.

Another promising approach for dry developable system is the introduction of an organometallic compound (e.g. containing silicon) in the exposed or the non-exposed areas. It thus resembles the trilayer and bilayer techniques described above, but now all lithographic functions are combined in one single layer. In order to introduce silicon selectivity in the exposed areas of a layer, a polymer with functional groups can be used that, upon exposure, are transformed into groups that react with a silicon containing gas. Thus poly(p-formyloxystyrene) is imagewise exposed to UV and treated with hexamethyldisilazane [24]. The liberated phenolic OH groups are silylated and upon development in an oxygen plasma a negative image of the mask is obtained.

Another system based on selective incorporation of silicon into exposed areas of the resist is called Desire [25]. After imagewise exposure of the novolak-diazo-oxide resist, the latent image is treated with hexamethyldisilazane at elevated temperature and subsequently developed in an oxygen plasma to give a negative image of the mask [25a]. The selective incorporation of silicon in the exposed areas is attributed to a difference in diffusion rate of the silicon containing gas into the exposed and unexposed areas [25b]. A mechanistic study [25c] on the rate of silylation of various polymers containing phenolic OH groups led to the conclusion that a top region of such a polymer layer is silylated completely after a relative short silylation time. Selective incorporation of silicon in the layer may arise from a difference in the rate by which the topregion progresses in depth.

## 3.2 Pattern Stabilization

After pattern delineation, the next step in a production process is e.g. etching (wet or plasma), ion implantation or metal deposition. Some of these techniques result in heating of the resist above 200 °C. Although novolak resins start to crosslink from a temperature of 120 °C, crosslinking cannot prevent thermal flow of a resist pattern because the flow occurs at a lower temperature already. Several solutions to this problem have been found, and are described below.

### 3.2.1 Deep UV Hardening

A convenient method for pattern stabilization is deep UV hardening. A novolak-based photoresist can thus be given an improved thermal flow stability up to 200 °C. It has been suggested that indene carboxylic acid is the active agent for UV hardening [26]. Later the same authors postulated that salt formation causes an improved thermal flow stability [27]. However, recently it was found that the light-sensitive component of the resist plays no role in UV hardening [28]. The hardening proceeds via a direct novolak excitation followed by crosslinking. This is confirmed by the fact that the hardening proceeds via a benzylic radical intermediate [29].

The UV light, which can induce resist hardening, is deep UV and mid UV (320 nm). As deep UV is very strongly absorbed by the novolak in the resist, crosslinking occurs only in a surface layer [30]. Although this superficial hardening appears to be sufficient to prevent thermal flow of patterns in thin resist layers (1.5 μm), a thoroughly hardened resist layer can be obtained by mid UV hardening (300 nm < λ < 320 nm) [28]. It should be emphasized that the required exposure dose for hardening photoresists is approximately 10–100 times higher than for imaging.

### 3.2.2 Plasma Hardening

If a novolak-based photoresist pattern is placed in a plasma under non-etching conditions (low power or inert gas), the thermal flow stability increases. Several gases can be utilized, e.g. $CF_4$ [31a], nitrogen [31b] and also oxygen [31c]. Although plasma treatment is a cumbersome method for stabilizing resist patterns towards thermal flow, it may be useful in processes where this treatment can be combined with subsequent plasma etching of the underlying substrate. The mechanism of this hardening effect is not known. Possibly reactive species from the plasma induce crosslinking of the novolak. It is also possible that deep UV radiation, evolving from the plasma, is partly responsible for the hardening of the photoresist pattern. Recently, fluorocarbon-like species have been detected in a surface-layer of $CF_4/O_2$ plasma hardened novolak-resist [31d].

### 3.2.3 Chemical Hardening

Patterns of novolak-based resists can be chemically hardened by treatment with formaldehyde in an acidic medium [32]. The reaction that takes place is an inter-molecular condensation. The resin becomes highly crosslinked by reaction at its para positions. Although the method results in a high thermal flow stability (300 °C), it is not attractive for implementation in a production process because of the hazardous nature of the chemicals involved.

### 3.2.4 Molding

A novolak-based resist pattern, which tends to flow at 120 °C, can hold its original shape when it is embedded and covered with an inert polymeric material. Then, it is possible to induce a thermal crosslinking of the novolak at 180–220 °C without distortion of the patterns. Two polymeric materials have been described, i.e. PMMA [33] and a fluorine bearing resin [34]. After the hardening bake, the mold is removed and the hardened patterns can be used in the next process step.

The PMMA-molding method has been developed to be implemented as a process-step in the PCM-system. Removal of the mold is achieved during development of the PMMA-patterns [33]. The fluorine-bearing resin material is coated out of an aqueous solution on top a resist pattern and baked. Removal of the mold is done by rinsing with water [34].

## 3.3 Erosion in Plasma Etch Systems

Apart from thermal flow of resist patterns also resist erosion can occur during plasma etching. Resist erosion refers to thickness loss of the resist pattern during the plasma etch process. The problem is most severe for process conditions required for anisotropic etching of the substrate. As in these processes ion bombardment is involved, it is difficult to avoid the phenomenon of resist erosion. However, polymers containing aromatic groups are more etch-resistant than aliphatic polymers. A study of the mechanism of the etching of polymers in oxygen and argon plasmas under reactive ion etching conditions, revealed that there is only a slight temperature dependence of the etch rate of novolak in an oxygen plasma [34a].

## 3.4 Lift-off Metallization

Metallization by a "lift-off" procedure involves the generation of resist patterns with undercut profiles. A metal is evaporated over the entire surface in such a way that a discontinuity exists between metal on the substrate and metal on top of the resist (see Fig. 8). Then the resist is dissolved and the metal on top of the resist thus lifts off.

In the first method published, PMMA was used as an E-beam resist [35]. Undercut profiles are easily obtained as a result of the non-linear energy absorption in the resist, i.e. the absorption reaches a maximum at about two-thirds of the beam penetration range. In optical lithography, the energy is highest at the top of the resist and thus results in positive sloping profiles in case of positive photoresists. However, undercut profiles can be obtained by means of a chlorobenzene soak before or after exposure [35]. The reason for this overhang formation is the extraction of residual casting solvent and low-molecular-weight resin species [36] and/or removal of photoactive compound [37]. This decreases the dissolution rate in the developer, especially in the non-exposed areas.

Alternative methods for production of overhanging resist profiles suitable for lift-off metallization are Image Reversal [38] and Promote [15]. These methods were described in Sect. 3.1.4 and 3.1.3 respectively. Negative acting resists can also be used to produce overhanging resist profiles. Thus the novolak-bisazide system is probably suited for lift-off applications [3]. Also a negative acting novolak-diazo-oxide resist was reported to produce negative profiles. This was achieved by development in organic solvents instead of aqueous base [38a].

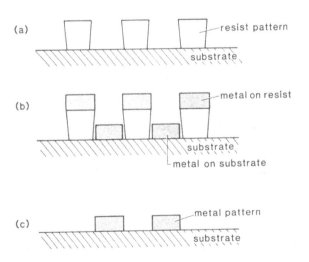

**Fig. 8a–c.** Schematic representation of the lift-off procedure; **a** a resist pattern with negatively sloping profiles is made either by the soak procedure or by image reversal; **b** metal is deposited on the patterned resist, so that a discontinuity exists between metal on the substrate and metal on the resist; **c** the metal on the resist has been lifted off by dissolution of the resist pattern

# 4 Discs for Optical Recording

The storage of audio and video signals by optical recording, as implemented in the Compact Disc and Laser Vision Disc, has been a breakthrough in the field of information storage. The information is stored on a disc in which submicron pits are arranged in spiral tracks and can thus be read out by a laser beam [39].

The manufacture of these discs starts with a glass substrate coated with a positive photoresist having a layer thickness of only 0.12 µm. This resist is the novolak-diazo-oxide resist, which is also used in IC manufacture. The information is „written" in the resist layer by a modulated laser beam, which is moved radially over the rotating glass substrate. Subsequently, the resist is developed and a silver layer is deposited on top. This metallized relief structure is utilized as a master from which a replication process starts. The first step in this process is electroplating of a nickel shell onto the silvered surface of the master. The nickel copy is than separated from the relatively soft resist on the master and contains the negative of the master surface structure. This negative nickel master can be used as a stamper in mass replication methods for audio discs or video discs. Audio discs are replicated by injection moulding, video discs by injection moulding or by the 2p process [39], an especially developed photopolymerization technique.

The holes that are made in the photoresist on the glass substrate have submicron dimensions (width 0.5 µm, length 0.5–2.0 µm). As the resist thickness used is very small (0.12 µm) these sizes can be realized without serious resolution problems with the above-mentioned positive photoresist.

The laser beam used to write the information into the photoresist, is an $Ar^+$-ion laser. Although the 458 nm line from this laser is not the most intense one, it is used in master fabrication because the photoresist is not sensitive at longer wavelengths.

# 5 Printed Circuits

Printed circuits are among the oldest field of application of photoresists. Negative resist types are preferred for their high chemical resistance and low cost. The chemical resistance must be high enough to withstand the plating and etching operations on the substrate. The disadvantage of negative systems, e.g. swelling of the exposed crosslinked parts of the coating during development, is not a serious limiting factor for the resolution because the pattern dimensions in printed circuits are not extremely small (down to 25 µm). Three different systems to produce patterns for application in printed circuitry are currently used and are described below.

## 5.1 Wet Photoresists

The first system, introduced as Kodak KPR resist in 1956, was based on poly(vinyl cinnamate) derivatives as resins and organic solvents like cellosolve acetate [1]. UV exposure brings about crosslinking, after which the unexposed parts can be developed with the original solvent of the resist. A second system, which was introduced for printed circuits in 1960, was based on derivatives of poly(isoprene) rubber as a resin and a bis-azide sensitizer. Upon UV exposure, crosslinking occurs and $N_2$ is evolved.

The unexposed parts can be developed with an aromatic solvent like xylene. This system loses much sensitivity due to reaction of the activated sensitizer (a nitrene) with oxygen instead of crosslinking the polymer. To avoid this, exposure is carried out in $N_2$ atmosphere.

The mentioned wet photoresist systems have gradually been losing their position in the printed circuit field since the introduction of dry-film resist systems in 1970.

## 5.2 Dry-Film Photoresists

Dry-film resists are ready-to-use systems where the photoresist itself is enclosed between two foils. The resists used in dry-films are mostly negative systems, based on acrylate chemistry (see Sect. 2.1).

Dry-films, as introduced in 1970 and based on a patent of Celeste [40], have largely replaced wet resist systems for the fabrication of printed circuits. Most pronounced advantages of dry-film photoresists over wet photoresists are the simplified handling and the diminished chance on pinhole formation. Furthermore, the availability of a range of well-defined thicknesses (25, 38 and 50 μm are mostly used) is especially important during electrolytic metal deposition.

In this process, a thick resist is required to avoid lateral growth of the metal pattern.

In applying dry-film photoresists, the bottom foil of the system is peeled of just before the photoresist is laminated on the substrate. The top foil of the system (polyester) remains in place until developing starts, so that it protects the resist system before and during exposure. During exposure, it also acts as a barrier to oxygen. Dry-film photoresists can be organic-solvent developable (1,1,1-trichloro-ethane) or aqueous base developable (1 % $Na_2CO_3$) [41]. The latter system is preferred because of the less hazardous nature of the developer.

Dry-films also have some disadvantages compared with wet photoresists, e.g. the high cost of the material itself and the processing equipment. Nevertheless, at the present time (1986) details between 50 and 250 μm are mainly produced with dry-film resists and it has been stated that 25 μm can be reached in the near future [42].

## 5.3 Screen Printing

Screen printing is a very old technique for making patterns on a substrate. Screen printing is the art of forcing ink through a stencil which is mounted on a tightly stretched screen. To print a pattern on a substrate, a rubber or plastic sqeegee is used to push a small puddle of ink across the stencil area on the screen. In printed circuitry, screen printing is the preferred technique to produce resist patterns on copper clad substrates, especially if the dimensions exceed 250 nm [43]. As the resist does not need to be a photoresist, mostly heat-curable materials are used [44].

The pattern to be reproduced in a screen-printing process is recorded in the stencil by a photolithographic technique [45]. The light-sensitive system of the stencil may be ammonium dichromate or a water-soluble aromatic azide with a water-soluble polymer [46]. After crosslinking by UV exposure, the system can be developed with water. The stencil thus obtained may be applied to a stainless steel screen material. Such a screen can print some 10,000 items [47].

# 6 Cathode Ray Tubes

Cathode Ray Tubes (CRT) contain two elements that are made by means of microlithography: the shadow mask and the TV screen. When a TV set operates, electrons emitted from three differently positioned electron guns pass through the holes of the shadow mask and excite the phosphor raster on the TV screen. As three different phosphors are present on the screen (for the colours green, blue and red), a full colour image is produced [48].

Both the shadow mask and the TV screen are manufactured by using photoresists based on a water-soluble polymer and ammonium dichromate as the light-sensitive crosslinking agent. The photoresist used for making shadow masks is based on flish-glue or some other gelatine. The process is straightforward and requires only one lithographic step (exposure and development) and a subsequent etch process.

On the other hand, the manufacture of the TV screen is a multi-step operation [48]. The lithographic procedure has to be repeated for each of the three colours separately and in some cases an extra step is necessary in order to generate a non reflecting matrix pattern around the colour dots [49]. The colour dots are produced on the screen by flow coating of a suspension of any of the three phosphors in an aqueous poly(vinyl alcohol)-ammonium dichromate solution. After drying, the coating is imagewise irradiated through the shadow mask. The same mask is used for delineation of the three phosphors, but the position of the light source during exposure is different for each case. These positions correspond to the positions of the three electron guns mentioned above. In all cases, development is achieved by spraying with water.

After delineation of the resist with the last phosphor, the polymer is removed from the phosphor dots by a high temperature bake. The polymer must disappear completely, otherwise no vacuum can be maintained in the CRT. Another important function of the bake at high temperature is that the phosphor dots are made to adhere very well to the screen surface.

The shadow mask, which was used to delineate the raster on the screen, is mounted together with the screen in one CRT to obtain a close correlation in each tube between mask and raster.

For the lithographic procedure as described for the TV screen, a coating is used with a very low polymer content: 1 gram of poly(vinyl alcohol) with 5 to 10 grams of phosphor powder. So, the light-sensitive system can only form a thin shell around the phosphor grains. That the suspensions nevertheless show a fairly good stability is partly attributable to adsorption of dichromate ions on the surface of the phosphor grains.

The quality of the image can be enhanced by several methods that give the patterns a sharper profile either by influencing the light intensity distribution [11] (see Sect. 3.1.1) or by aftertreatment of the developed pattern. Spraying with acetone immediately after development [50] gives a sharper profile, which is ascribed to rapid displacement of water out of the swollen image.

# 7 Miscellaneous Products

Apart from the applications discussed above, there are several other products where microlithography is involved. Among these products are liquid crystal displays, liquid crystal television and colour filters for image sensors. The manufacture of most of these products does not require resist properties additional to those which already· exist in most conventional photoresists. Accordingly, a selection of these resists is normally made.

## 7.1 Liquid Crystal Displays

Liquid Crystal Displays (LCD) contain a thin layer of liquid crystal material sandwiched between two substrate glasses containing electrode patterns (e.g. indium oxide) [51]. These patterns are made by a photolithographic procedure, followed by etching of the electrode material. Screen printing is also used, but only for large geometries. This technique was discussed in Sect. 5.3. As LCD's have become very popular in such products as watches and calculators, this field is an important consumer of photoresist materials.

## 7.2 Liquid Crystal Television

Flat arrays of liquid crystal displays are very attractive for use as a television screen because they reduce weight and volume considerably as compared with the cathode ray tube.

In order to address each LCD on the panel, amorphous silicon thin film transistors have been used [52], which are made by a conventional lithographic procedure. Although the latest reported size is "only" 12.5 cm (5 inch), it is a promising development in a field where the cathode ray tube has enjoyed a monopoly position since the introduction of television.

## 7.3 Colour Filters for Image Sensor Applications

The replacement of the conventional image tube by a solid-state image sensor in colour TV cameras is an important step forward in the registration of video signals. To obtain a three-colour signal, the image sensor has to be provided with a colour filter. This can be done directly on the silicon wafer, which contains many sensors, by subsequent delineation of three gelatin dichromate layers and intermediate dyeing [53]. Thus the first gelatin dichromate layer is patternwise irradiated and developed. Then the stripes are dyed red in a solution of e.g. a water-soluble acid dye in diluted acid. The red stripes are now covered with a barrier layer (e.g. nitrocellulose), which prevents the red stripes from being dyed during further processing.

By repeating this process twice using a green and a blue dye, green and blue stripes respectively, are formed, next to the first red stripes. Gelatin is a useful resin for this purpose because it can be dyed very easily as described above. On the other hand, the resolution is limited because of swelling during development and dyeing, and this has already led to a search for alternative methods. These include dyeing with heat transfer dyes and plasma etching of coloured layers.

# 8 Polymers in Photoresists

After having discussed the photochemistry involved in a number of photoresists and their technological applications, we will consider in some more detail the role of polymers in photoresist systems. The most familiar polymers in photoresist are poly(vinyl alcohol), poly(vinyl cinnamate), poly(isoprene) and novolak. These resins are manufactured on an industrial scale according to general procedures, but the exact conditions, applied in the production process are mainly proprietary to the manufacturers.

The polymer influences many properties of the resist system. These properties are divided into two categories; physical and functional. Thus, viscosity, solid content, solvent compositions and molecular weight distribution of the polymer are important physical properties. The functional properties include sensitivity, spectral response, contrast, adhesion to the substrate, development conditions, thermal flow, plasma resistance, and planarization capability.

In order to get reproducible lithographic results in any production process, it is important that the functional properties of the applied photoresist be kept constant from one to the other lot. Obviously, the functional properties of a resist are correlated with its physical properties. In this section, some of these correlations will be discussed for both negative and positive photoresists. Finally, some speciality polymers for miscellaneous applications will be described.

## 8.1 Negative Photoresists

The polymers-poly(vinyl alcohol), poly(vinyl cinnamate) and poly(isoprene)-are very well-known in negative photoresists (see Sect. 2.1). As these resists are insolubilized by crosslinking, the sensitivity of the resist is closely related to the molecular weight of the polymer. A high molecular weight results in high sensitivity because already a few crosslinks give total insolubilization. On the other hand, a high molecular weight of the resin limits its solubility and the choice of solvents. It also strongly increases the viscosity of a solution of the polymer.

Another important property of polymers for negative photoresists is the molecular weight distribution. A narrow distribution gives the sharpest contrast between irradiated and non-irradiated areas of the photoresist.

Apart from these general polymer aspects in negative photoresists, there are also some specific ones.

Poly(vinyl alcohol), for example, can best be used at its maximum water-solubility (for 88% hydroxyl groups) where it also has a low tendency to crystallize [54]. Poly(isoprene) shows optimal results at 90% bicyclic structures and 10% uncyclized poly(isoprene) [1].

As negative photoresists based on the above-mentioned polymers exhibit serious swelling during development, their resolution is limited to approximately 5 μm. Therefore, some negative resist formulations based on novolak have been developed recently [3,5]. As novolak is mostly used in positive photoresists, the properties of this resin are discussed in the next section.

## 8.2 Positive Photoresists

Most positive photoresists are based on novolak. This resin is derived from a cresol mixture, formaldehyde and an oxalic acid catalyst. A typical ratio for the cresol mixture is 70% meta, 20% para and 10% ortho cresol.

This mixture may result in a novolak resin with an $M_n$ (number average molecular weight) of 500–900 with a polydispersity ($M_w/M_n$) of 40–70 [29]. Although one would expect a better resolution of a resist based on a novolak with a narrower molecular weight distribution, no such effect was observed. By measuring the contrast values of photoresists, based on different fractions of a novolak (separated by gel permeation chromatography), it was concluded that low molecular weight fractions ($M_w < 12000$) do lead to a resist with a diminished resolution, but if these fractions are present in an unfractionated novolak, no detrimental effect is observed [29]. On the other hand, an improvement of the contrast value was observed when the ratio of meta to para (m/p) was reduced from 10/0 to 4/6 at a molecular weight of approximately 12,000 [55]. The position of the methylene bond in meta cresol novolak was also found to be important. It turns out that a relative abundance of methylene bonds on ortho positions to the OH group gives a photoresist with a higher contrast.

Apart from effects on resolution, the ratio of meta-, para- and ortho-cresol in the novolak also influences thermal flow and sensitivity; o-cresol lowers the softening point of the novolak, m-cresol gives an increased resist sensitivity [56].

A very important property of novolak is that the resolution of photoresists based on this resin is not limited by swelling during development in aqueous base. Recently, it was argued that some swelling, at least superficial, may occur [57]. This enables penetration of ions into the layer which can explain the observation that the rate of dissolution decreases, when the cation of the aqueous base developer increases in size.

## 8.3 Miscellaneous

Many polymers have been designed for the formulation of photoresists with special properties, e.g. high thermal flow stability, resistance to oxygen plasma and high sensitivity. As these polymers may become important in future resist applications, some of them are discussed below.

### 8.3.1 Photoresists with High Thermal Flow Stability

Ordinary novolak, as discussed in Sect. 8.2, has a low thermal flow stability. During the bake treatment (120 °C), which is normally applied to developed resist patterns, thermal flow leads to deformation. Although this problem can be solved by hardening the pattern by chemical or physical means (see Sect. 3.2), many phenolic polymers with high $T_g$ have been studied as replacements for the novolak resin. For example poly(vinyl phenol) ($T_g$ 160–180 °C) has been utilized in a commercially available photoresist [58]. Also copolymers of N-(p-hydroxyphenyl) maleimide and various olefins have been described [59].

### 8.3.2 Photoresists that are Resistant to Oxygen Plasma

For the bilayer system described in Sect. 3.1.7 many silicon-containing polymers have been designed [22]. Upon exposure of a layer of these polymers to an oxygen plasma, an upper part of the layer is converted to $SiO_2$ which protects the underlying material from being etched. Most of these polymers are only suitable for deep UV or E-beam irradiation. Important exceptions are modified novolaks which can be used in photoresist in the conventional way. Thus p-trimethylsilylmethylphenol and cresol were subjected to co-condensation with formaldehyde to give a silicon-containing novolak resin [60]. Recently, a negative Si-containing photoresist was formulated as a mixture of poly(allyldimethylsilyl-α-methylstryrene) and a bisazide [61].

### 8.3.3 Highly Sensitive Photoresists

Although the novolak-diazo-oxide system is satisfactory in many processes, a considerably effort is being made to find resist formulations that show higher sensitivity. It has been demonstrated [5,62] that by using photocatalytic systems the sensitivity can be enhanced considerably as compared to the diazo-oxide system. In these systems the difference in dissolution rate is achieved by catalytic cleavage of the inhibitor [5] or liberation of a phenolic OH group [62] in the irradiated areas. The catalyst is a Brönsted acid which is formed after photolysis of an onium salt or a polyhalogen compound. The systems are still aqueous base developable to ensure that no swelling occurs.

## 9 Summary and Outlook

In many technologies, relief images in photoresists are utilized to define structures in the underlying substrate or in a material which is to be deposited. Generally, the resist structures have only a temporary function, i.e. after etching or deposition the resist is removed. Nevertheless the photoresist plays an important role in these technologies because the image quality in the photoresist eventually determines the accuracy with which the pattern on the mask is transferred to the substrate or to the deposited material.

Negative photoresists based on a polymer resin and a crosslinking agent are mostly used in CRT and PCB applications. As high resolution is not a prerequisite here, the emphasis is on adhesion to the substrate, chemical resistance and some other properties related to the process.

Positive photoresists based on novolak and a diazo-oxide are mostly used in applications where high resolution is required, e.g. in IC and optical recording technologies. The search for high resolution resist systems, especially for application in ICs, has led to many interesting concepts, including modified single-layer systems and intriguing multilevel schemes.

With respect to single layer systems (conventional, ImRe, Promote and dry developable resists), we expect that computer programs that can simulate latent image formation and development processes will become of great importance in the near future. A better understanding of the mechanism which underlies the dissolution or etching of resist layers in a developer or plasma respectively is also crucial in order to achieve the ultimate resolution of single layer photoresists. In this respect, well chosen and perfectly defined polymers are required.

Although multilevel schemes are more complex than single layer systems, the former will probably grow in importance because high resolution capability can be reached in a very straightforward way. These systems often require tailor-made polymers and photosensitive materials, thus giving rise to a lot of research in this field.

A further improvement of resolution can be found in deep UV or X-ray lithography. This development requires special resist formulations that are sensitive (and sufficiently transparent) to these kinds of radiation. Although much research is going on in this field at present, this effort will probably increase if deep UV or X-ray projection tools become more readily available.

# 10 References

1. DeForest, W. S.: Photoresists, Materials and Processes, McGraw-Hill, New York 1975
2. Grimm, L., Hilke, K. H., Scherner, E.: J. Electrochem. Soc. *130*, 1767 (1983)
3. Koibuchi, S., Isobe, A., Makino, D., Iwayanagi, T., Hashimoto, M., Nonogaki, S.: SPIE Proc. *539*, 182 (1985)
4. Spak, M., Mammoto, D., Jain, S., Durham, D.: SPE-Regional Technical Conference: "Photopolymers, Principles, Processes and Materials", Ellenville, N.Y., 28–30 October 1985, Technical Papers, 247
5. Steppan, H., Buhr, G., Vollmann, H.: Angew. Chem. *94*, 471 (1982)
6. Pacansky, J., Lyerla, J. R.: IBM J. Res. Developm. *23*, 42 (1979)
7. Hinsberg, W. D., Willson, C. G., Kanazawa, K. K.: SPIE Proc. *539*, 6 (1985)
8. "Introduction to Microlithography", Thompson, L. F., Willson, C. G., Bowden, M. J., Editors, ACS Symposium Series 219, Washington DC (1983), Chapter 3
9. "Introduction to Microlithography", Thompson, L. F., Willson, C. G., Bowden, M. J., Editors, ACS Symposium Series 219, Washington DC (1983), Chapter 2
10. U.S. Patent 3511653 (1970) granted to American Cyanamid Co.
11. U.S. Patent 3965278 (1978) granted to U.S. Philips Corp.
12a. Griffing, B. F., West, P. R.: Solid State Technol. *28*, (5) 152 (1985)
12b. West, P. R., Davis, G. C., Griffing, B. F.: J. Imag. Sci. *30*, 65 (1986)
13. Halle, L. F.: J. Vac. Sci. Technol. *B3*, 323 (1985)
14. Vollenbroek, F. A., Nijssen, W. P. M., Kroon, H. J. J., Yilmaz, B.: Microelectron. Eng. *3*, 245 (1985)
15. Vollenbroek, F. A., Spiertz, E. J., Kroon, H. J. J.: Polym. Eng. Sci. *23*, 925 (1983)
16. Gijsen, R. M. R., Kroon, H. J. J., Vollenbroek, F. A., Vervoordeldonk, R.: SPIE Proc. *631*, 108 (1986)
17. Lin, B. J.: Solid State Technol. *26*, 105 (1983)
18a. Wijdenes, J., Geomini, M. J. H. J.: SPIE Proc. *539*, 97 (1985)
18b. de Granpre, M. P., Vidusek, D. A., Legenza, M. W.: SPIE Proc. *539*, 103 (1985)
19. Moran, J. M., Maydan, D.: J. Vac. Sci. Technol. *16*, 1620 (1979)
20. "Introduction to Microlithography", Thomson, L. F., Willson, C. G., Bowden, M. J., Editors, ACS Symposium Series 219, Washington DC (1983), Chapter 6
21. Hatzakis, M., Paraszczak, J., Shaw, J.: Proc. Microcircuit Engineering Conference, Lausanne 1981, p. 386
22. Onhishi, Y., Suzuki, M., Saigo, K., Saotome, Y., Gokan, H.: SPIE Proc. *539*, 62 (1985)
23. Taylor, G. N., Wolf, T. M.: Proc. Microcircuit Engineering Conference, Lausanne 1981, p. 381, and references therein
23a. Tsuda, M., Yabuta, M., Oikawa, S., Yokota, A., Nakane, H.: Proc. Microcircuit Engineering Conference, Cambridge 1983, p. 371

24. MacDonald, S. A., Ito, H., Hiraoka, H., Willson, C. G.: SPE-Regional Technical Conference: "Photopolymers, Principles, Processes and Materials", Ellenville N.Y., 28-30 October 1985, Technical Papers, 177
25. Coopmans, F., Roland, B.: SPIE Proc. *631*, 34 (1986)
25a. Eur. Pat. 0184567 granted to U.C.B. (1986)
25b. Roland, B., Coopmans, F.: Solid State Devices and Materials *18*, 33 (1986)
25c. Visser, R. J., Schellekens, J. P., Reuhman, M. E., van Yzendoorn, L. J.: to be published in: SPIE Proc. Advances in Resist Technology (1987)
26. Hiraoka, H., Pacansky, J.: J. Vac. Sci. Technol. *19*, 1132 (1981)
27. Hiraoka, H., Pacansky, J.: J. Electrochem. Soc. *128*, 2645 (1981)
28. Spiertz, E. J., Vollenbroek, F. A., Verhaar, R. D., Dil, J. G.: Proc. Microcircuit Engineering Conference, Berlin 1984, 527
29. Pampalone, T. R.: Solid State Technol. *27* (6), 115 (1984)
30. Allen, R., Foster, M., Yen, Y.: J. Electrochem. Soc. *129*, 1379 (1982)
31a. Ma, W. H. L.: SPIE Proc. *333*, 19 (1982)
31b. Moran, J. M., Taylor, G. N.: J. Vac. Sci. Technol. *19*, 1127 (1981)
31c. U.S. Patent 3920483 (1975)
31d. Verkerk, M. J., Witjes, A. J., Veenvliet, H.: to be published in: J. Mat. Sci., 1986
32. Moreau, W. M.: Proc. Microcircuit Engineering Conference, Cambridge 1983, 321
33. Lin, B. J., Chao, V. W., Petrillo, K. E., Yang, D. J. L.: SPE-Regional Technical Conference: "Photopolymers, Principles, Processes and Materials", Ellenville N.Y., 28 30 October 1985, Technical Papers, 75
34. Grünwald, J. J., Spencer, A. C.: SPIE Proc. *631*, 62 (1986)
34a. Visser, R. J., de Vries, C. A. M.: ISPC 8th International Symposium on Plasma Chemistry, Tokyo 1987
35. Hatzakis, M., Canavello, B. J., Shaw, J. M.: IBM J. Res. Developm. *24*, 452 (1980)
36. Halverson, R. M., MacIntyre, M. W., Matsiff, W. T.: ibid. *26*, 590 (1982)
37. Mimura, Y.: J. Vac. Sci. Technol. *B4*, 15 (1986)
38. Moritz, H.: Proc. Microcircuit Engineering Conference, Berlin 1985, 45
38a. Yamashita, Y., Kawazu, R., Itoh, T., Kawamura, K., Ohno, S., Kobayashi, K., Asano, T., Nagamatsu, G.: Microelectron. Eng. *3*, 305 (1985)
39. "Principles of Optical Disc. Systems", G. Bouwhuis, Editor, Adam Hilger Ltd., Bristol 1985
40. U.S. Patent 3,469,982 (1969) granted to J. R. Celeste
41. Fullwood, L.: Insulation/Circuits, January 1982, 47
42. Wopschall, R. H.: Solid State Technol. *29* (6), 153 (1986)
43. "Handbook of Printed Circuit Manufacturing", R. H. Clark, Editor, van Nostrand and Reinhold, New York 1985
44. Schueler, S. D., Green, W. J.: Electronics, April 1984, 17
45. Roffey, C. G.: Photopolymerization of Surface Coatings, Wiley, New York 1982
46. Ger. Offen. DE 1622763 granted to Kalle (1959)
47. Nersesian, R. M.: Appendix 6 in "Handbook of Printed Circuit Manufacturing", R. H. Clark, Editor, van Nostrand and Reinhold, New York 1985
48. Scharner, E., Grimm, L., Hilke, K. J.: Chemie in unserer Zeit *9* (6), 163 (1975)
49. B.P. 1,180,195, Neth. Appl. 6,804,370 and Neth. Appl. 7,407,985
50. Neth. Appl. 7,007,777
51. Bahadur, B.: Mol. Cryst. Liq. Cryst. *109*, 3 (1984)
52. Yamano, M., Takesada, H.: J. Non-Crystalline Solids 77 & 78, 1383 (1985)
53. U.S. Patent 4,339,514 and Eur. Patent 90,865 granted to Polaroid
54. Molyneux, P.: "Water-soluble Synthetic Polymers: Properties and Behaviour", Vol. I, CRC Press, Boca Raton 1983
55. Hanabata, M., Furuta, A., Nemura, Y.: SPIE Proc. *631*, 76 (1986)
56. Eur. Pat. 0070624 granted to Hunt (1983)
57. Arcus, R. A.: SPIE Proc. *631*, 124 (1986)
58. Ger. offen DE 3309222 (1983) granted to Shipley Co.
59. Turner, S. R., Arcus, R. A., Houle, C. G., Schleigh, W. R.: SPE-Regional Technical Conference: "Photopolymers Principles Processes and Materials", Ellenville N.Y., 28-30 October 1985, Technical Papers, 35

60. Tarascon, R. G., Shugard, A., Reichmanis, E.: SPIE Proc. *631*, 40 (1986)
61. Saigo, K., Watanabe, F., Ohnishi, Y.: J. Vac. Sci. Technol. *B4*, 3, 692 (1986)
62. Willson, C. G., Ito, H., Fréchet, J. M. J., Tessier, T. G., Houlihan, F. M.: J. Electrochem. Soc. *133*, 181 (1986)

Editor: K. Dušek
Received March 23, 1987

# Electrochemistry and Electrode Applications of Electroactive/Conductive Polymers

A. F. Diaz[1], J. F. Rubinson[2] and H. B. Mark, Jr.[2]

This report describes some of the recent work on the electrochemical and electrode applications of polymers which are electroactive and can be switched to an electrically conductive state, as well as the inherently conductive $(SN)_x$. The materials fall into two general categories. There are the polymer films which can be prepared *in situ* by the electrochemical polymerization of aromatic compounds, and there are the polyenes such as polyacetylene and polythiazyl. Many of the electrode applications being considered are based on the electroactive/conductive properties of the films such as display devices, and storage batteries. Some applications make use of the conductive property of the materials such as protective coatings against corrosion, and other applications make use of the possibility for molecular selectivity such as chemically selective electrodes and sensors.

1 Introduction . . . . . . . . . . . . . . . . . . . . . . . . . . . 114

2 Polyaromatic Polymer Films . . . . . . . . . . . . . . . . . . 115
    2.1 Polyaniline . . . . . . . . . . . . . . . . . . . . . . . . 115
    2.2 Polypyrrole . . . . . . . . . . . . . . . . . . . . . . . 123
    2.3 Polythiophene . . . . . . . . . . . . . . . . . . . . . . 129
    2.4 Other Polyaromatic Films . . . . . . . . . . . . . . . . 130

3 Linear Polyene Polymers . . . . . . . . . . . . . . . . . . . 131
    3.1 Polyacetylene . . . . . . . . . . . . . . . . . . . . . . 131
    3.2 Polythiazyls $(SN)_x$ . . . . . . . . . . . . . . . . . . . . 134

4 Other Organic Electrode Materials . . . . . . . . . . . . . . 136

5 Conclusions . . . . . . . . . . . . . . . . . . . . . . . . . . 136

6 References . . . . . . . . . . . . . . . . . . . . . . . . . . . 136

[1] IBM Almaden Research Center, 650 Harry Road, San Jose, California 95 120
[2] Department of Chemistry, University of Cincinnati, Cincinnati, Ohio 45 221

Advances in Polymer Science 84
© Springer-Verlag Berlin Heidelberg 1988

# 1 Introduction

A tremendous amount of interest has developed in the last six years on the electrochemical characteristics and the electrode properties of polymer films which can be
switched into a conductive state [1]. It is the electroactivity, the property of these
materials to switch between two states, coupled with the fact that the two states have
widely differing conductivities, which makes "conductive polymers" attractive for
many technological applications Eq. (1). This is also the characteristic which makes
this area of science and technology particularly suited for electrochemists. In this
manuscript, it is often referred to the switching properties of the polymers. These
electroactive polymers can be driven between two oxidation states by varying the
applied voltage. Under controlled conditions, the process is thermodynamically
reversible and the two states of the polymer can be produced repeatedly by cycling
the voltage. This is known as the switching process. Many applications are based on
this property and the applicability of the material is determined by two principle
reaction characteristics: the switching maximum rate and the reversibility of the reaction. The latter is often measured in terms of the number of switching cycles accomplished without loss of charge density, optical density (when applicable) and chemical
degradation. Besides the switching/conductive properties, the fact that the materials
are electrochromic, can be structurally modified, and are not susceptible to photocorrosion, make them a serious alternative to metals for many technologies. These
are important considerations which should not be overlooked when pursuing the
development of a technology based on these materials because other considerations,
such as cost and stability, may often favor the use of a metal. These materials have
been considered for displays, charge storage, electrodes with chemical selectivity,
protective coatings for photoelectrodes against corrosion, ion gates, time release
electrodes for chemicals, and molecular transistors. These materials are far from ideal
and are often limited by their chemical and mechanical stability, and switching rates.
The relative importance of these limitations varies with the application, thus a specific
application often needs to be defined when considering improvements or optimization
of these materials.

$$\text{Neutral, insulating polymer} \rightleftharpoons \text{Cationic, conductive polymer-anion composite} \tag{1}$$

In this review, we discuss the electrode/electrochemical properties and technological
applications of $\pi$-conjugated polymers. It is organized by material and the materials
are grouped in two classes, the polyaromatic polymer films such as the polymer derivatives of aniline, pyrrole, and thiophene, and the linear polyenes such as polyacetylene,
polythiazyl. The report does not intend to review the preparation procedures or the
physical properties of these materials since they have been summarized elsewhere [1, 2].
It is also limited to work published in the scientific literature without attempting to
include the numerous patents which have appeared in the last few years.

# 2 Polyaromatic Polymer Films

The materials discussed in this section are polyaromatic polymer/anion composite films prepared electrochemically by the polymerization of the derivatives of pyrrole, aniline, thiophene, azulene and benzene. The films are normally used directly as grown on the metal substrate where they have good adhesion and electrical contact. The switching properties of these materials are described with more detail in the first section which describes polyaniline. However, it should be understood that these characteristics are generally common to all of the materials.

## 2.1 Polyaniline

In 1862, Letheby reported the formation of a dark green precipitate from the electrochemical oxidation of aniline in aqueous solution [3]. This material soon became known as "aniline black", and has been of intermittent interest to electrochemists [4-12] as is outlined by Mohilner, Adams and Argersinger [4]. The interest has been both in the mechanic aspects of the reaction which produces "aniline black" [5, 6] as well as in the properties of the resulting material. Prior to 1980, the material at hand was primarily a powdery precipitate consisting of lower molecular weight oligomers including the octamer which could on occasion be solubilized. These studies were performed with the powdery deposit directly on the electrode or with pressed pellets. Polyaniline is electroactive Eq. (2) and switches in aqueous solutions in the potential range 0.1 to 0.8 V anodic of the hydrogen electrode [7], producing broad oxidation/reduction waves in the voltammogram. The reaction is accompanied by a change in the conductivity of the material by a factor of $10^6$. It was quickly recognized that this material could be useful in technological applications such as a carbon electrode [8], redox membranes with permselective properties [7] and with sensitivity to pH [7], charge storage [9, 10] and electronic switches [11].

$$\left.\left[\bigcirc\!\!-\!NH_2-\bigcirc\!\!-\!NH_2\right]_n \rightleftharpoons \left[\bigcirc\!\!-\!NH=\bigcirc\!\!=NH\right]_n^{2\oplus} + 2H^\oplus + 2e^\ominus \quad (2)\right.$$

In 1980, a procedure was reported for the preparation of polyaniline as a thin and coherent film, by the electrooxidation of aniline in aqueous acid solution using a platinum electrode and with cycling of the voltage between $-0.2$ and $+0.8$ V *versus* SCE [12]. Since the original report, good quality films of polyaniline have also been prepared on other electrode materials, including Si [13], GaAs [13], GaP [13], Cd-chalcogenides [13] and carbon [14]. The general electroactive characteristics of the films are consistent with those of the powdery materials described in the earlier reports. They switch in aqueous acid solution in the region $-0.2$ to $+0.8$ V, and the voltammograms have different forms depending on the preparation conditions and the treatment of the film. This is seen in Fig. 1 which shows voltammagrams of films prepared in two different laboratories. The amount of faradaic charge involved in the oxidation reaction is 0.16–0.29 Faradays/repeating unit [12, 17], and is often accompanied by a large capacitive current. This is a characteristic of all of the electropolymerized films which are electroactive and which switch from insulating to electrically

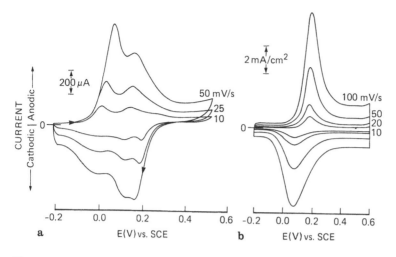

**Fig. 1a and b.** Cyclic voltammogram of polyaniline film on platinum in aqueous 0.1 M sulfuric acid solution **(a)** and 1 M HCl solution **(b)** using a sodium chloride calomel reference electrode (reprinted with permission from Ref. [12, 18])

conductive. As seen in Fig. 1, the reaction is complicated, involving the interconversion of benzenoid to quinoid structures, and a hysteresis appears both in the current and the optical absorption response to the voltage sweep [18]. In addition, the peaks do not have the expected shapes for a surface localized reaction. For example, the anodic peak at 0.18 V, although sharp, is not quite symmetrical, and the corresponding cathodic peak is poorly defined. The peak positions depend on the thickness of the film and will shift anodically by 20–40 mV in the thicker films.

The fact that polyaniline is conveniently switched in aqueous acid solutions, and even in the presence of chloride and fluoride salts [18, 19], is important to technological applications because it avoids the use of organic solvents. The electroactivity of the film is not destroyed during storage in ambient conditions, in either the oxidized or neutral form. On the other hand, prolonged storage does make the material more resistant to oxidation and there is a *ca.* 100 mV anodic shift in the switching potential. In contrast with polyaniline, the other electroactive polymers in this class are not stable to moisture, air and halides such as chloride and fluoride.

The material is electrochromic and the switching reaction produces a color change. Thus one of the first applications considered is electrochromic display devices [15,16]. This technology requires that the material have a good optical contrast ratio, stability and a rapid switching speed. The film is yellow in the cathodic region (−0.2 V) and changes to green, blue and black as the potential is swept anodically to 0.8 V *versus* SCE. The absorption changes at three wavelengths are shown as a function of the applied voltage in Fig. 2. In aqueous HCl solution, the neutral film (−0.2 V) has a sharp optical absorption peak at 305 nm and no absorption in the visible. The oxidized film (0.6 V) has a broad absorption beginning at 500 nm and a peak at 740 nm. The color changes have been attributed to changes in the amount of electrolyte salt in the film [18, 19]. In aqueous sulfuric acid solution, the optical absorption spectra show a broad peak at 420 nm for the neutral film and at 630 nm for the blue film [12].

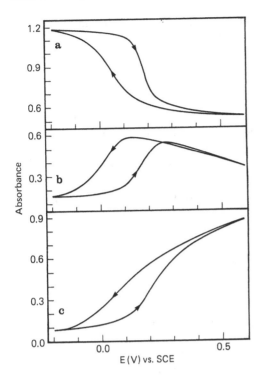

**Fig. 2a–c.** Change in absorbance at 305 nm (a), 420 nm (b) and 740 nm (c) of polyaniline film-coated OTE with changing electrode potentials in 1 M HCl. Deposited amount of polyaniline: 48 mC/cm², and sweep rate of 5 mV/s (reprinted with permission from Ref. [18])

The reaction is coulombically reversible and stable to switching when the anodic potential limit is kept below 0.4–0.7 V [12, 16]. In fact Tamura and co-workers report that polyaniline can survive $10^6$ switching cycles when the anodic potential limit is below 0.4 V [16]. The switching times are in the order of milliseconds [15], and the rate depends on the applied voltage limits and the thickness of the film since the reaction is limited by the diffusion of ions in the film [17, 18]. The apparent diffusion constant for the process is $10^{-10}$ to $10^{-8}$ cm²/s [20].

Energy storage is an application which was recognized quite early [9] and which makes use of the switching property of polyaniline. Reports on this topic are much more complete and serious efforts have been made to determine the practical limitations associated with using polyaniline as electrodes in batteries. The two key issues pertinent to the practical development of a rechargeable cell are reversibility which depends on the voltage limits and current density. Several reports by MacDiarmid and co-workers have appeared describing the use of polyaniline as electrodes in batteries [21–23]. The material has been studied both in film form and as a pressed pellet of the powder. Polyaniline film electrodes have been tested as cathodes in a cell containing aqueous zinc chloride/ammonium chloride solution, pH 4 and using a zinc metal anode. The cell is reversible Eq. (3) and the highest reversibility was obtained using current densities in the range 0.05–0.1 mA/cm². A coulombic capacity of 88% is reported for 200 charge/discharge cycles [23]. The constructed cell is stable to storage showing a decay in the cell voltage from 1.35 to 1.20 V during 30 days. The coulombic recovery at 0.1 mA/cm² was 98% [23]. The power density for such a battery was estimated equal to 184 Watt h/kg, and is comparable to the value for the lead-acid battery,

186 Watt h/kg. Similar results are reported using pressed pellets of polyaniline powder which had been produced by a chemical method [23].

$$ \left[\hspace{-4pt}\begin{array}{c}\end{array}\hspace{-4pt}\right. \!\!\!-\!\!\bigcirc\!\!-NH_2-\!\bigcirc\!\!-NH_2 \left.\hspace{-4pt}\begin{array}{c}\end{array}\hspace{-4pt}\right]^{2\oplus}_{/n} + Zn + HCl \rightleftharpoons \left[\hspace{-4pt}\begin{array}{c}\end{array}\hspace{-4pt}\right. \!\!\!-\!\!\bigcirc\!\!-NH=\!\bigcirc\!\!=NH \left.\hspace{-4pt}\begin{array}{c}\end{array}\hspace{-4pt}\right]^{2\oplus}_{/n} Cl^{\ominus} + ZnCl_2 \quad (3) $$

An improvement in the performance of these films was reported by Genies and co-workers using polyaniline prepared and studied in ammonium fluoride-hydrogen fluoride eutectic salt mixtures [24, 25]. The films were prepared with 100% yield and had the same electrochemical characteristics as the films prepared in aqueous sulfuric acid solutions. In the eutectic mixtures, the switching reaction is highly reversible, surviving 2000 switching cycles with a faradaic yield of 100%. For comparison, in aqueous sulfuric acid electrolyte, faradaic yields of 100% were only observed during the first 100 cycles. Genies gives a brief report of another cell composition for poly-aniline. The film was used as a cathode in aqueous HF solutions and in nonaqueous propylene carbonate/lithium perchlorate electrolyte with a LiAl anode [25]. The open circuit voltage for the latter is 3.7 V.

Polyaniline films prepared in neutral aqueous solutions have been considered as liquid junction photosensors for solar energy conversion [26]. The film employed in this application is different from the material described above. It was grown on a platinum electrode in a pH 6 solution, conditions known to produce polymer which lacks electroactivity [17, 28]. In the electrochemical experiment, a 1 cm$^2$ film on a plati-num electrode was mounted in a cell containing 0.2 M LiClO$_4$. The electrode was irradiated with light in the visible range using a 500 W xenon lamp. The photoresponse depends on the applied voltage and at $-0.3$ V, a photocurrent of 27 μA/cm$^2$ could be maintained for several hours. As can be seen in Fig. 3, the response is reversible and can be switched on and off by turning the light on and off. Also seen is the fact that the photoresponse is slow.

Polyaniline films have also been considered as protective coatings for photoelec-trodes [13], and for the surfaces of stainless steel [27] and Fe [28]. The stability of semicon-ductor electrodes is greatly improved when coated with polyaniline. This is a very important finding since the stability of semiconductor electrodes is a serious problem

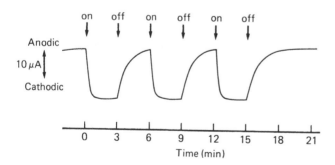

**Fig. 3.** Current change induced by irradiation of a polyaniline film on platinum in aqueous 0.1 M lithium perchlorate with on-off switching of visible light (30 mW/cm$^2$). Applied potential: $-0.3$ V *versus* Ag-AgCl (reprinted with permission from Ref. [26])

in many applications. For example, Noufi [13] studied polyaniline on silicon electrodes for use both as anodes and cathodes. One of the cell reactions employed was the $Fe^{2+/3+}$ couple which was found to be more reversible on polyaniline than on platinum metal (Fig. 4). Using the polyaniline coated n-silicon electrode as a photoanode, in a dilute sulfuric acid solution pH 1 containing $Fe^{2+/3+}$, photocurrents in the range of 6 mA/cm$^2$ were produced. Currents maintained for 70 hours show a 30% reduction while with the uncoated electrodes, the current decreases to zero within seconds (Fig. 5). Similar results were observed when using polyaniline coated p-silicon electrodes as photo-cathodes in a cell containing $[Fe(CN)_6]^{3-}$ and poised at $-0.75$ V. Polyaniline films have also been considered as protective coatings for stainless steel surfaces because they significantly reduce the rate of weight loss due to anodic corrosion [27]. The polyaniline films were grown on 5 cm$^2$ stainless steel sheets type 410 and 430 in aqueous perchloric acid solutions. The film coated surfaces were then immersed in 1 M sulfuric acid and the time required for the metal to become activated at the open circuit potential was between one and 17 hours depending on the pretreatment of the metal. This corresponds to an average penetration rate due to corrosion of 25 µm/year. For comparison, bare stainless steel type 410 becomes active within minutes under the same conditions and after 64 hours at room temperature, the average penetration rate due to corrosion was calculated to be equal to 31,000 µm/year. Mengoli [28] used nonelectroactive polyaniline films prepared in alkaline solutions to passivate Fe sheets against anodic corrosion. This work reports the largest area electrodes, 25 cm$^2$, yet to be reported for the use in the polymerization of aniline.

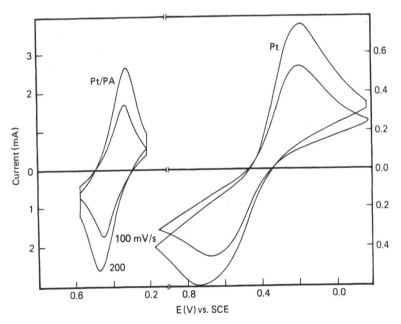

**Fig. 4.** Cyclic voltammogram of 0.1 M $Fe^{2+/3+}$ in 0.1 M sodium sulfate/sulfuric acid solution, pH 1 with a bare and a polyaniline-coated Pt electrode; sweep rate is 100 and 200 mV/s (reprinted with permission from Ref. [13])

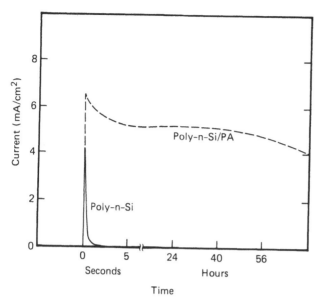

**Fig. 5.** Photocurrent stability of n-Si/polyaniline and bare n-Si photoanode measured with the $Fe^{2+/3+}$ reaction in 0.1 M sodium sulfate/sulfuric acid solution, pH 1; light intensity is 90 mV/cm² and potential is 0.1 V versus platinum (reprinted with permission from Ref. [13])

In the early reports on polyaniline, several applications were described for the use of the material as electrodes in aqueous solutions. Polyaniline was proposed as a general purpose carbon electrode [8] and as a pH sensitive permselective membrane for the pH range 0 to 3 [7]. More recently, the applications focus on the chemical selectivity which the material can impart towards redox species in solution, either by promoting the desired reaction or by minimizing the undesired ones. These selectivities make use of the characteristics of the film such as promoted electron transfer, switching properties, and the selective diffusion of the redox species across the film to the underlying electrode. In the majority of the reports, the principle of the application is demonstrated, however, the limitations of the material, such as stability, efficiency, etc., are not fully determined.

Bockris [29] demonstrated that polyaniline could be used to reduce carbon dioxide. This reaction, which is normally difficult to accomplish in the absence of metal ion catalysts, is promoted on a boron doped p-type silicone electrode coated with polyaniline. The reaction was carried out in an aqueous perchloric acid solution under a carbon dioxide atmosphere and with irradiation using a 150 mW/cm² xenon lamp. Photocurrents of *ca.* 10 mA/cm² were observed and formic acid and formaldehyde were produced with a combined faradaic yield of 20–28 % (see Eq. (4)). Oyama demonstrated a chemical selectivity based on redox potentials using the switching behavior of the film [14]. By switching the electrical conductivity of the film, the film is made to perform like a redox diode. Since the film is conducting in a limited potential range, it will oxidize solution species having oxidation potentials which are anodic of this region and reduce those species having reduction potentials which are anodic of this region. Finally, a selectivity based on size exclusion was demonstrated with electro-

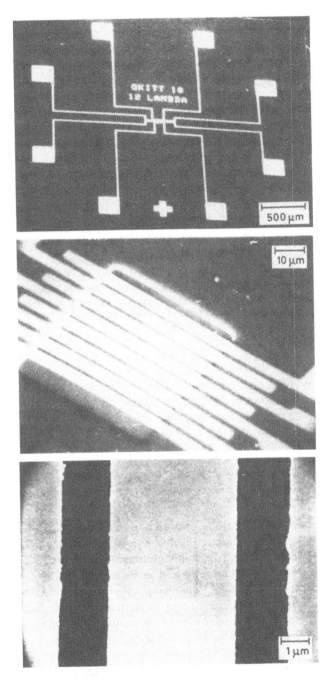

**Fig. 6.** Scanning electron micrograph showing a completed chip containing one fabricated eight-electrode array (top), a view of the area of the eight-microelectrode array exposed to solution for functionalization and electrochemical characterization (middle), a close view of one of the 4.4 micron gold electrodes separated by 1.7 micron from the other two (reprinted with permission from Ref. [30])

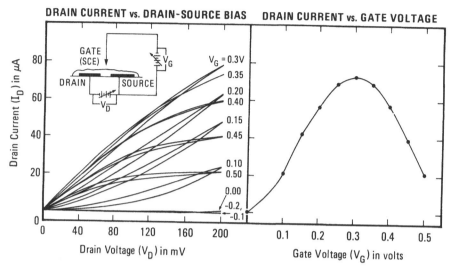

**Fig. 7.** Characteristic of the device represented by Scheme using a polyaniline film (*ca.* 5 microns thick) in 0.5 M sodium hydrogen sulfate solution. The $I_d$ *versus* $V_d$ plots were measured by varying $V_d$ at 10 mV/s from 0 to +0.2 and back to 0.0, and at different $V_g$ values. The plot of $I_d$ versus $V_g$ on the right is for $V_d$ equal 18 V (reprinted with permission from Ref. [30])

inactive films of polyaniline on metal electrodes which provide a physical barrier to the approach of ions. This film was shown to discriminate between protons and metal ions in aqueous solutions [20]. The resulting current for the reduction of protons was greater than for the highly charged ions such as $Fe^{2+}$ and $Eu^{3+}$ by a factor of 6 and > 6000, respectively. The authors suggest the use of these coated electrodes as pH sensors.

$$CO_2 \rightarrow HCOOH + H_2C(OH)_2 \tag{4}$$

More recently, Wrighton and co-workers have described a novel electrode construction which results from the combination of the electroactive properties of polyaniline and microelectrodes [30]. Films were prepared in configurations of microelectrode arrays consisting of small area electrodes, 4.4 μm wide and 5–10 μm thick, which are in close proximity, 1.4 μm separation [30]. The electron micrographs for such a circuit are shown in Fig. 6. These electrodes were prepared employing photoresist/metallating procedures which are fairly standard in the microelectronics industry. The microelectrodes were used in an electrochemical cell which was constructed as shown in Fig. 7, and were in contact with an aqueous 0.5 M sodium hydrogen sulfate solution. This film/microelectrode configuration was shown to function as a transistor-like triode device. The electrode configuration with 2 microelectrodes detects changes in the resistance of the film between the electrodes by measuring the response in the drain current ($I_d$) at a given drain voltage ($V_d$) as the "gate" voltage ($V_g$) is adjusted. As seen in Fig. 7, the drain current tracks the drain voltage and the magnitude of the response is adjusted by the gate voltage. With polyaniline, the currents can be made to increase to 80 µA (200 mV drain voltage) when the gate voltage is increased to 0.3 V. The electrode is quite sensitive and a 5–10 µm film can be switched with *ca*. 1 microcoulomb. The film is stable to the input signal and to current flow from source to drain for up to 16 hours. An important property of the polyaniline film is the $I_D$ response which increases and decreases as $V_G$ increases towards more anodic potentials. In another configuration using three electrodes, the microelectrode device has been considered as a redox sensor. The outer electrodes are *source* and *drain*, and the center electrode is the gate which is switched when in contact with the redox system in solution. The redox system is the input signal.

## 2.2 Polypyrrole

Like "aniline black", "pyrrole black" has also been reported as a black intractable solid resulting from the oxidation of pyrrole [31, 32], and much of the early work focused on the chemical characterization and electrical properties of the material. Unlike "aniline black", the early work did not include electrochemical studies and the switching properties of the material did not become known until recently. Thus, there was little insight into the possible technological applications of this material. In 1980 the electrochemical generation of coherent films of polypyrrole on a platinum electrode was demonstrated [33]. Among other things, this permitted the investigation of the electrochemical characteristics of the film, and more generally, the study of conducting polymers in film form instead of pressed powder pellets [32]. In contrast with the polyaniline films which are normally quite thin, relatively thick films of polypyrrole can be prepared by electropolymerization. Furthermore, the properties of the film vary

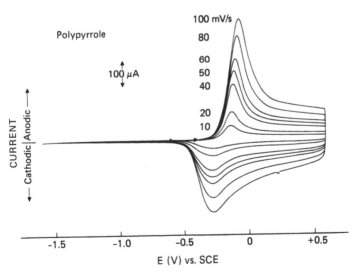

**Fig. 8.** Cyclic voltammograms of a polypyrrole tetrafluoroborate film (20 nm thick) on a platinum surface measured in 0.1 M tetraethylammonium tetrafluoroborate/acetonitrile solution (reprinted with permission from Ref. [36])

with the thickness, dividing the proposed applications into two categories, depending on whether the film employed is thin or thick. In this section, the applications employing thin films of polypyrrole which are electroactive are discussed separately from the thicker films which are mainly conductive and poorly electroactive.

Thin films of polypyrrole (less than 0.1 μm thick) are electroactive (Eq. (5)), and can be switched between the neutral and the oxidized state at about 0.1 V *versus* SSCE [33]. In aprotic solvents and in the absence of oxygen, the reaction is coulombically reversible and the film can be switched between the two oxidation states repeatedly without decay of the electroactivity. In the presence of oxygen, the electroactivity of the film decays slowly, and the film settles in a conducting form. The oxidation reaction introduces 0.25–0.3 charges per pyrrole ring along the chain and is accompanied by the incorporation of counter anions into the film [34, 35]. As observed with the polyaniline films, the switching reaction is not simple, (see Fig. 8) [35, 36], and is accompanied by structural reorganizations in the polymer plus the limited diffusion of ions in and out of the film. The diffusion constant is *ca.* $10^{-10}$ cm$^2$/sec. [34] for films which are *ca.* 1 μm thick. In addition, the electrical characteristics of the film change drastically. The electrical conductivity changes by *ca.* eight orders of magnitude [35] and the ionic impedance varies by a factor of ten [37].

$$\left(\!\!\begin{array}{c} \text{O} \\ \text{N} \\ \text{H} \end{array}\!\!\!\!\!\begin{array}{c} \text{H} \\ \text{N} \\ \text{O} \end{array}\!\!\right)_n + A^\ominus \;\rightleftharpoons\; \left(\!\!\begin{array}{c} \text{O} \\ \text{N} \\ \text{H} \end{array}\!\!\!\!\!\begin{array}{c} \text{H} \\ \text{N} \\ \text{O} \end{array}\!\!\right)_n^{\oplus}\!\! A^\ominus + e^\ominus \qquad (5)$$

Since the films are electrochromic, they have been proposed for display applications [38]. The oxidation reaction is accompanied by a color change in the film from yellow (neutral) to blue-black (oxidized). The optical contrast ratio in the visible spectrum depends on the thickness of the film and is about 1000, between the neutral and the oxidized form. The switching speed also depends on the film thickness and is in the order of milliseconds. Considering these characteristics plus the stability of the film, these materials are marginally competitive with organic electrochromic materials.

The redox/electrochromic behavior of the polypyrrole films was also studied in aqueous 0.5 M KCl solutions by Tamura [38]. The film switches with a peak at −0.3 V versus SCE, and the shape of the voltammogram was different depending on whether the electrolyte was chloride or sulfate. The optical absorption spectra are similar to the one observed in acetonitrile, and the switching response in the voltage range −0.85 to 0.45 V was found to be rapid for the anodic pulse and slower for the cathodic pulse. The material and color changes were stable for 1000 switching cycles.

Reports from two laboratories propose the use of these materials for optical storage. In a study by Inganas, reflectance changes resulting from the switching of a poly(N-methylpyrrole) film on single crystal n-type silicon were studied [39]. The film was switched in acetonitrile, and the reflectance response was measured from the illumination of the surface at a 45° angle to a light detector. The response is not linear with the amount of charge passed and again found to be slower in the cathodic direction of the write/erase cycle. In another study [40], it was shown that the switching property. of the film combined with the photoactive nature of the silicon substrate could be used to write on the film using a laser beam. Films were deposited on a silicon electrode

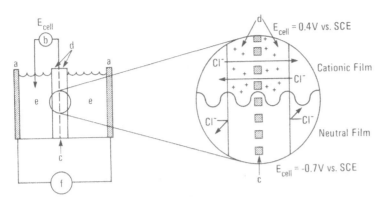

**Fig. 9.** Experimental setup for impedance measurements with electrochemical control of membrane impedance: platinized platinum electrodes (a); constant voltage power supply, (b); gold minigrid electrode (c); polypyrrole film, (d); 1 M KCl solution (e); constant current ac circuit, (f). At right is a microscopic view of membrane, illustrating effect of membrane potential on ionic resistance (reprinted with permission from Ref. [37])

and mounted in a cell containing propylene carbonate electrolyte. A spot on the film could be switched by irradiation with a He—Ne laser. The write resolution was poor and the size of the spot was ten times greater than the area of incident light. In addition, the switching rate was slow, *ca.* 1 second for a 90 nm thick film.

Thin films of polypyrrole grown on a gold minigrid sheet have been proposed by Murray and co-workers as an "ion gate" to control the flow of ions between two electrolytic solutions [37, 41, 42]. This application arises because the mobility of ions is different in the oxidized and the neutral film. For this application, a 15 μm polypyrrole film was prepared by electropolymerization on a gold minigrid electrode and the electrode/membrane was mounted in a cell containing aqueous KCl as shown in Fig. 9. Stepping the potential applied to the electrode/membrane from 0.4 to −0.7 V produced a change in the impedance from 132 to 1280 ohms. The response was slow, taking minutes for a significant change and up to an hour to equilibrate.

Polypyrrole films have also been suggested for use as time release electrodes. Making use of the migration of ions in and out of the film during the switching reaction, Miller and co-workers have shown that ions such as ferrocyanide and glutamate can be released into the solution in a controlled manner. In particular, the possibility of having controlled release of glutamate in aqueous solutions pH 6.95 is of interest in the area of neuroscience [43].

There is also interest in using switching films deposited on an array of microelectrodes which when mounted in an electrochemical cell functions like a transistor [30, 44]. For this application, an eight gold microelectrode array, 3 μm wide, 140 μm long, 0.12 μm thick, and separated by 1.4 μm, was connected by a coating of electropolymerized polypyrrole. The electrochemical characteristics of this electrode are analogous to those described with polyaniline and the reader is referred to Sect. 2.1 for details.

Thicker films of polypyrrole, 0.2–2.0 μm thick, are of interest for a different set of applications. The switching characteristics of the thicker films are much less pronounced due to the slower migration of ions in and out of the film. These films, when left as a coating on the platinum electrode, are conductive and will drive solution reactions. In aprotic solvents, the cyclic voltammogram of solution species such as ferrocene resembles those measured on platinum [45-47]. There is little or no change in the thermodynamic or kinetic characteristics of the reaction. In aqueous solutions, the film produces large background currents which are larger in the region anodic of the switching potential. As a result, the voltammograms of redox species are distorted and resemble quasi-reversible reactions. The voltage limits of the film/electrode are established by the reduced conductivity of the film at voltages cathodic of $-0.5$ V, and the loss of adhesion at voltages anodic of *ca.* 1.5 V. The anodic stability of the film is also less in aqueous solutions than in aprotic solutions because the gas evolution resulting from the electrooxidation of water causes the film to blister off the electrode. The stability of the film in any solvent is also limited by exposure to alkali or strong oxidants such as chlorine and bromine [46].

Energy storage is another area where these films have been considered [48-58]. Making use of their conductive properties, polypyrrole has been shown to work quite well as a protective layer against photocorrosion for photoelectrodes. Photocorrosion is probably the single biggest limitation in the use of semiconductors in photochemical solar cells and is one area where conducting polymers may successfully find an application. The films meet many of the material requirements for this application because they are stable to corrosion, they can be prepared in relatively thin form, and they are sufficiently conductive to drive solution redox reactions. In addition, this application requires that the films have good electrical and adhesive contact with the substrate. Films of polypyrrole have been deposited on gallium arsenide and silicon semiconductor electrodes and tested for photoelectroactivity. The studies relevant to the use of polypyrrole films as coatings on electrodes for photochemical devices were recently included in a review by Frank [48]. In this article, the authors present a detailed analysis of the technical problems plus the characteristics of the polypyrrole film associated with these devices. We therefore present only a few examples of these studies and the reader is referred to Frank's article for a more complete discussion of this application.

Noufi, Frank and co-workers have reported that the photocorrosion of single crystal n-gallium arsenide [47] and polycrystalline n-silicon [49] electrodes can be suppressed when used in aqueous solutions containing ferrous/ferric ions. The coated n-silicon electrode had a short circuit current of 2.9 mA/cm$^2$, a power efficiency of 3%, and functioned for 122 hours of continuous irradiation (143 mW/cm$^2$) with only a 30% reduction in the photocurrent. For comparison, the current decays to zero within a minute with a uncoated electrode [49]. These results are shown in Fig. 10. The n-gallium arsenide electrode showed a similar photocurrent response. Adhesion of the film to the semiconductor surface is a problem with these electrodes and is much more prevalent with the gallium arsenide, where the film peels off the electrode in aqueous solutions.

The adhesion of polypyrrole films on semiconductor surfaces has been improved by pretreating the semiconductor surface in two different ways. In a report by Skotheim and co-workers [50-52], the adhesion was improved by precoating a single crystal of n-silicon with 0.5 nm of platinum prior to the electropolymerization of pyrrole. This pretreated multilayer electrode showed an improved stability with only a 7% reduction in photocurrent due to the presence of the platinum. These electrodes were

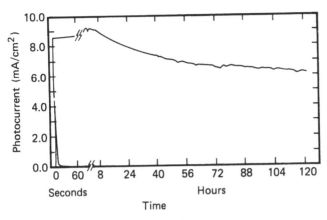

**Fig. 10.** Photocurrent-time behavior of naked (secs) and polypyrrole covered (hrs) polycrystalline n-type Si electrodes illuminated with tungsten-halogen light at 143 mW/cm² under short-circuit conditions. Solution containing 0.15 M FeSO₄, 0.15 M FeNH₄(SO₄) 12 H₂O and 0.1 M Na₂SO₄ in water at pH 1 was stirred continuously (reprinted with permission from Ref. [49])

stable during six days of continuous use with no decay in the short circuit photo-current ($9$ mA/cm²) while electrodes with platinum and no polypyrrole, or poly-pyrrole and no platinum, were stable for only a few minutes. In a separate study, treating the silicon oxide surface of the n-silicon electrode with N-(3-trimethoxy-silyl)propylpyrrole prior to the anodic polymerization of pyrrole also improved the adhesion of the film [53]. With this treatment, the film is presumably anchored covalently to the oxide surface. These electrodes were used in aqueous ferrous/ferric solutions for 25 hours with no decay in the photocurrent, while untreated electrodes lost 70% of the photocurrent after 18 hours.

These films have also been used to improve the charge transfer efficiency of photo-voltaic cells employing solid state electrolyte [54]. Polypyrrole was used to modify the interface between n-Si and polyethylene oxide-KI/I₂ solid electrolyte in a thin film photoelectrochemical cell (see Fig. 11). This cell has an open circuit potential of 320 mV when irradiated with 100 mW/cm² tungsten-halogen light. The photocurrent response and the stability of the cell was improved by an order of magnitude when the n-Si surface was precoated with 1–2 nm of platinum.

Polypyrrole films containing selected metal particles, metal ions or other com-ponents have been used for electrocatalysis or electrodes with chemical selectivity. For example, tantalum electrodes coated with a polypyrrole film and then with a 7.2 nm film of platinum will catalyze the electrooxidation of water to produce oxygen [55]. This multilayered electrode performs like a platinum electrode and is stable to the reaction conditions. For comparison, the uncoated tantalum electrode readily forms an insulating oxide layer and self-passivates, and a tantalum electrode coated with polypyrrole but no platinum is a poor catalyst for the oxygen generating reaction. Another study demonstrated the use of polypyrrole films containing ruthenium oxide to catalyze the photooxidation of water to produce oxygen [56]. For this application, $[RuO_4]^{2-}$ anions were incorporated into a film grown on a n-GaP electrode by anion

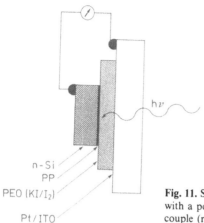

hν

n-Si
PP
PEO (KI/I₂)

Pt/ITO

**Fig. 11.** Schematic configuration of a photoelectrochemical cell with a polymer solid electrolyte and an iodide/triiodide redox couple (reprinted with permission from Ref. [54])

exchange and subsequently reduced to $RuO_2$. With this electrode, oxygen evolution occurs 200 mV less anodic than the voltage observed with a film not containing ruthenium oxide. The electrode has stability problems since the film blisters and separates from the GaP electrode in a few minutes. In a separate study, polypyrrole films supported on cadmium sulfide were coated with ruthenium oxide-silver paint [57]. The catalyst coated electrode successfully improved the stability of the cadmium sulfide against photodegradation in aqueous solutions and afforded the photooxidation of water to produce oxygen.

Polypyrrole films have also been considered for storage battery applications. This application makes use of the switching properties for energy storage and the conductive properties to minimize heating. Mengoli and co-workers describe a cell employing a polypyrrole anode for the bromine/bromide and the iodine/iodide couples with a zinc/zinc bromide cathode [58]. The cell was found to be reasonably stable during 20 to 30 charge/discharge cycles when the charging was 50 % of capacity, and the recovered charge was 90 % of the original. With capacity charging, the cell was not stable.

Finally, two reports have appeared in which polypyrrole-modified electrodes have shown electrochemical selectivity. In one report, polypyrrole coated electrodes are reported to selectively permit the $Fe^{2+/3+}$ reaction to occur in the presence of the thionine/leucothionine couple in dilute sulfuric acid [59]. The authors proposed the use of this selectivity to design a simple photogalvanic cell which avoids the requirement for a compartment separator. In another study, polypyrrole films were used to improve the yield of chiral products produced by electrochemical asymmetric oxidation [60]. In this work, poly(L-valine) coated electrodes were used to generate optically active phenylcyclohexylsulfoxide from the corresponding sulfide (Eq. (6)). The yield of optically active product was improved by 60–100 % when the poly(L-valine) layer was coated on the polypyrrole film (on Pt) compared to a coating directly on platinum.

$$C_6H_5-S-C_6H_{11} \rightarrow C_6H_5-\overset{\overset{\displaystyle O}{\|}}{S}-C_6H_{11} \qquad (6)$$

## 2.3 Polythiophene

Polythiophene films are prepared by anodic polymerization of either the monomer, dimer or their β-substituted analogs [61-68]. The films have the same compositional, electrochemical/electrochromic characteristics as the films described in the previous sections. Polythiophene is oxidized to a level of 0.02–0.22 charges/thiophene ring (Eq. (7)), and the rate of switching depends on the voltage limits used and is in the order of milliseconds [61, 62]. These films are reasonably stable in ambient conditions and gradually lose conductivity and electroactivity during a couple of weeks of exposure [62, 66, 67]. The coulombic reversibility of the switching reaction combined with the electrochromic properties suggests that these materials could be used for display devices [65] and electrodes for light weight batteries [66, 67].

$$\left(\!\!\!\!\begin{array}{c}\text{S}\\\text{S}\end{array}\!\!\!\!\right)_n + A^\ominus \;\rightleftharpoons\; \left(\!\!\!\!\begin{array}{c}\text{S}\\\text{S}\end{array}\!\!\!\!\right)_n^{\oplus} A^\ominus + e^\ominus \qquad (7)$$

Polythiophene has been considered as an electrode in energy storage cells. Like polyaniline and polypyrrole, it offers the convenience that films with good electrochemical properties can be readily acquired using a relatively straightforward preparation procedure [61, 68, 69]. The films were prepared on a platinum electrode using 1 M lithium perchlorate in propylene carbonate. In a cell containing the same electrolyte and a lithium counter electrode, the film oxidizes at 3.5 V anodic of lithium and has an open circuit potential near 2.8 V. The film could be switched with a coulombic efficiency of 95% when the anodic potential was maintained below 3.8 V of lithium. The reaction introduces 0.16–0.20 charges/thiophene ring. With higher anodic potentials the efficiency drops off.

The polymer from polythienothiophene, a thiophene analog, has also been considered as an electrode for a storage cell in conjunction with a lithium electrode in a propylene carbonate-lithium perchlorate electrolyte [70]. In contrast with polythiophene, this polymer can be both oxidized and reduced with a 2 V separation (Eq. (8)).

$$\left(\!\!\!\!\begin{array}{c}\text{S}\\\text{S}\quad\text{S}\end{array}\!\!\!\!\right)_n + A^\ominus \;\rightleftharpoons\; \left(\!\!\!\!\begin{array}{c}\text{S}\\\text{S}\quad\text{S}\end{array}\!\!\!\!\right)_n^{\oplus} A^\ominus + e^\ominus . \qquad (8)$$

Derivatives of polythiophene have also been considered as supports for catalysts used in solar energy cell [71]. Aggregates of platinum and silver were electrodeposited on poly(3-methylthiophene). The aggregates are 3–200 nm in size and are deposited along the strands of the polymer in amounts up to 20–50% the weight of the polymer. The deposited clusters are found to catalyze the reduction of protons in aqueous HCl solutions. Current densities as high as 200 mA/cm² were observed and the electrode was stable for 170 hours. For comparison, electrodes prepared by electrodepositing silver on gold produced currents which were half the size. Thiophene polymers containing Cu(II) are reported to have electrocatalytic properties. These materials were prepared chemically and they promote the evolution of oxygen when immersed in aqueous solutions containing platinum and cerium salts [72].

## 2.4 Other Polyaromatic Films

A variety of other aromatic compounds have been successfully electropolymerized and found to have the general electroactive properties described for the polyaniline, polypyrrole and polythiophene films. Films have been prepared from azulene [73, 74], furan [63], indole [63], carbazole [74], pyrene [74], triphenylene [74] and phenylene [75] (see Scheme 1). In principle, all of these films could be used as electrodes in energy

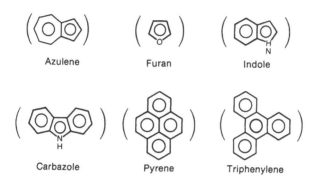

Azulene          Furan          Indole

Carbazole          Pyrene          Triphenylene

*Scheme I.*

storage cells, however only the reports on polyphenylene describe such studies. Electroactive polyphenylene has been prepared as a film by electropolymerizing benzene in liquid HF [75] and as a powder by a chemical method [76, 77]. The polymer can be switched with coulombic reversibility in aprotic solvents [75-7], and the material can function both as the anode and the cathode in rechargeable batteries. The open circuit voltage for a cell using the polymer anode/polymer cathode configuration in $LiAsF_6$/ propylene carbonate electrolyte is 3.3 V, and is 4.4 V with the polymer cathode/lithium anode configuration [77, 78]. The charging stoichiometry for the polymer used as shown in Eq. (9) can be varied from 0 to 0.5 charges/aromatic unit, and has a gravimetric charge storage capacity equal to 0.15 Ah/g with 0.1 charge/repeating unit charge level. The polymer/lithium electrode combination was found to be more stable than the polymer/polymer combination.

$$(C_6H_4)_n^+ AsF_6^- + Li \rightleftharpoons (C_6H_4)_n + LiAsF_6 \qquad (9)$$

The polyazulene polymer films [73, 74], are conducting in the oxidized forms and have conductivities in the range 0.01 to 1 S/cm [73, 74]. The polymer can be switched in aprotic solutions with coulombic reversibility, and the reaction is accompanied by a color change. The level of oxidation for the polymer is approximately 0.3 charges/ aromatic unit. At present, we are not aware of any reports describing technology development activity for the other materials.

Finally, an asbestos membrane reinforced with polyphenylene sulfide was used in an electrolysis cell, though not as an electrode but as a cell compartment separator. The cell was used to produce hydrogen by the electrolysis of alkaline solutions at high temperatures. For this application, polyphenylene sulfide is impregnated into an asbestos card, then sulfonated with sulfur trioxide (Eq. (10)). It was found necessary to sulfonate the polymer in order to reduce the rigidity of the composite membrane. Unfortunately, sulfonation affects the stability of the composite. The material is still more stable than plain asbestos and will survive boiling KOH solution for 140–180 hours.

$$\left(\!\!\left\langle\!\!\bigcirc\!\!\right\rangle\!\!-\!\!S\!\!-\!\!\left\langle\!\!\bigcirc\!\!\right\rangle\!\!-\!\!S\!\!\right)\!\! + SO_3 \longrightarrow \left(\!\!\left\langle\!\!\bigcirc\!\!\right\rangle\!\!\overset{S}{\underset{SO_3H}{\Big|}}\!\!\left\langle\!\!\bigcirc\!\!\right\rangle\!\!-\!\!S\!\!\right)\!\! \qquad (10)$$

## 3 Linear Polyene Polymers

### 3.1 Polyacetylene

Of the materials included in this report, the greatest number of application related studies have been carried out with polyacetylene. There is a large number of patents which have been issued for the use of polyacetylene in secondary storage batteries and displays applications which make use of the electroactivity of the material. The electroactive nature of this material was first demonstrated with thin films casted on a platinum surface [82]. The voltammogram is shown in Fig. 12. The films can be switched between the neutral and the oxidized form at about 0.7 V *versus* SSCE in acetonitrile (Eq. (11)). The reaction is accompanied by a color change from light yellow (neutral) to dark blue (oxidized) and a change in the electrical properties from insulating to

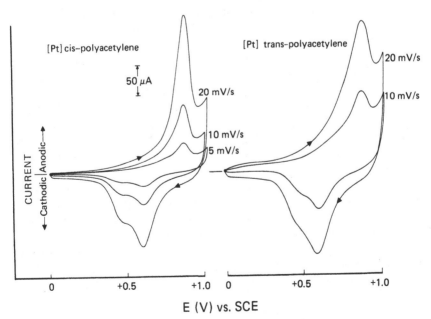

**Fig. 12.** Cyclic voltammograms of polyacetylene films on platinum measured in acetonitrile containing 0.1 M tetraethylammonium tetrafluoroborate (reprinted with permission from Ref. [82])

conducting. The electrochemical stoichiometry of the reaction is approximately 0.01 charges/repeating unit along the chain, and the reaction is accompanied by a large capacitive current. In the absence of air and moisture, the reaction is coulombically reversible and the material can be cycled repeatedly. The material is inherently sensitive to the atmosphere and any exposure immediately reduces the electroactivity and eventually destroys all the active properties of the polymer.

$$(CH)_m + A^- \rightleftharpoons (CH)_m^+ A^- + e^- \tag{11}$$

Although the switching reaction is coulombically reversible, it is not simple. It is accompanied by structural reorganizations of the polymer and there are limitations to the diffusion of ions and electrolyte into the polymer strands. The peak heights in the voltammogram scale linearly with sweep rate, but the peaks are not completely symmetrical and the positions of the oxidation and reduction peaks do not coincide. The current response for the first scan of a freshly prepared film is always larger than for the subsequent scans. Similar results are obtained with both the neutral cis and trans polymers [82].

MacDiarmid and co-workers [83] have reported that polarization of $(CH)_x$ films at 9 V in various electrolytes can produce either n or p-type doping. This anodic doping can be accomplished with various tetraalkylammonium electrolyte salts, such as perchlorate, trifluoromethylsulfonate, and arsenic hexafluoride. The oxidized films can reach charge levels of 0.05–0.10 charges/carbon unit, and electrical conductivities approaching $1000$ S cm$^{-1}$. The cathodic doping is best accomplished using lithium electrolyte salts. This possibility for both n and p-type electrochemical doping has led to the development of light weight, rechargeable batteries using $(CH)_x$ as the cathode and/or anode [84–86]. The advantage being sought in going to these materials is a higher charge to mass ratio since the materials are lighter than metals. Cells using lithium perchlorate doped $(CH)_x$ have been found to have power densities as high as ca. 30 kW/kg. The cells have coulombic efficiencies of 100–86% and energy efficiencies of 81–68% [86]. In addition, these materials are expected to be free of some of the problems associated with the discharging/charging cycle such as growth of dendrites. Prior efforts to produce lightweight cells have involved the use of lithium metal, which tends to produce dendrites during the recharging step [87]. These batteries have been shown to be rechargeable with as many as 20 charge-discharge cycles with little change in the charging or discharging characteristics. Also, if left standing for as long as 48 hours, there is no spontaneous discharge [84].

Shirakawa and co-workers [88] described the electrode properties of polyacetylene in aqueous media as resembling those of a metal electrode, except for the effect of the presence of oxygen. Hydrogen gas evolution was observed when the voltages were swept to values more negative than $-1.0$ V versus Ag/AgCl. On the other hand, the evolution of oxygen was never observed even with voltage sweeps to very positive values. The subsequent scans to negative potentials did, however, show increased cathodic currents. This is postulated to be a result of the reduction of some form of adsorbed oxygen.

Polyacetylene was also investigated for possible use as a photoelectrode [88, 89, 90–92]. In 0.1 M acetate (pH 5.6), a dark potential of 65 mV versus Ag/AgCl was measured. Within 30 seconds after illumination, a steady state value of 190 mV was achieved. This process was reversible, *i.e.*, if the light was turned off, the dark potential returned to its original value. The photopotential (potential when illuminated minus dark potential) varies with pH, from approximately 50 mV at pH 1 to a fairly constant level of approximately 90 mV at pH 5–8. Cathodic photocurrent was found to vary with wavelength, exhibiting a maximum at about 620 nm. Measurable photocurrent was observed even at wavelengths of 700 nm [88]. In another study, a photochemical cell was described in which polyacetylene is the active photoelectrode in contact with sodium polysulfide as the electrolyte, as shown in Fig. 13. When illuminated at a level of 1 sun, an open circuit voltage of 0.3 V and a short circuit of 40 μA/cm² were obtained. The quantum efficiency of the cell is shown in Fig. 14. The system was found to be limited by the high resistance, small effective electrode area and solution absorbance [90].

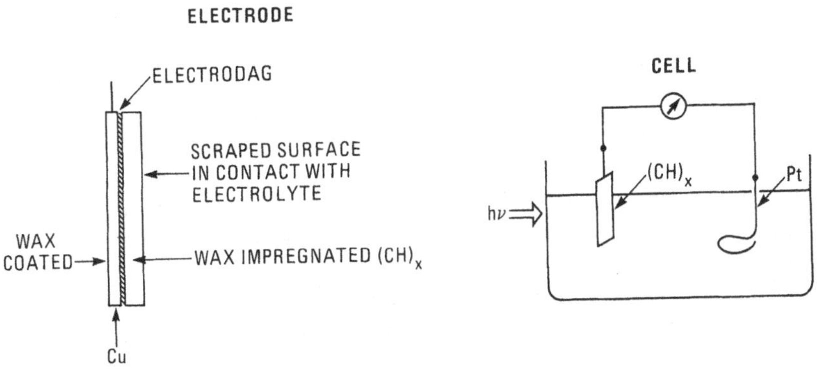

**Fig. 13.** Diagram of photoelectrochemical cell (reprinted with permission from Ref. [90])

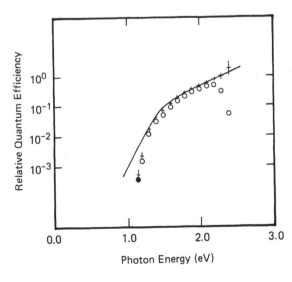

**Fig. 14.** Relative quantum efficiency of the polyacetylene photoelectrochemical cell. The open circles represent data taken with light incident through the highly colored electrolyte. The data corrected for light absorption by the electrolyte (x), and photoconductivity response normalized to the photoelectrochemical cell data at 2 eV (———), (reprinted with permission from Ref. [90])

A different cell design was described by Yamase and co-workers [91] using trans-$(CH)_x$ films in the presence of methylviologen ($MV^{2+}$). The cell is not active in the dark and is activated with light. The light-induced reaction appears to involve the $MV^{2+}$ as a primary acceptor of the electron from $(CH)_x$. The quantum yield is quite low for the process, $10^{-3}$, and it is postulated to result from the limited lifetime of the generated hole. In addition, the spectral response was unusual and did not match the excitation spectrum.

Solar energy conversion using an undoped p-$(CH)_x$—ZnS heterojunction has been demonstrated. The cell has an open circuit voltage of 0.8 V, and the photoresponse of the diode-like junction shows a peak absolute efficiency at 3.1 eV of about 0.3 %. This low value is due to inefficient carrier collection at the point of contact [92].


## 3.2 Polythiazyls (SN)$_x$

The use of polythiazyl, $(SN)_x$, as an electrode was first reported in 1977 [93]. In contrast with the polymer films described above this material is not grown electrochemically and must be mounted on a support with good electrical contact for electrode uses. The material is fibrous in nature and has been used as a single crystal, a composite paste and a film. In general, polythiazyl shows good electrode characteristics, and the behavior of electrodes is similar both with the fibers parallel to the exposed face and with the ends perpendicular to the face, except with regard to surface modification with some metal cations. As an electrode, the material does not influence the kinetics or thermodynamics of most solution redox systems. Its working range and background currents are very much dependent on factors such as pH [94, 95] and the identity of solution species [94–102]. The negative breakdown potential in acidic and near neutral aqueous solution is in the range —1.3 to —1.9 V versus SCE and depends on the pH and the electrolyte [93, 94]. Hydroxide media extend the range to about —2.0 V [100]. The most dramatic difference is in borate medium (pH 9.4) where there is no appreciable breakdown as far negative as —2.8 [101].

Oxygen is strongly chemisorbed on $(SN)_x$ surfaces. Fresh surfaces initially produce $O_2$ at the anodic breakdown, but OH radicals produced in prolonged cycling to anodic breakdown voltages result in surface hydroxides. An irreversible peak then appears at 0.4 V versus SCE [96]. The positive breakdown region is not affected to any great extent by the pH of the medium [95]. Studies of the background currents in aqueous media for polythiazyl electrodes showed that two distinct and different types of current-voltage curves are observed which are a function of the preparation of the $(SN)_x$ [104]. Thus, surface chemical anisotropy of the electrode is thought to arise from surfaces that are predominantly nitrogen or predominantly sulfur. This phenomenon occurred only at the perpendicular surfaces.

The negative breakdown potential is also influenced by the cation in solution where the working range for parallel electrodes is extended in the order Li < Na < K < Cs. In calcium nitrate electrolyte, the breakdown current was beyond —0.2 V for the parallel electrode and at —1.5 V for the perpendicular electrode [94]. Transition metal ions produce a dramatic effect. When chromium(III) [94] or iron(III) [88] ions were present in solution, or when the electrodes were dipped in concentrated solutions

(1 M) of these salts, background currents could be decreased by as much as a factor of ten, with an accompanying extension of the working range. Lead and mercury ions react to form persistent deposits of $(SN)_x$ when a crystal is dipped in a solution of the metal salt [94]. These deposits can be seen under an electron microscope. On a parallel surface, the deposit tends to occur at the cracks in the material [101]. Palladium and silver also form deposits on $(SN)_x$, but if a treated electrode is cycled in 0.05 M $NaClO_4$, the metal is slowly lost from the surface [97].

The strong interactions with metal ions extend to the use of metal-modified electrodes in electrocatalysis. Catalysis has been demonstrated with four systems. Chromium treatment results in as much as a 200 mV positive shift in the reduction peak for $IO_3^-$ in acetate buffer [97]. This has been compared to the necessity for prior oxidation of the platinum electrode surface [103]. Ruthenium pretreatment of $(SN)_x$ electrodes results in a catalytic current for the $I^-/I_2$ couple in phosphate buffer, pH 7.6. These electrodes also photoelectrochemically reduce protons to hydrogen at $-0.05$ V versus SCE in dilute sulfuric acid solution [102]. Molybdate treated electrodes have been used to electrochemically reduce acetylene at potentials of 1.5 V versus SCE in borate and hydroxide solutions. Iron treated electrodes show some ability to facilitate this reaction, but the rate is slower than with the molybdate treated electrodes [95].

The material was also studied in nonaqueous electrolytes such as acetonitrile [105, 106], propylene carbonate [106, 107], methylene chloride and ethanol [106]. In acetonitrile, the working range is wide, extending about 2 V in both the anodic and the cathodic regions [106]. The initial scan with a fresh electrode shows the appearance of a reduction peak for $(SN)_x$ which is postulated to be due to superficial degradation of the material to $S_7N^-$. The behavior of the silver reduction reaction is very similar to that in aqueous solution. $(SN)_x$ electrodes unmodified and modified with bromide or lithium ions were studied in propylene carbonate containing $KPF_6$, $MClO_4$ and MBr (where M is Li, Na, K, $Et_4N$) electrolyte. The formation of alkali metal compounds of $(SN)_x$ in the cathodic region was proposed [107].

Polythiazyl paste electrodes are also well behaved, and do not influence the kinetics or thermodynamics of solutes undergoing outer sphere electron transfer reactions. They are resonably stable to continuous use and to storage [107]. An $(SN)_x$-graphite electrode was studied in aqueous solution and has been found to function like a pH electrode in the pH range 0.3 to 5. Below pH 5, oxidative potentials are postulated to result in the formation of $(SNO)_x$ [108]. In both nonaqueous and aqueous media, the paste electrodes prepared using Apiezon RM grease function well in the region $-0.5$ to 0.9 V versus SCE [109, 110]. These electrodes are easily fabricated and the surface is easily renewed by wiping. $(SN)_x$ film electrodes were studied in aqueous solution, and were found to behave like the crystal electrodes. The background currents and the peak separations are, however, much larger. The acetylene reduction reaction is catalyzed on these film electrodes. The utility of $(SN)_x$ film electrodes is, however, limited by the poor mechanical properties. The films tend to flake with use in solutions [111].

$(SN)_x$ is a very electronegative species [112] and, therefore, has been utilized as an $(SN)x$-GaAs polymer-semiconductor solar cell [113]. The high electronegativity of the material increases the Schottky barrier heights and thus, provides a means of raising the open circuit voltage by as much as 40 % above that found for cells where the surface contact with the semiconductor is a metal or metal oxide.

# 4 Other Organic Electrode Materials

Electrode applications have also been considered with other organic materials which are categorically different from the two classes of materials described in the previous sections. Probably the most important example, from a technology applications stand point, are the perylene-iodine complexes which have been developed successfully and are currently used as long-life pacemakers. The material is used as a cathode in a cell containing lithium iodide solid electrolyte and a metallic lithium anode. The cell has an open circuit potential of 2.81 V and a projected decay of 2 V after 12 years. This type of cell which does not liberate gases, offers the advantage that leaks and separator failures are minimized [114]. In addition, solid state cells have been reported using perylene-iodine cathodes with silver powder anode and $KAg_4ICN$ solid electrolyte [115], and using the iodine adducts of polyvinylpyridine [116], poly-2,5-thienylene, and poly(N-vinyl-2-pyrrolidinone), and phenothiazine [117].

Electrode applications have also been considered for some donor-acceptor complexes. Tetracyanoquinodimethane-tetrathiafulvalinium (TCNQ-TTF) was used as pressed pellet electrodes in aqueous solutions [118-120], and in acetonitrile [121]. In both solvents, the working range is restricted by the redox chemistry of the TTF-TCNQ salt. The complex is stable in water for periods of several weeks, and in electrolytic solutions with most electrolytes except chlorides and bromides. The cyclic voltammograms for the $[Fe(CN)_6]^{3-/4-}$ and Cu(II)/Cu(I) reactions measured with these electrodes are similar to those measured with a platinum electrode. TCNQ complexes with other donors were also probed for electrode applications [120]. The films of copper phthalocyanine (CuPc) on gold substrates were investigated for photoelectrochemical response in both near IR [122] and the visible regions [122-125]. The electrodes were immersed in 0.1 M aqueous sodium sulfate solution and irradiated with near IR light from a 150 W halogen lamp [122]. The photocurrent in the near ir has a maximum at 1100 nm and a shoulder at 1020 nm and generally coincides with the optical absorption spectrum of the material indicating that it is closely associated with an intrinsic photoexcitation process [122]. In the visible region, the photocurrent follows the absorbance spectra and seems to involve a singlet to triplet excitation. Both anodic and cathodic photocurrents were obtained depending on the applied potential.

# 5 Conclusions

In conclusion, some very important experiments have been performed in attempting to establish the electrode characteristics of the electroactive/conductive polymers. Continued work with these materials will require a parallel effort in improving their properties, if they are to successfully impact a technology.

# 6 References

1. For a convenient update in this field, see papers in Conferences such as Conductive Polymers Seymour RB (ed) Plenum Press, New York 1981; Proceedings from the International Conference on Low Dimensional Conductors, Mol. Cryst. Liq. Cryst. Cryst., 1982, Vols. 77–79; Davidson T

(ed) Polymers in Electronics. ACS Symposium Series 242, based on symposium in the Division of Organic Coatings and Plastics Chemistry, 185th Meeting, American Chemical Society, Seattle, Washington, March 20–25, 1983. International Society of Electrochemistry, 35th Meeting, Berkeley, California, August 5–10, 1984; Symposium on Materials on Emerging Technologies in the Division of Industrial and Engineering Chemistry, 190th Meeting, American Chemical Society, Chicago, Illinois, September 8–13, 1985

2. Diaz AF, Bargon J (1986) in: Skotheim TA (ed) Electrochemical Synthesis of Conducting Polymers; in Handbook of Conducting Polymers. Marcel Dekker Inc., New York, p 81
3. Lethely H (1862) J. Chem. Soc. *15*: 161
4. Mohilner DM, Adams RN, Argensinger Jr., WJ (1962) J. Amer. Chem. Soc. *84*: 3618 (see references to earlier work on the electrochemistry of aniline)
5. Brietenbach M, Heckner KH (1973) J. Electroanal. Chem. Interfac. Electrochem. *43*: 267
6. Dunsch L (1975) ibid. *61*: 61
7. Messina R, Sarazin C, Yu LT, Buvet R (1976) J. Chim. Phys. *73*: 9
8. Yasui T (1935) Bull. Chem. Soc. Japan *10*: 306
9. Jozefowicz LT, Yu LT, Perichon J, Buvet R (1969) J. Polym. Sci. C *22*: 1187
10. Cristofini F, de Surville R, Jozefowicz M, Yu LT, Buvet R (1969) C. R. Acad. Sci., Paris, Ser. C *268*: 1346
11. Langer J. (1978) Solid State Comm. *26*: 839
12. Diaz AF, Logan JA (1980) J. Electroanal. Chem. *111*: 111
13. Noufi R, Nozik AJ, White J, Warren LF (1982) J. Electrochem. Soc. *129*: 2261
14. Oyama N, Ohnuki Y, Chiba K, Ohsaka T (1983) Chem. Soc. Japan, Chem. Let., 1759
15. Diaz AF; U. S. Patent No. 4,304, 465, December 8, 1981
16. Kobayashi T, Yoneyama H, Tamura H (1984) J. Electroanal. Chem. *161*: 419
17. Kitani A, Izumi J, Yano J, Hiromoto Y, Sasaki K (1984) Bull. Chem. Soc. Japan *57*: 2254
18. Kobayashi T, Yoneyama H, Tamura H (1984) J. Electroanal. Chem. *177*: 281
19. Kobayashi T, Yoneyama H, Tamura H (1984) ibid. *177*: 293
20. Ohnuki Y, Matsuda H, Ohsaka T, Oyama N (1984) ibid. *158*: 55
21. MacDiarmid AG, Chian JC, Halpern M, Huang WS, Krawczyk JR, Mammone RJ, Mu SL, Somasiri NLD, Wu W (1984) Polym. Prepr. Chem. Soc., Div. Polym. Chem. *25*: 248
22. MacDiarmid AG, Chiang JC, Halpern M, Huang WS, Mu SL, Somasiri NL, Wu W, Yaniger SI (1985) Mol. Cryst. Liq. Cryst. *121*: 173
23. MacDiarmid AG, Mu SL, Somasiri NLD, Wu W (1985) Mol. Cryst. Liq. Cryst. *121*: 187
24. Genies EM, Tsintavis C (1985) J. Electroanal. Chem. *195*: 109
25. Genies EM, Syed AA, Tsintavis C (1985) Mol. Cryst. Liq. Cryst. *121*: 181
26. Kaneko M, Nakamura H (1985) J. Chem. Soc., Chem. Comm., 346
27. DeBerry DW (1985) J. Electrochem. Soc. *132*: 1022
28. Mengoli G, Munari MT, Bianco P, Musiani MM (1981) J. Appl. Polym. Sci. *26*: 4247
29. Aurian-Blajeni B, Taniguchi I, Bockris JOM (1983) J. Electroanal. Chem. *149*: 291
30. Paul EW, Ricco AJ, Wrighton MS (1985) J. Phys. Chem. *89*: 1441
31. Dall'Olio A, Dascola Y, Varacca V, Bocchi V (1968) Compt. Pend. C *267*: 433
32. Gardini GP (1973) Adv. Heterocycl. Chem. *95*: 15 and references therein
33. Diaz AF, Castillo JI (1980) J. Chem. Soc., Chem. Comm., 397
34. Genies EM, Bidan G, Diaz AF (1983) J. Electroanal. Chem. *149*: 101
35. Diaz AF, Kanazawa KK (1982) Miller JS (ed) in: Extended Linear Chain Compounds, Vol. 3. Plenum Pub. Corp., p 417
36. Diaz AF, Castillo JI, Logan JA, Lee WY (1981) J. Electroanal. Chem. *129*: 115
37. Burgmayer P, Murray RW (1982) J. Amer. Chem. Soc. *104*: 6139
38. Kuwabata S, Yoneyama H, Tamura H (1984) Bull. Chem. Soc. Japan *57*: 2247
39. Inganas O, Lundstrom I (1984) J. Electrochem. Soc. *131*: 1129
40. Yoneyama H, Wakamoto K, Tamura H (1985) ibid. *132*: 2414
41. Burgmayer P, Murray RW (1983) J. Electroanal. Chem. *147*: 339
42. Burgmayer P, Murray RW (1984) J. Phys. Chem. *88*: 2515
43. Zinger B, Miller LL (1984) J. Amer. Chem. Soc. *106*: 6861
44. White HS, Kittlesen GP, Wrighton MS (1984) ibid. *106*: 5375
45. Diaz AF, Vasques Vallejo JM, Martinez Duran A (1981) IBM J. Res. and Devel. *25*: 42
46. Bull RA, Fan FF, Bard AJ (1982) J. Electrochem. Soc. *129*: 1009

47. Noufi R, Tench D, Warren LF (1980) ibid. *127*: 2310
48. Frank AJ (1982) Electrically Conductive Polymer Layers on Semiconductor Electrodes, in: Energy Resources Through Photochemistry and Catalysis. The authors wish to thank A. Frank for making available a preprint
49. Noufi R, Frank AJ, Nozik AJ (1981) J. Amer. Chem. Soc. *103*: 1849
50. Skotheim T, Lundstrom I, Perjza J (1981) J. Electrochem. Soc. *128*: 1625
51. Skotheim T, Petersson LG, Inganas O, Lundstrom I (1982) ibid. *129*: 1737
52. Inganas O, Skotheim T, Lundstrom I (1982) Physica Scripta *28*: 863
53. Simon RA, Ricco AJ, Wrighton MS (1982) Amer. Chem. Soc. *104*: 2031
54. Skotheim TA, Inganas O (1985) J. Electrochem. Soc. *132*: 2116
55. Cooper G, Noufi RJ, Frank AJ, Nozik AJ (1982) Nature *295*: 578
56. Noufi RJ (1983) J. Electrochem. Soc. *130*: 2126
57. Frank AJ, Honda K (1982) J. Phys. Chem. *86*: 1933
58. Mengoli G, Musiani MM, Tomat R, Valcher S, Pletcher D (1985) J. Appl. Electrochem. *15*: 697
59. Murthy ASN, Reddy KS (1983) Electrochim. Acta *28*: 473
60. Komori T, Nonaka T (1983) J. Amer. Chem. Soc. *105*: 5690
61. Diaz A (1981) Chemica Scripta *17*: 145
62. Waltman RJ, Bargon J, Diaz AF (1983) J. Phys. Chem. *87*: 1459
63. Tourillon G, Garnier F (1982) J. Electroanal. Chem. *135*: 173
64. Diaz AF, Bargon J, Waltman RJ (1982) Proceedings of the Symposium in Membranes and Ionic and Electronic Conducting Polymers, Electrochem. Soc.
65. Garnier F, Tourillon G, Gazard M, Dubois JC (1983) J. Electroanal. Chem. *148*: 299
66. Waltman RJ, Diaz AF, Bargon J (1984) J. Electrochem. Soc. *131*: 1452
67. Tourillon G, Garnier F (1983) ibid. *130*: 2042
68. Tourillon G, Garnier F (1985) Mol. Cryst. Liq. Cryst. *121*: 305
69. Kaufman JH, Chung TC, Heeger AJ, Wudl F (1984) J. Electrochem. Soc. *131*: 2092
70. Mastragostino M, Scrosati B (1985) ibid. *132*: 1259
71. Tourillon G, Garnier F (1984) J. Phys. Chem. *88*: 5281
72. Czerwinski A, Laguren-Davidson L, Van Pham C, Zimmer H, Mark Jr. HB (1985) Anal. Lett. *18*: 2395
73. Bargon J, Mohmand S, Waltman RJ (1983) Mol. Cryst. Liq. Cryst. *93*: 279
74. Bargon J, Mohmand S, Waltman RJ (1983) IBM J. Res. Develop. *27*: 330
75. Rubinstein I (1983) J. Electrochem. Soc. *130*: 1506
76. Elsenbauer RL, Shacklette LW, Sowa JM, Chance RR, Ivory DM, Miller GG, Baughman RH (1982) Polymer Reprints, (1), ACS Meeting, p 132
77. Shacklette LW, Elsenbaumer RL, Chance RR, Sowa JM, Ivory DM, Miller GG, Baughman RL (1982) J. Chem. Soc., Chem. Comm., 361
78. Shacklette LW, Toth JE, Murthy NS, Baughman RH (1985) J. Electrochem. Soc. *132*: 1529
79. Giuffre L, Montoneri E, Modica G, Ho B, Tempesti E (1982) Adv. Hydrogen Energy *3*: 319
80. Montoneri E, Giuffre L, Tempesti E, Maffi S, Modica G (1984) Int. J. Hydrogen Energy *9*: 571
81. Giuffre L, Montoneri E, Modica G, Ho B, Temepesti E (1984) ibid. *9*: 179
82. Diaz AF, Clarke TC (1980) J. Electroanal. Chem. *111*: 115
83. Nigrey PJ, MacDiarmid AG, Heeger AJ (1979) J. Chem. Soc., Chem. Comm., 594
84. MacInnes D, Druy MA, Nigrey PJ, Nairns DP, MacDiarmid AG, Heeger AJ (1981) ibid. 317
85. Nigrey PJ, MacDiarmid AG, Heeger AJ (1982) Mol. Cryst. Liq. Cryst. *83*: 1341
86. Kaneto K, Mayfield N, Nairns DP, MacDiarmid AJ (1982) J. Chem. Soc., Far. I *78*: 3417
87. Epelboin I, Froment M, Garreau M, Thurnin J, Warin D (1980) J. Electrochem. Soc. *127*: 2100
88. Shirakawa H, Ikeda S, Aizawa M, Yoshitake J, Suzuki S (1981) Synth. Met. *4*: 43
89. a) Snow A, Brant P, Weber D, Yang NL (1979) J. Polym. Sci., Polymer Lett. Ed. *17*: 263; b) Tomkiewicz Y, Schutz TD, Brown HB (1979) Phys. Rev. Lett. *43*: 1532
90. Chen SN, Heeger AJ, Kiss Z, MacDiarmid AG, Gau SC, Peebles DL (1980) Appl. Phys. Lett. *36*: 96
91. Yamase H, Harada H, Ikawa T, Ikeda S, Shirakawa H (1981) J. Chem. Soc., Japan *54*: 2817
92. Ozaki M, Peebles DL, Weinberger BR, Chiang CK, Gau SC, Heeger AJ, MacDiarmid AG (1979) Appl. Phys. Lett. *35*: 83

93. Nowak RJ, Mark Jr HB, MacDiarmid AG, Weber DJ (1977) J. Chem. Soc., Chem. Comm., 9
94. Nowak RJ, Kutner W, Mark Jr HB (1978) J. Electrochem. Soc. *125*: 232
95. Rubinson JF, Behymer TD, Mark Jr HB (1982) J. Amer. Chem. Soc. *104*: 1224
96. Czerwinski A, Voulgaropoulos AN, Johnson JF, Mark Jr HB (1979) Anal. Lett. *12A*: 1089
97. Nowak RJ, Kutner W, Rubinson JF, Voulgaropoulos A, Mark Jr HB (1981) J. Electrochem. Soc. *128*: 1927
98. Voulgaropoulos A, Nowak RJ, Kutner W, Mark Jr HB (1978) J. Chem. Soc., Chem. Comm., 244
99. Mark Jr HB, Nowak RJ, Kutner W, Johnson JF, MacDiarmid, AG (1978) Bioelectrochem. Bioenergetica Acta *5*: 215
100. Rubinson JF (1981) Ph.D. Univ. of Cincinneti
101. Voulgaropoulos AN (1980) Ph.D. Univ. of Cincinneti
102. Mark Jr HB, Voulgaropoulos A, Meyer CA (1981) J. Chem. Soc., Chem. Comm., 1021
103. Anson FC (1959) J. Amer. Chem. Soc. *81*: 1554
104. Voulgaropoulos AN, Mark Jr HB (1980) Anal. Lett. *13*: 959
105. Bernard C, Tarby C, Robert G (1980) Electrochim. Acta *25*: 435
106. Nowak RJ, Joyal CL, Weber DC, Venezky DL (1980) paper presented at the Spring meeting of the Electrochemical Society
107. Tarby C, Bernard C, Robert G (1980) Electrochim. Acta *26*: 663
108. Beaudoin S, Bernard C, Vallot R, Robert G, Yu LT (1978) C. R. Acad. Sci. Paris *286*: 217
109. Nowak RJ, Joyal CL, Weber DC (1983) J. Electroanal. Chem. *143*: 413
110. Shenoy KP, Mulligan KJ, Mark Jr HB (1983) J. Electrochem. Soc. *130*: 2391
111. Rubinson JF, Behymer TD, Mark Jr HB, Nowak RJ (1983) ibid. *130*: 121
112. Scranton RA, Nooney JB, McCaldin JO, McGill TC, Mead CA (1976) Appl. Phys. Lett. *29*: 47
113. Cohen MJ, Harris Jr JS (1978) ibid *33*: 812
114. Schneider AA, Greatbatch W, Read R (1974) paper presented at the 9th International Power Resources Symposium, Brighton, U. K.
115. Louzos DV, Garland WG, Mellors GW (1973) J. Electrochem. Soc. *120*: 1151
116. Yamamoto T, Kuroda S, Yamamoto A (1982) Inorg. Chim. Acta *65*: Li75 and references therein
117. Matsumoto T, Matsunage Y (1981) Bull. Chem. Soc. Japan, 648
118. Jaeger CD, Bard AJ (1979) J. Amer. Chem. Soc. *101*: 1690
119. Wallace WL, Jaeger CD, Bard AJ (1979) ibid. *101*: 4840
120. Jaeger CD, Bard AJ (1980) ibid. *102*: 5435
121. Sandman DJ, Zoshi GD, Samuelson L, Burke WA (1982) Syn. Metals *4*: 249
122. Minami J (1980) J. Chem. Phys. *72*: 6317
123. Minami J, Watanabe T, Fujishima A, Honda K (1979) Ber. Bunsenges., Phys. Chem. *83*: 476
124. Giraudeau A, Fan FRF, Bard AJ (1980) J. Amer. Chem. Soc. *102*: 5137
125. Jaeger CD, Man FRF, Bard AJ (1980) ibid. *102*: 2592

Editors: G. Henrici-Olivé and S. Olivé
Received March 16, 1987

# Polymer-Coated Electrodes:
# New Materials for Science and Industry

Masao Kaneko
Institute of Physical and Chemical Research, Wako-Shi, Saitama, 351, Japan

Dieter Wöhrle
Organische und Makromolekulare Chemie, Universität Bremen, Bibliothekstr. NW 2,
D-2800 Bremen 33, FRG

Polymer modified electrodes form a new field that combines polymer chemistry with electrochemistry and which aims at developing new materials for electrocatalysis, electronics, photoelectrochemistry and photoelectronics. This review describes the methods used to prepare polymer coatings on a carrier electrode and discusses various ways of controlling electron processes. The subjects treated include electron transport through the polymer layer, catalysis at the interface, charge storage for polymer batteries, display devices and polymer-based electronics. A central topic is electron pumping in the polymer layer by photophysical and photochemical processes activated by visible light. Several prospective applications are mentioned.

1 Introduction . . . . . . . . . . . . . . . . . . . . . . . . . 143

2 Electron Cycles and Their Artificial Control . . . . . . . . . . . . . . 143

3 Preparation of Polymer-Coated Electrodes . . . . . . . . . . . . . . 145
   3.1 Method A: Polymer Coating . . . . . . . . . . . . . . . . . . 147
      3.1.1 A1: Casting of Polymer Solutions . . . . . . . . . . . . . . 147
      3.1.2 A2: Coating and Adsorption Process . . . . . . . . . . . 148
      3.1.3 A3: Dip Coating . . . . . . . . . . . . . . . . . . . . 150
      3.1.4 A4: Spin Coating . . . . . . . . . . . . . . . . . . . . 151
      3.1.5 A5: Electro- and Photodeposition . . . . . . . . . . . . . 152
      3.1.6 A6: Impregnation Process . . . . . . . . . . . . . . . . 152
      3.1.7 A7: Sublimation Technique . . . . . . . . . . . . . . . . 153
      3.1.8 A8: Covalent Binding . . . . . . . . . . . . . . . . . . 153
   3.2 Polymerization on Electrodes . . . . . . . . . . . . . . . . . 153
      3.2.1 B1: Electrochemical Polymerization . . . . . . . . . . . . 153
      3.2.2 B2: In situ Synthesis . . . . . . . . . . . . . . . . . . 162
      3.2.3 B3: Plasma Polymerization . . . . . . . . . . . . . . . . 163
      3.2.4 B4: Glow Discharge Polymerization . . . . . . . . . . . . 164
   3.3 Multilayer-Coated Electrodes . . . . . . . . . . . . . . . . . 164

4 Electron Processes at Polymer-Coated Electrodes . . . . . . . . . . . 164
   4.1 Electron Transport and Electrochemistry . . . . . . . . . . . . . 165
   4.2 Redox Reactions . . . . . . . . . . . . . . . . . . . . . . 171
   4.3 Catalysis . . . . . . . . . . . . . . . . . . . . . . . . . 172

Advances in Polymer Science 84
© Springer-Verlag Berlin Heidelberg 1988

**5 Charge Storage and Electric Functions at Polymer-Coated Electrodes** . . . . 180
   5.1 Charge Storage . . . . . . . . . . . . . . . . . . . . . . . . . . 180
   5.2 Display Devices . . . . . . . . . . . . . . . . . . . . . . . . . . 187
   5.3 Permeation and Transportation of Ions and Molecules . . . . . . . . 189
   5.4 Polymer-Based Electronics . . . . . . . . . . . . . . . . . . . . . 192

**6 Electron Pumping by Photophysical Processes** . . . . . . . . . . . . . 196
   6.1 Semiconducting Polymer Systems . . . . . . . . . . . . . . . . . 196
       6.1.1 Solid-State Devices . . . . . . . . . . . . . . . . . . . . . 198
       6.1.2 Liquid-Junction Devices . . . . . . . . . . . . . . . . . . . 201
   6.2 Inorganic Semiconductor/Polymer Membrane Systems . . . . . . . . 204
       6.2.1 Stabilization and Photoelectrochemical Catalysis
             by Polymer Coating . . . . . . . . . . . . . . . . . . . . . 204
       6.2.2 Sensitization by Dye Coating . . . . . . . . . . . . . . . . . 208
   6.3 Miscellaneous . . . . . . . . . . . . . . . . . . . . . . . . . . . 211

**7 Electron Pumping by Photochemical Processes** . . . . . . . . . . . . . 212
   7.1 Photoredox Systems . . . . . . . . . . . . . . . . . . . . . . . . 213
   7.2 Non-Photoredox Systems . . . . . . . . . . . . . . . . . . . . . . 215

**8 Future Prospects** . . . . . . . . . . . . . . . . . . . . . . . . . . . 219

**9 References** . . . . . . . . . . . . . . . . . . . . . . . . . . . . . . 220

# 1 Introduction

Polymer-coated electrodes are attracting more and more attention because polymer coatings can lead to the development of new types of electronic and photoelectronic devices. By coating an electrode with a polymer membrane the electrode is provided with various properties. There have been intensive studies of fundamental aspects of polymer-coated electrodes [1-18].

Polymers can be coated on electrodes by various methods. A simple method is to cast the polymer solution, which forms a film after drying. Active species are very often incorporated into this coated polymer layer by adsorption. Electropolymerization or other methods of inserting various monomers can directly give a polymer coating on electrodes. Other ways of preparing polymer coatings directly from monomers also exist.

Polymer-coated electrodes exhibit various functions for which electron transfer is an essential step. Polymer-coated electrodes can control electron processes according to their design, as will be described in this review article. Since electron cycles and their control are essential and most important processes in nature, they are briefly discussed in Sect. 2. The processes concerned are pumping, transport, storage and utilization of electrons.

The preparation of polymer-coated electrodes is reviewed in Sect. 3. Electron transport, a fundamental behaviour of polymer-coated electrodes, is described in Sect. 4, which includes the electrochemistry, redox reactions and catalyses of polymer coating. Charge storage and electric properties of polymer-coated electrodes are presented in Sect. 5, together with various applications. Electron pumping driven by light energy is described in Sects. 6 (photophysical processes) and 7 (photochemical processes), where the classification is according to the principles of the processes concerned.

In order to understand the way polymer-coated electrodes function, a few results on the modification of carriers by electrochemically active, low molecular weight compounds are included (for more details see [21]). The electrochemistry of polyacetylene is also mentioned. Coatings of pyropolymers obtained by high temperature treatment of electrochemically inactive and insulating saturated polymers are omitted from this article, as are graphite and its modification.

# 2 Electron Cycles and Their Artificial Control

The main energy source of biological life, as well as human community life, is solar radiation. The solar electromagnetic energy, most of which lies in the visible light region, is converted into chemical energy by photosynthesis in plants. The energy cycles in nature are represented by the electron processes shown in Fig. 1. These cycles are the basis of artificial ones [19,20].

The electrons extracted from water are pumped up in the photosynthesis system to a high energy level, and then they reduce carbon dioxide to give carbohydrates. The carbohydrates are a kind of reservoir of electrons whose energy has been raised by solar radiation. Dinitrogen is also involved in such processes, but it is omitted in this brief survey. The plants thus produced are almost all energy sources for animals:

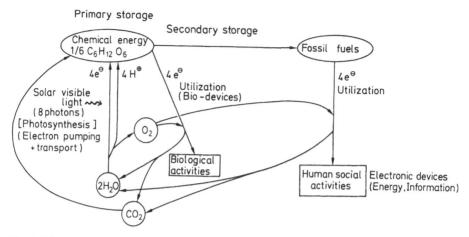

**Fig. 1.** Electron cycles in nature [Photosynthesis]

humans utilize the energy directly from plants and indirectly via animals. When electrons are extracted from water in photosynthesis, dioxygen is liberated as a by-product. In the respiratory system of animals, the electrons that were stored in the plants in a high energy state are transferred to dioxygen. By this process energy is liberated and used for biological activities. Water and $CO_2$ are the products of these processes, the electrons returning to the water molecule.

Oils and coals, which are the fossil fuels derived from plants, are the secondary reservoir of high-energy electrons and supply the majority of the energy required for human community activities. The electrons stored in the fossil fuels are transferred to dioxygen on cumbustion and liberate energy, again producing $H_2O$ and $CO_2$. In the important electrical systems in human activities and industrial use, the electrons are driven by the energy that is generated on burning the fuels.

The energy cycles in nature and in human activities are thus represented by electron processes. The main processes are pumping, transportation, storage, and utilization of electrons. An important objective of modern science and technology is to control these electron processes. The use of polymer-coated electrodes is one of the most promising approaches since these polymer-coatings provide electrodes with designed properties and lead to direct processes with electrons. Among these electron processes, transport, storage and utilization are the most fundamental ones, so they are described in Sects. 4 and 5 before electron pumping (Sects. 6 and 7).

In Sect. 4 electron transport at polymer-coated electrodes is described. Electrons are transported by electron-conductive polymers, and also by electron exchanges occurring between redox groups incorporated into polymer layers. Redox reactions of the incorporated groups also bring about chemical reactions and catalysis. Charge storage by redox reactions or by doping of polymer films is utilized in the construction of polymer batteries, as is described in Sect. 5. This section also includes display devices and polymer-based electronics. These electron processes and their applications are summarized in Table 1.

Artificial electron pumping induced by light energy is achieved by two main processes: photophysical and photochemical ones. Principal processes of electron pump-

**Table 1.** Electron processes at polymer-coated electrodes and their applications

| Process | Mechanism | Applications |
|---|---|---|
| Electron transport | Electron conduction<br>Electron exchange | Electron-conductive films<br>Redox reactions<br>Catalyses<br>Electronics |
| Charge storage | Redox reactions<br>Doping-undoping | Secondary batteries<br>Display devices<br>Permeation and transportation<br>of ions and molecules<br>Electronics |

ing induced by light energy are excitation by light, successive charge separation and transport of the separated charges. In order to realize these processes, photophysical and photochemical processes are used; the details are shown in Table 2. Photophysical processes utilize mainly semiconductor junctions, while photochemical processes use electron transfer reactions of light-absorbing compounds. All these electron processes are controlled by electric/photoelectric devices, in which polymer-coated electrodes play an important role.

# 3 Preparation of Polymer-Coated Electrodes

Mainly during the last ten years, several preparative methods of forming polymer structures containing electrochemical and/or chemical reaction centres at electrochemical interfaces have been developed. Two main pathways must be considered. *Method A* uses polymers which are coated mainly by adsorption or covalently to the carrier electrode. The electroactive group either may be inserted into an inert polymer matrix by an ion exchange process (electrostatic binding) or may be part of the polymer (covalent binding). *Method B* refers to the direct formation of the polymer on the carrier electrode from the corresponding monomer by a one-step procedure. The polymer contains the electroactive groups as a part of the polymer chain or network. In comparison to method A electrodes the method B ones often exhibit higher electronic conductivity leading to a different electrochemical response.

Various types of carriers are used for the polymer coatings: metals such as platinum (Pt) or gold (Au), carbon (C) in the form of graphite or glassy carbon, or inorganic semiconductors such as $SnO_2$, Si or GaAs. The layer thickness of the coatings varies a great deal, extending from $\sim 10$ nm to a few micrometers. The number of electroactive centers in the films coated on the carriers is of the order of $10^{-10}$–$10^{-6}$ mol cm$^{-2}$. The apparent concentration of electroactive sites in the polymer is often as high as 0.1–5 M. Compared with coatings of monolayers of low molecular weight compounds the electrochemical response is larger and easier to observe. The type and also thickness of the layer may influence the rate, intercalation of ions, chemical reactivity and velocity of the electrochemical process. This must be considered in preparing polymer-modified electrodes.

Due to the several hundred papers that have been published in the field of polymer-

**Table 2.** Electron pumping processes induced by light energy

| Principal processes | Photophysical processes | | Photochemical processes |
|---|---|---|---|
| | Band mechanism | Exciton mechanism | |
| Excitation by irradiation (Interaction of photons and electric field of a substance, and the subsequent change of the electron energy from the ground to excited states) | Electron transition from valence band to conduction band | Exciton generation | Electron transition from the ground to excited states |
| Charge separation | | Ion — pair formation by ionization of exciton, applied potential or reaction with impurities | Electron transfer reaction of the excited state with acceptor or donor |
| Transport of separated charges | Transport of separated charges through the gradient of electric field | | Successive electron transfer reaction with the second donor or acceptor |
| Output    (1) Electricity | Electron transport between electrodes and separated charges | | |
|             (2) Chemical energy | Catalytic reduction and oxidation reactions by the separated charges | | |
| Representative schemes | Semiconductor $\xrightarrow{h\nu}$ $e^- + h^+$ | | $D + P + A \xrightarrow{h\nu} D^+ + P + A^-$ |

$E_f$: Fermi level; P: Photoreaction centre; D: Donor; A: Acceptor

modified electrodes, the following survey can only consider a few typical examples, mainly taken from recent papers. Further references are given in these papers or in two review articles [21, 22].

## 3.1 Method A: Polymer Casting

A polymer is coated onto the carrier (area ~0.1–2 cm²) from a solution. This is elegantly simple. The systems in contact with an electrolyte can very often be electrochemically oxidized or reduced completely with good kinetic facility. A disadvantage of the electrostatic binding of redox centres is that instability due to desorption from the film may occur in the long term. This may be prevented by covalent binding of the redox centres.

### 3.1.1 A1: Casting of Polymer Solutions

A few microliters of a dilute solution containing the dissolved polymer and dissolved redox-active part are spread onto the carrier followed by evaporation of the solvent. In order to obtain smooth films the removal of the solvent must be carefully optimized. This method has the advantage that the concentration of the redox active parts is known.

**Examples**
a) C/poly(4-vinylpyridine) with $Fe(CN)_6^{3-}$ [23].

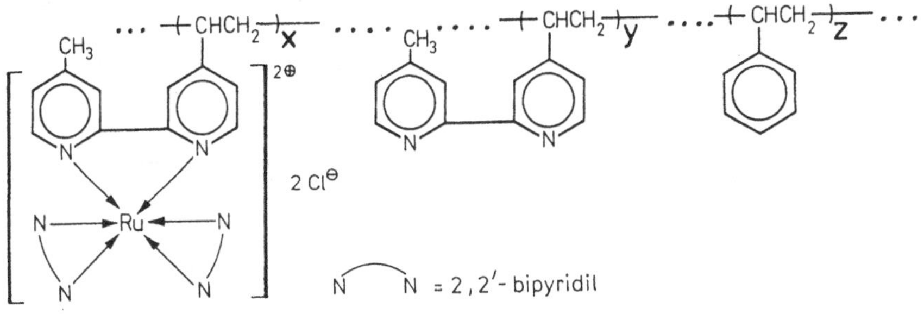

$1$

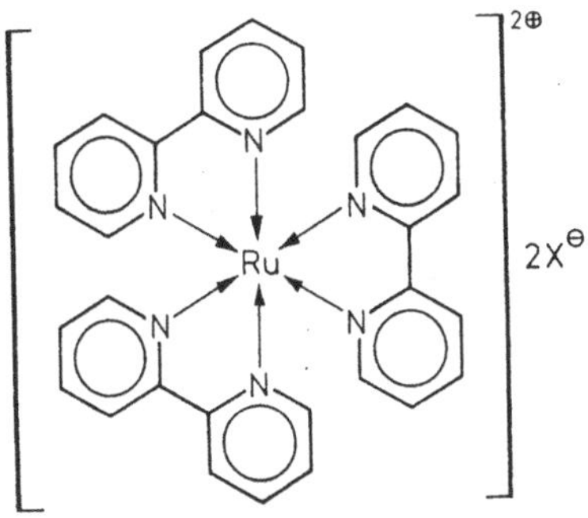

$2 \left( Ru (bpy)_3^{2\oplus} \right)$

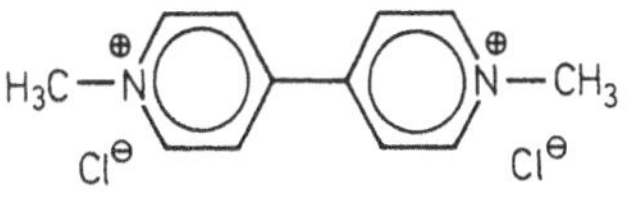

$3 \quad MV^{2\oplus}$

b) C or Si-doped GaAs/covalently polymer bound $Ru(bpy)_3^{2+}$ (*1*) ($\sim 0.3$–$0.6$ mol $l^{-1}$, thickness $\sim 0.15$–$0.3$ μm) [24,25].

c) C/Nafion ($\sim 10^{-4}$ g cm$^{-2}$) with $Ru(bpy)_3^{2+}$ (*2*) ($\sim 4 \times 10^{-9}$ mol cm$^{-2}$) and $MV^{2+}$ (*3*) ($\sim 4 \times 10^{-9}$ mol cm$^{-2}$) [26].

d) C/intermolecular complexes of poly(viologens) (*4, 5*) with poly(*p*-styrenesulfonate) or Nafion (each $\sim 10^{-7}$–$10^{-10}$ mol cm$^{-2}$) [6,27–29].

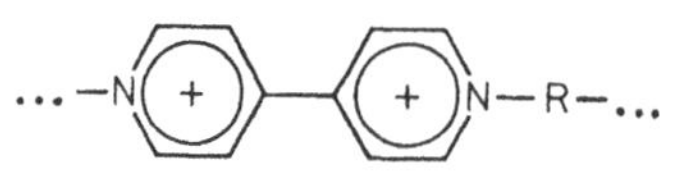

4

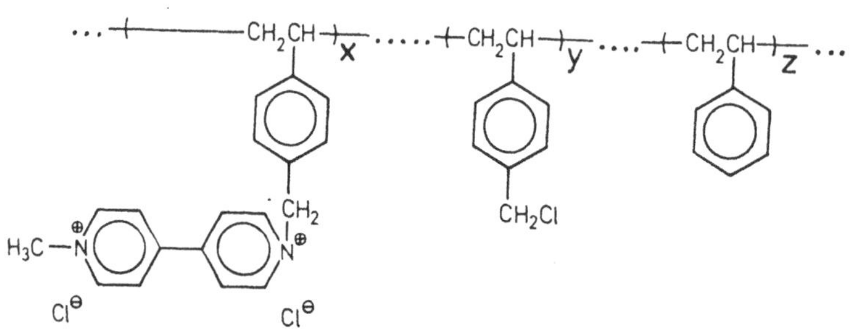

5

e) C or $SnO_2$/polymer covalently bound anthraquinone (*6*) (thickness $\sim 3$–$0.01$ μm) [30]. Films were also made by dispersing fine particles of a pigment in a polymer binder. For this a solution of a polymer containing the dispersed material is coated on electrodes.

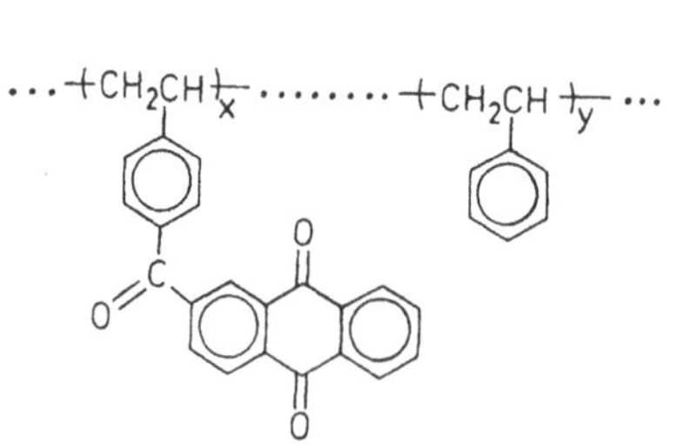

6

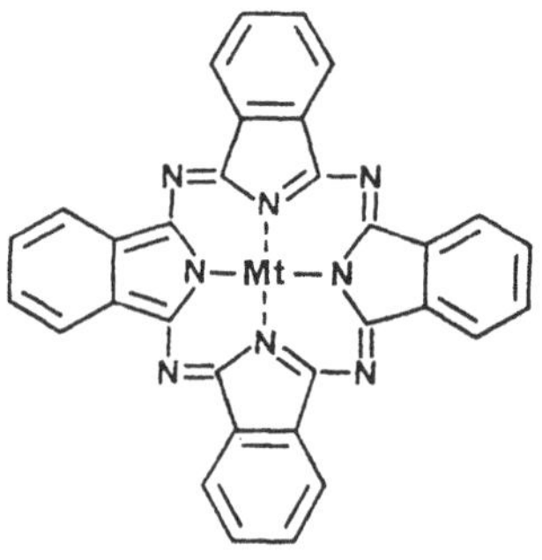

7 (Pc) with Mt=2H, Cu, Fe...

**Examples:**

a) $SnO_2$/metal-free phthalocyanines (*7*) and polyvinylcarbonate or others (weight ratio $\sim 2:1$) suspended in $CH_2Cl_2$ (thickness of film $\sim 5$–$0.5$ μm) [31,32].

b) $In_2O_3$/metal-free phthalocyanine and various polymers (weight ratio $\sim 1:1$) suspended in DMF or DMAc (thickness of films $\sim 1$–$3$ μm) [33].

## 3.1.2 A2: Coating and Adsorption Process

First a layer of an electrochemically inert polymer is coated onto the electrode from a solution (organic solvent). Then the coated electrode is placed in contact with an

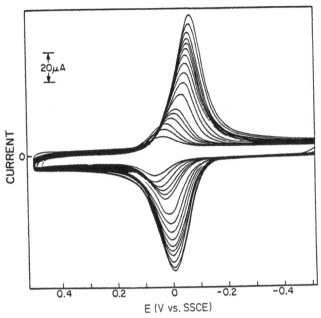

**Fig. 2.** Cyclic voltammograms recorded in $Fe(CN)_6^{3-}$ solution with a graphite electrode coated with benzyl-quaternized polyvinylpyridine. Supporting electrolyte, 0.2 M $CF_3COONa$ at pH 10; scan rate, 200 mV s$^{-1}$

aqueous electrolyte containing a soluble redox-active compound for a few minutes to several hours. The adsorption of the redox-active compound onto the polymer matrix is observed in its cyclic voltammogram as a time-dependent increase of the redox wave due to the adsorbed species (Fig. 2) [34]. The active compound is adsorbed onto the polymer matrix. In the case of poly(4-vinylpyridine) coatings the electrode is in contact with a solution of low pH. The pyridine moieties are protonated, becoming a polyelectrolyte. The counter anions on the polycation polymer are exchanged with redox anions in solution. The assembly is removed from the solution and can be used in another solvent that does not contain an external dissolved redox-active compound. The amount of redox-active compound inside the matrix is determined by electrochemical or chemical methods.

**Examples:**

a) C/poly(4-vinylpyridine) ($\sim 5 \times 10^{-7}$ mol cm$^{-2}$) $\leftarrow$ $Mo(CN)_8^{3-}$, $IrCl_6^{2-}$, $W(CN)_8^{4-}$, $Fe(CN)_6^{3-}$ (each $\sim 3 \times 10^{-8}$ mol cm$^{-2}$) [5,35].

b) C or n-type $SnO_2$/Nafion or polyfluoropolycarboxylate (thickness $\sim 0.3$–3 µm) $\leftarrow$ $Ru(NH_3)_6^{3+}$, $Ru(bpy)_3^{2+}$ (2) or $Os(bpy)_3^{2+}$ [36,37].

c) C/Nafion $\leftarrow$ covalently polymer bound $Ru(bpy)_3^{2+}$ (1) or covalently polymer bound methylviologen$^{2+}$ (5) [38].

$$\left[ \begin{matrix} S \\ S \end{matrix} \right\rangle = \left\langle \begin{matrix} S \\ S \end{matrix} \right]$$

8

d) Pt/Nafion (thickness $\sim 1$ μm) ← tetrathiafulvalene$^+$ (8) [39].

e) C/Polycarbonate    and    N-(4-ethoxybenzylidene)-4'-n-butylamine    thickness
   $\sim 0.1$ μm) ← $Fe(CN)_6^{3-}$ [40].

Stabilization of the adsorbed species was achieved by electrostatic encapsulation using bipolar films on electrodes. After $Fe(CN)_6^{3-}$ was incorporated into the film of cationic poly-$p$-xylylviologen, anionic sulfonated polysulfone film was coated on its top to produce bipolar film coating [41]. Metal interlayers were grown in polyimide film coated on an electrode. The metal ions such as $Ag^+$, $Cu^+$ and $Au^+$ incorporated into the polyimide film were reduced electrochemically to give metal interlayers [42]. Clay-coated electrodes [43−46] are also useful for incorporating cationic complexes. Sodium ions of the clay are exchanged, for example, with $Fe(bpy)_3^{2+}$ or $Os(bpy)_3^{2+}$.

## 3.1.3 A3: Dip Coating

The carrier electrode is dipped into a solution of a polymer and subsequently dried. Higher concentration and longer soaking time yield thicker coatings. In general, the redox-active compound, as in the case of A1 or A2, is added to the polymer solution or is diffused into the polymer film.

**Examples:**

a) C/poly(4-vinylpyridine) ($< 10^{-7}$ mol cm$^{-2}$, thickness $\sim 20$ μm) ← Ru(III) EDTA (concentration in the film $> 10^{-8}$ mol cm$^{-2}$; EDTA = ethylenediaminetetraacetic acid) [47].

Figure 3 shows the time dependence of the coordinative attachment of Ru(III) EDTA from the solution to the PVP coating [47]. A limiting value is reached after $\sim 2000$ s [40% of the pyridine groups coordinating to Ru(III)]. Figure 4 makes it clear that the fraction of pyridine groups coordinating to Ru(III) EDTA depends on the thickness of the PVP film (30% saturation at $7.6 \times 10^{-8}$ mol cm$^{-2}$ and

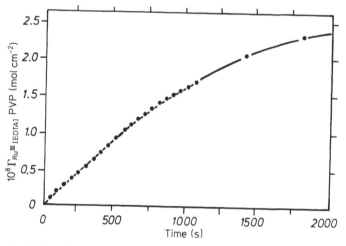

**Fig. 3.** Time dependence of the attachment of Ru(III)EDTA to a graphite electrode coated with $6 \times 10^{-8}$ mol cm$^{-2}$ pyridine groups in polyvinylpyridine. $2 \times 10^{-4}$ M Ru(III)EDTA dissolved in an aqueous solution containing 0.2 M CF$_3$COONa

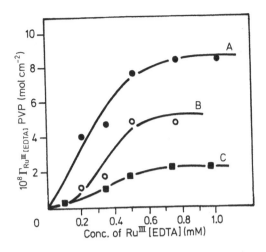

**Fig. 4.** Coordination of Ru(III)EDTA to graphite electrodes coated with increasing quantities of polyvinylpyridine. Ru(III)-EDTA dissolved in aqueous solution containing 0.2 M $CF_3COONa$ at pH 7.1. Surface concentration of pyridine groups in polyvinylpyridine: A, $1.2 \times 10^{-6}$; B, $3.3 \times 10^{-7}$; C, $7.6 \times 10^{-8}$ mol cm$^{-2}$

$7\%$ saturation at $1.2 \times 10^{-6}$ mol cm$^{-2}$ pyridine surface concentration). Therefore, in contrast to a solution reaction, only some parts of the pyridine groups exhibit coordination to the metal complex.

b) C/Nafion (thickness $\sim 7$ μm) ← Ru(bpy)$_3^{2+}$ (2) [48].

c) SnO$_2$/poly(4-vinylpyridine) (thickness $\sim 0.1$ μm) ← anionic dye rose bengal (concentration in the film $\sim (2-10) \times 10^{-9}$ mol cm$^{-2}$, $< 1$ M] [49].

d) Pt/poly(4-nitrostyrene) [88].

### 3.1.4 A4: Spin Coating

Thin films of polymers are formed by dropping a small amount of a solution of a polymer onto the carrier electrode, and then spinning it at several thousands r.p.m. As in the case of *A1* or *A2*, the redox-active compound is added to the polymer solution or diffused afterwards from a solution. During centrifugal spinning the evaporation of the solution leads to an increase of the concentration of the polymer causing an increased viscosity and the formation of a solid film. Only in the case of Newtonian fluids does the solution of the hydrodynamic equations lead asymptotically to a uniform layer thickness that is independent of the liquid profile at the start of the rotation [50,51]. The thickness of the films depends on the rotation speed, the evaporation rate and the initial viscosity [52,53].

**Examples:**

a) C or Pt/various alkylated poly(4-vinylpyridines) partly cross-linked (thickness $\sim 0.06-1.5$ μm). K$_3$/K$_4$ Fe(CN)$_6$ and other anions [54,55].

$$...-CH_2CH-...$$

9

b) Pt or SnO$_2$ on quartz/tetrathiafulvalene-polymer (9) (thickness ∼4–0.005 μm) [56].
c) Pt on quartz/tetracyanoquinodimethane-polymer (10) (thickness ∼1–0.03 μm) [57].
d) Pt/substituted linear polysilanes (thickness ∼0.06 μm) [58].

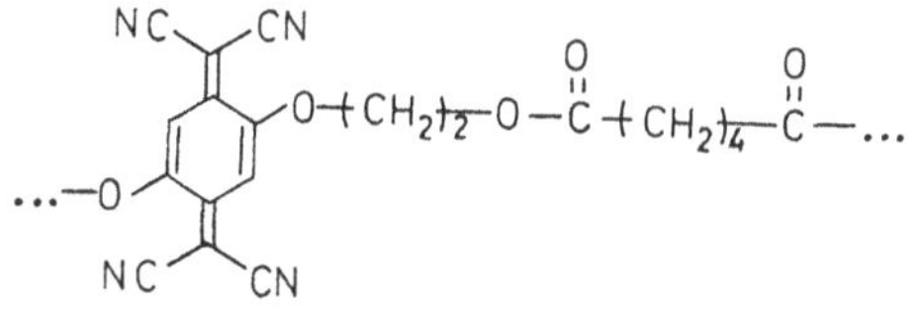

10

### 3.1.5 A5: Electro- and Photo Deposition

The solubility of polymers depends on their ionic state. Electrooxidized polymers such as poly(vinylferrocene) (11) were found to deposit on the anode due to the poor solubility of the oxidized polymer in the solvent used [59,60]. Photooxidized poly(vinylferrocene) was precipitated from methylene chloride solution by photooxidation [61].

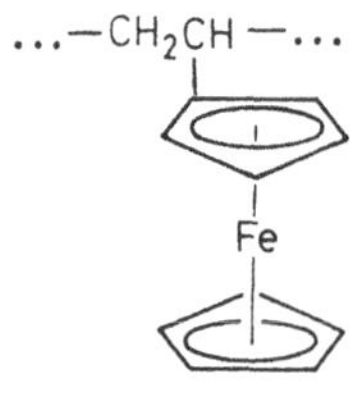

11

### 3.1.6 A6: Impregnation Process

Catalysts for dioxygen reduction are often obtained by reprecipitation of an active chelate such as Fe(II) or Co(II) containing phthalocyanines, porphyrins or tetra-aza(14)-annulenes on carbon materials. Polymeric Fe(II)phthalocyanine (12) was dissolved in concentrated sulfuric acid [62,63]. After addition of a carbon substrate the phthalocyanine was precipitated by pouring the mixture on ice and the precipitate was filtered.

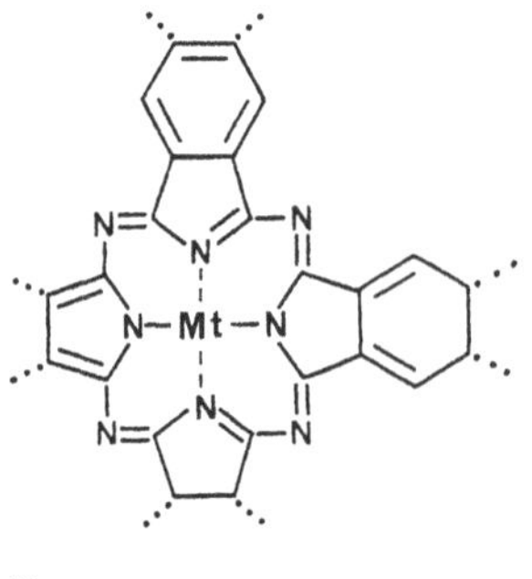

12 (polyPc) with Mt=Cu, Fe...

### 3.1.7 A7: Sublimation Technique

Poly(thiazyl) films were obtained on Mylar by sublimation from $(SN)_x$ crystals or films [64].

### 3.1.8 A8: Covalent Binding

In general, the surface of carrier electrodes such as $SnO_2$, Pt or C may be modified chemically. At the reactive groups obtained it is possible to bind polymers and/or redox-active compounds covalently.

**Examples:**
a) $SnO_2$—OH + alkylaminosilane(monolayer) + poly(acrylchloride) + hydroxy-methylferrocene ($\sim 7 \times 10^{-10}$ mol cm$^{-2}$) [65].
b) Pt or $SnO_2$ surface-OH + $N,N'$-bis(trimethoxysilyl)propyl-4,4'-bipyridinium-(derivative of methylviologen) coupling by cathodic polymerization ($\sim 10^{-9}$ mole $\times$ cm$^{-2}$) [66]

## 3.2 Method B: Polymerization on Electrodes

Starting from several monomers, polymer coatings on carrier electrodes are obtained by various techniques. The monomers are activated by a physical or chemical method. The polymers prepared are often insoluble due to cross-linking during polymerization, therefore the determination of real structure by, for example, elemental analysis, IR, Raman or electronic spectra, ESR, ESCA, neutron scattering or SEM is not very easy. Smooth to rough films are obtained. The polymer films often exhibit high electronic conductivity.

### 3.2.1 B1: Electrochemical Polymerization

The number of papers describing polymer coating by electrochemical polymerization of monomers dissolved in an organic solvent or water in the presence of conducting salts has increased in recent years [67]. The main polymerization is the oxidative one which occurs at the anode. The monomers used are aromatic heterocyclic, benzoid or nonbenzoid molecules. The polymers are obtained in an oxidized, high conductivity state (polymeric cations) containing incorporated counterions from the conducting salt. The films can be electrochemically cycled between the oxidized, conducting state and the neutral, insulating state. Some monomers were polymerized by a reductive electrochemical procedure. Most of the polymers described in this section can be prepared as amorphous powders by a normal chemical procedure. The advantage of electrochemical polymerization is that in a one-step procedure conducting films are obtained on a carrier electrode. Besides the preparation of pure polymer films, work is in progress in some laboratories to prepare composites with saturated polymers in order to improve the ease of processing and mechanical properties of the films.

The polymerization is carried out under inert gas usually in a three-electrode (working electrode, counter electrode, reference electrode), one- or two-compartment cell. When the electrode process is known, a two-electrode cell (working electrode, counter electrode) is also employed. The polymer coating occurs at the working electrode which may have a surface area of several square centimetres. Various materials, e.g. Pt, C, Au, SnO$_2$, Si and GaAs are used. The concentrations of the monomer and of the supporting electrolyte are each of the order of $<0.1$ mol l$^{-1}$. The electrochemical reaction was mainly carried out potentiostatically but also sometimes galvanostatically. The thickness of the films is $<1$ mm. Thicker films can be peeled off from the electrode surface to yield free-standing electrical conducting films.

The electrochemical polymerization often proceeds with an electrochemical stoichiometry of $\sim 2$ electrons/polymerized unit. A charge in excess of $\sim 2$ is consumed in the doping or partial oxidation of the polymeric film which is formed on the carrier electrode. In the oxidative polymerization anions of the electrolyte are incorporated into the polymer matrix.

As a typical example of electrochemical polymerization, cyclic voltammograms (CV) of aniline during the polymerization are shown in Fig. 5, where (a) is the CV under acidic, (b) neutral, and (c) basic reaction conditions [68]. For all the figures the first scan shows the oxidation of aniline at around 1 V giving a polyaniline (PAn) film on the electrode. For the PAn formed under acidic conditions (Fig. 5a) the waves based on doping and undoping of the films are observed between ca. 0–0.5 V for the 2nd and 10th scans, indicating that the film is electroactive. For the PAn formed under both neutral and basic conditions (Figs. 5b and c), the scans after the first cycle show no distinguishable peak, indicating that the films formed under these conditions are electroinactive.

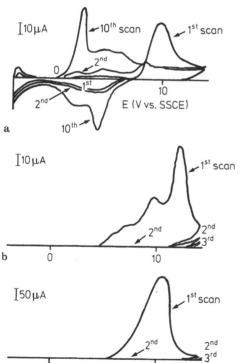

Fig. 5a–c. Cyclic voltammograms recorded during the oxidation of aniline (0.1 M) on platinum electrodes (0.0078 cm$^2$). **a** 0.5 M Na$_2$SO$_4$, H$_2$SO$_4$ (pH 1.0); **b** 0.5 M Na$_2$SO$_4$, 0.2 M Na$_2$HPO$_4$/NaH$_2$PO$_4$ (pH 7); **c** 0.5 M NaClO$_4$, 0.3 M pyridine. Scan rate, 50 mV s$^{-1}$

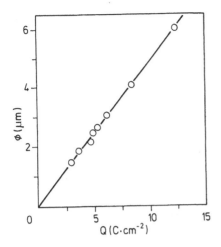

**Fig. 6.** Yield of polypyrrole (PP, *14*) vs amount of electricity that has flowed. Anode, $SnO_2$ glass

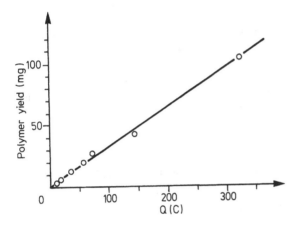

**Fig. 7.** Correlation between the thickness of films and the amount of the charge that has passed during the anodic oxidation of *N,N*-dimethylaniline. Polymerization carried out at 1.0 V vs SCE in 0.5 M $Na_2SO_4$ at pH 1.0 containing 0.1 M dimethylaniline

The electrochemically induced polymerization is easy to control. The polymer yield is proportional to the amount of electricity that has flowed (Fig. 6)[69]. A linear correlation also exists between the thickness of the films prepared and the amount of charge that has passed during anodic oxidation (Fig. 7)[70]. The morphology of the polymer film depends on factors such as the solvent, the current density and the anion of the electrolyte. Therefore, in addition to the structural units of the polymers, the morphology must also be considered in order to get reproducible physical properties of the films.

Table 3 lists some potentials of monomers and polymers. As in the case of PAn the polymers are easier to oxidize — reversibly — than the inserted monomers — irreversibly — (for PAn see Fig. 5). Also, polymers not prepared by an electrochemical technique, e.g. the phthalocyanines *12*, polyacetylene and graphite, can be oxidized or reduced reversibly. Not all organic compounds are able to be polymerized by an electrochemical technique on a carrier electrode. In the case of oxidative polymerization the neutral monomers are first oxidized at the anode irreversibly. The cation radicals $R^+$ formed are unstable and rather reactive. A high concentration

**Table 3.** Cyclic voltammetric data of some monomers and polymers (V vs SCE)

| Monomer | Peak potential | Polymer | Peak potential | |
|---|---|---|---|---|
| | $(E_{pa})$ R$^+$/R | | P$^+$/P | P/P$^-$ |
| Pyrrole [71 b)] | 1.20 | PP (14) | 0.27 | |
| N-Methylpyrrole [71 b)] | 1.12 | Poly(N-methylPP) | 0.26 | |
| Thiophene [71 b)] | 2.06 | PT (17) | 0.96 | |
| 3-Methylthiophene [71 b)] | 1.86 | Poly(3-methylPT) | 0.72 | |
| Pyrene [71 b)] | 1.33 | Poly(pyrene) | 1.1 | |
| Octacyanophthalocyanines (26) [246)] | 0.08 −0.06 (R/R$^-$) | PolyPc (12) [121 b, c)] | | −0.09 (at pH 0) −1.07 (at pH 14) |
| | | Polyacetylene (PA, 28)[145)] | 0.72 | −0.75 |
| | | Poly(p-phenylene) (PPP, 24) [145)] | 1.45 | −1.50 |
| | | Graphite [145)] | 1.64 | −1.78 |

of cation radicals is continuously maintained by the steady-state diffusion of R from the solution. The following situation can be explained in terms of the stability of the radical cation intermediate in the region of the electrode surface. The radical R$^+$ can react with the electrode surface ($k_p$), diffuse away from the electrode ($k_d$, relatively stable R$^+$), or give side reactions with the solvent S or the anions X$^-$ in the vicinity of the electrode surface ($k_s$) (Eq. (1 a)) [71)]. Whether polymerization occurs or not depends on the nature of R$^+$. The fraction $f_p$ of cation radicals which take part in the electro-polymerization is given in Eq. (1 b).

$$R \xrightarrow[E_{pa}]{k_1} R^+ \begin{cases} \xrightarrow{k_p} \text{oxidized polymer} \to \text{unoxidized polymer} \\ \xrightarrow{k_d} \text{radical cation diffusing away from the electrode} \\ \xrightarrow{k_s} \text{side reaction as nucleophilic substitution product} \end{cases}$$

($E_{pa}$: electrochemical peak oxidation potential)　(1 a)

$$f_p = \frac{k_p}{k_p + k_d + k_s([S] + [X^-])} \tag{1 b}$$

Figure 8 shows the potential range where some heterocycles polymerize [71)]. The cathodic cutoff for the polymerization (around 1.2 V) occurs when the stability of the radical cation is enhanced (intrinsically or via a substituent). When $k_d$ becomes greater than $k_p + k_s([S] + [X^-])$ diffusion of R$^+$ from the electrode results in the production of soluble products. The anodic cutoff (around 2.1 V) occurs when $k_s([S] + [X^-]) \gg (k_p + k_d)$. Then R$^+$ becomes unstable and reacts with the solvent or anions. Between around 1.2 and 2.1 V good conditions for the electropolymerization of such monomers exists where $k_p \gg k_d + k_s([S] + [X^-])$. The influence of substituents in pyrroles, thiophenes, indoles, azulenes, fluorenes, and pyrenes on whether electropolymerization of the monomers or other reactions can occur has been discussed in detail including consideration of electronic or steric effects [71)].

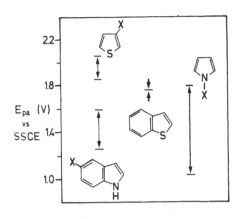

**Fig. 8.** Electrochemical potential range for polymerization of 1-substituted pyrroles, 3-substituted thiophenes, 5-substituted indoles and thianaphthene (V vs SCCE)

Two mechanisms for electropolymerization have been discussed [72a, b]. Cyclic voltametric studies support the hypothesis that the stepwise chain growth proceeds via radical-ion/radical-ion association (RR-route, Eqs. (2–4)). First, two radical

*RR-Route*

$$2 \text{ A} \xrightarrow[E_{pa}]{-2 e^-} A^{\ddagger} + A^{\ddagger} \rightarrow A^+ - A^+ \rightarrow A - A + 2 \text{ H}^+ \tag{2}$$

e.g.

$$\tag{3}$$

$$A^{\ddagger} + [A - (A)_n - A]^{\ddagger} \rightarrow [A - (A)_n - A - A]^{2+} \rightarrow (A)_{n+3} + 2 \text{ H}^+ \tag{4}$$

cations formed at the anode dimerize. After elimination of two protons a neutral dimer results which is oxidized again. Chain growth results from the reactions of oxidized monomers, dimers, trimers and oligomers which react with each other or by the addition of lower molecular weight cation radicals to form higher molecular weight products (oxidative polycondensation). One example is shown in Eq. (4). With increasing molecular weight the products become more and more insoluble and separate at the anode. Chain growth may occur in solution and at the polymer formed at the electrode. The electrochemical stoichiometry is around 2.1–2.5 electrons/monomer, where 2 electrons/monomer are involved in the polymer formation (under elimination of 2 protons/monomer). The excess charge (around 0.1–0.5 electrons/monomer) accounts for the concomitant oxidation of the polymer, which has a lower oxidation potential than the monomer. Due to the proton elimination the solution becomes increasingly acidic [71 b]. The number of eliminated protons was used to try to calculate the molecular weight of poly(3,4-dimethylpyrrole), which is around $10^5$. The mechanism of the electropolymerization cannot be compared with a normal cationic polymerization, a cationically initiated electrochemical polymerization of vinyl monomers [72 c], where an initial charge is needed only for one polymer chain.

The radical-ion/substrate association (RS-route) (Eqs. (5, 6)) describes the reaction of a cation radical with a neutral monomer. Chain growth occurs by the addition of different high molecular weight neutral molecules to radical cations.

*RS-Route*

$$A \xrightarrow[E_{pa}]{-e^-} A^{+\cdot} \xrightarrow{+A} A^+ - A^{\cdot} \xrightarrow{-e} A{-}A + 2\,H^+ \tag{5}$$

$$A^{+\cdot} + [A{-}(A)_n{-}A] \rightarrow [A{-}(A)_n{-}A{-}A]^{+\cdot} \xrightarrow{-e} (A)_{n+3} + 2\,H^+ \tag{6}$$

*Polymers from pyrroles*

Oxidation at the anode results in the formation of blue to black oxidized materials (thickness ~ 1 mm — 0.61 µm) of high conductivity. The films show a yellow colour in the uncharged low conductivity state upon undoping. The mechanical properties of 1 mm thick polypyrrole films (obtained on platinized electrodes in propylene carbonate partly in the presence of water) were reported [67a]. Polypyrrole (PP) was obtained in two different morphologies [67b]. Electrochemical polymerization in an aqueous solution of $H_2SO_4$ results in a compact structure of aggregated spheres (diameter 3–4 µm) of PP-$HSO_4$. The films have smooth surfaces and a density of 1.5 g/cm³. PP-$BF_4$ consists of a rough film (aggregation of 1 µm particles into hollow bowls of 30 µm diameter) with a density of 0.37 g/cm³. PP-$HSO_4$ has a surface area which is 15 times less than that of PP-$BF_4$.

**Examples:**

a) Pt/pyrrole, $Et_4NBF_4$ in acetonitrile [14, 73].
b) $SnO_2$ or C or Ti-Pt coated/pyrrole, $MgClO_4$ in acetonitrile [74].
c) Pt/pyrrole or 1-methylpyrrole, $KNO_3$ and others in water at different pH values [75].
   For pyrrole, polymerization in weak acidic or neutral solution has a fast rate at 0.65–0.80 V (vs SCE) for the reaction of Eq. (7).

$$n\,py - 2.5\,ne^- \rightarrow \{py{-}\overset{+}{p}y\}_{n/2} + 2\,nH^+ \tag{7}$$

d) Pt or C/pyrrole, molten salt $BuPyAl_2Cl_7$ [76].
e) Si/$SiO_2$/Au/pyrrole or 1-methylpyrrole, $Bu_4NClO_4$ in acetonitrile [77].
f) Pt/pyrrole, $H_2SO_4$ or $KPF_6$, $NaBF_4$, $NaClO_4$ in $H_2O$ [78].
g) Pt or $SnO_2$/pyrrole, $Et_4NClO_4$ in acetonitrile [79,80].
h) n-Si/Pt coated/pyrrole, $NaBF_4$ in acetonitrile [81].
i) n-Si or p-Si/pyrrole in propylene carbonate (36 mC cm⁻² for thickness 90 nm) [82].
j) Pt/various para-substituted 1-phenylpyrroles, Na, $Et_4BF_4$ in acetonitrile [83,84].
k) n-type Si/1-methylpyrrole, $Et_4NClO_4$ in acetonitrile under illumination (tungsten) [85].
l) Pt/viologen substituted pyrrole (*13*), $Bu_4NClO_4$ in acetonitrile [86].

*13*

Polypyrrole was modified by incorporating phthalocyanine tetrasulfonic acid [89, 90], $RuO_2$ [91] and several metals [92]. More flexible films were obtained by the synthesis of composites from polypyrrole and polyvinylchloride, polyvinylalcohol and poly-ether/polyester elastomer [67c,93–95].

The polymers from pyrrole mainly contain structural units *14* of 2.5-linkage.

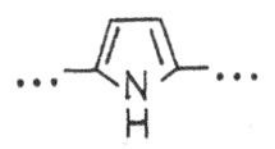

*14* (PP)

*Polymers from Benzannelated Pyrroles*

Unsubstituted or substituted carbazoles and indoles were polymerized to polymers containing the structural units *15* and *16*.

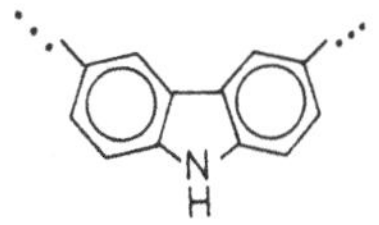

*15* 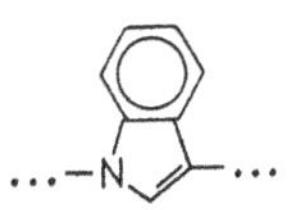                           *16*

**Examples:**

a) Pt/carbazol, Bu$_4$NClO$_4$ in DMF [96].
b) Pt/*N*-phenylcarbazole, Bu$_4$NPF$_6$ in nitromethane [72a, b].
c) Pt/1-vinylcarbazole, Bu$_4$NClO$_4$ in CH$_2$Cl$_2$ [97].
d) Pt/3-halogenated *N*-vinylcarbazole [98].
e) Pt/substituted and unsubstituted indoles, Bu$_4$NClO$_4$ in acetonitrile [71].

*Polymers from Thiophenes*

As in the case of pyrrole, highly conducting blue to black coloured polymers were obtained from various thiophenes by anodic polymerization. The polymers turned red in the unoxidized state. The polymerization of thiophene in organic solvents begins at around 2.1 V vs NHE, while with the dimer and tetramer of thiophene the oxidation of the monomers is easier (around 1.4 and 1.1 V) [72a]. The polymers prepared from thiophene are insoluble in organic solvents. Work is now in progress in some laboratories to polymerize thiophene derivatives which have a long aliphatic chain in position 3 of the thiophene ring. In this way soluble polymers are obtained by electrochemical polymerization. These polymers are better suited to structural investigation and processing.

The homogeneity of the polymers decreases when the film thickness is increased [99]. Poly(thiophenes) can be grafted as homogeneous thin films up to about 200 nm. Defects appear for thicknesses of 0.5–1 μm. When the thickness is increased to a few micrometers powdery deposits with fibrillar structure (diameter of fibrils in the conducting doped state ~800 Å in the undoped state ~200 Å) are observed. In thick layers the doping is inhomogeneous, except for CF$_3$SO$_3^-$, which gives the highest doping level of 50%. The specific conductivities of the doped poly(thiophenes) depend on the layer thickness. The polymerization of furane was also reported.

**Examples:**

a) Pt/thiophene, $LiClO_4$ in acetonitrile [100].

b) Pt or $SnO_2$/thiophene, $Bu_4NBF_4$ and others in nitrobenzene, benzonitrile or acetonitrile.

c) Pt or $SnO_2$/thiophene, $KPF_6$ in benzonitrile [101].

d) Pt/thiophene or various substituted thiophenes, $Et_4NBF_4$ in acetonitrile [102].

e) Pt/stirred emulsion of 3-methylthiophene, $NaClO_4$ in water [103].

f) $n$-GaAs coated with Pt/3,4-alkyl-substituted thiophenes, $LiClO_4$ in acetonitrile under tungsten illumination [104].

g) $In_2O_3$/dithienylalkylenes, $Bu_4NPF_6$ in acetonitrile, THF or nitrobenzene [105].

h) Pt or $SnO_2$/dithienothiophene, $Bu_4PF_6$ or $ClO_4^-$ in $CH_2Cl_2$ [106]

The structures *17–19* are postulated for some unoxidized polymers.

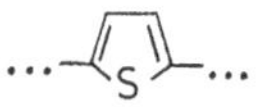

| *17* (PT) | *18* (n = 1–3) | *19* |
|---|---|---|
| poly(thiophene) | poly(dithienylalkylene)s | poly(dithienothiophene) |

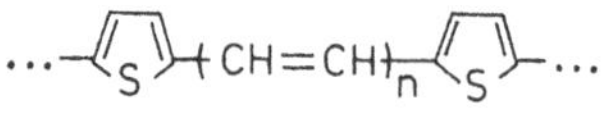

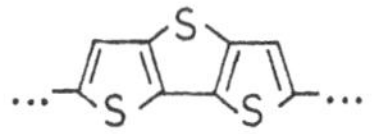

*Polymers from Anilines and Phenols*
Films of highly conducting polyanilines and insulating polyphenylene-oxides are prepared by the anodic polymerizations described before (see Fig. 5).

**Examples:**

a) C or Pt/aniline, $Na_2SO_4$ in $H_2SO_4$ (pH 1) or $PO_4^{3-}$ in $H_2O$ (pH 7) or $NaClO_4$ in pyridine, acetonitrile [68, 107, 108].

b) C or $In_2O_3$/N,N-dimethylaniline $Na_2SO_4$ in $H_2O$ (pH 1 and 13); incorporation of $Fe(CN)_6^{3-}$ studied [70].

c) Pt/phenol, aniline, 1,2-diaminobenzenes, various electrolytes and solvents [109].

d) C, Pt, Fe or Cu/8-quinolinols, NaOH in $CH_3OH$ [110].

e) C, Pt, Au/5-hydroxy-1,4-naphthoquinone (also 2-methyl derivative), NaOH in methanol [111].

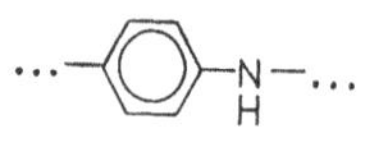

| *20a* (PAn) | *20b* (PAn) |
|---|---|

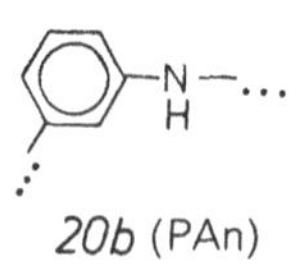

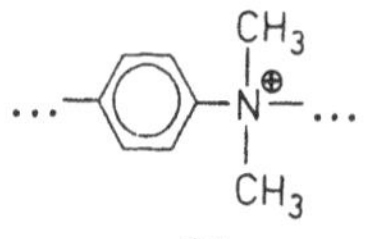

| *21* | *22* (PPO) | *23* |
|---|---|---|

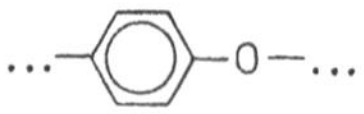

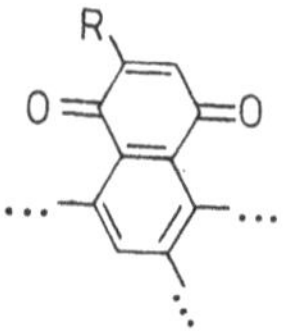

It is postulated that the polymers contain the structural units *20–23*. Polyaniline can be interconverted chemically or electrochemically between four forms by proton (acid-base) or proton and electron (oxidation-reduction) reactions. Form D is the only conducting version (Eq. (8)) [67].

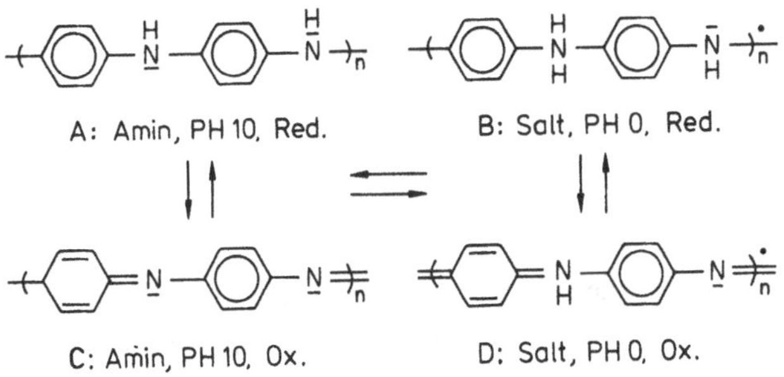

A: Amin, PH 10, Red.    B: Salt, PH 0, Red.

(8)

C: Amin, PH 10, Ox.    D: Salt, PH 0, Ox.

*Polymers from Aromatic Hydrocarbons*

In principle, many aromatic hydrocarbons such as benzene [72a, 112] can be polymerized anodically to dark-colored polymers. Real structure determination is difficult. Non-benzoid hydrocarbons such as azulene are also polymerized [113]. Polyvinylferrocene is anodically precipitated on carbon from $Bu_4BF_4$ in $CH_2Cl_2$ [108].

**Examples:**

a) Pt/benzene in aqueous HF [113].
b) Pt/pyrene, triphenylene, fluorene or fluoranthene, $Et_4BF_4$ in acetonitrile [114].
   Depending on the reaction conditions the structures *24* of polymers obtained from benzene are postulated.

*24* (PPP)

*Polynuclear Complexes*

High molecular weight polynuclear complexes are easily coated on electrodes as thin films by electrodeposition [115–117]. As a typical example, Prussian blue, which has the composition $Fe_4^{3+}[Fe^{II}(CN)_6]_3^{4-}$, is coated on Pt or graphite electrodes by applying $-0.5$ V (vs. SCE) at the electrode that is dipped in the aqueous mixture of $Fe(CN)_6^{3-}$ or $Fe^{3+}$. Ruthenium purple $(Fe_4^{3+}[Ru^{II}(CN)_6]_3^{4-})$ and osmium purple $(Fe_4^{3+}[Os^{II}(CN)_6]_3^{4-})$ were also coated as thin films.

*Polymers from Other Monomers*

Oxidative electropolymerization of Ni-complexes of the unsubstituted and substituted macrocycle $Me_4(Bzo)_2[14]tetraeneN_4$ in various organic solvents at 1.4 V vs SCE led to 185 "monolayers" of the corresponding polymer *25* on the carrier electrode [118].

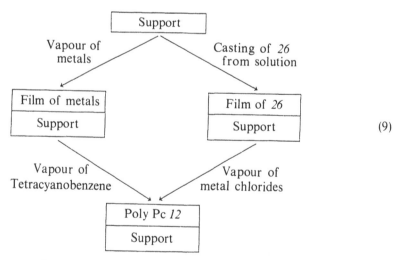

*25*

**Examples:**

a) Reductive electropolymerization of, for example, $[Os(bpy)_2(vp)_2](PF_6)_2$, $[Ru(bpy)_3](PF_6)_2$ and others in acetonitrile with $Et_4NClO_4$ (as electrolyte) on Pt results in layers of $<1\,\mu m$ thickness of the corresponding polymers (bpy: 2,2'-bipyridine, vp: 4-vinylpyridine) [119a].

b) Reductive electropolymerization of 4-phenylchinoline and benzamidazobenzophenanthroline [119b, c].

### 3.2.2 B2: In situ Synthesis

The surface of carriers can react chemically with low molecular weight compounds at higher temperatures to form polymer coatings. Thin films of polymeric phthalocyanines (*12*) were prepared on various supports as shown in Eq. (9) [120, 121].

One route uses the reaction of metal films with tetracyanocompounds. A wide range of supports is available: pure metals like Cu, Ni, Fe or Ti; or metal films on suitable carriers, e.g. Ti/Cu, ITO-glass/Cu, quartz/Cu, or carbon/Cu. Reaction of the metal surface at 300°–500 °C with gaseous 1,2,4,5-tetracyanobenzene gave coatings of structurally uniform polymers *12* with cyano end groups (thickness $<1\,\mu m$).

Cu films of 1.5 nm thickness led to smooth films of *12* (Mt=Cu) with a thickness of 48 nm and a specific gravity of 0.56 g cm$^{-3}$. Increasing the thickness of the Cu films >20 nm led to a more than proportional enhancement of the layer thickness and a decrease of specific gravity. Filaments were observed. The large surface area is one prerequisite for good electrochemical activity. The morphology of the films depends on the thickness of the metal film, the kind of metal and the tetracyanocompound. Investigations of the mechanism of film growth show that Cu atoms diffuse through the existing film of the polymer in order to react with tetracyanobenzene. Another route is the reaction of films of phthalocyanineoctacarbonitrile (*26*) (the monomer of the poly-

*26*

mers *12*) with metal chlorides. At first very smooth films of the monomer *26* were cast onto supports in layer thicknesses of 10–100 nm. Reaction of these films with gaseous metal chlorides such as FeCl$_2$ at 400 °C results in the reaction of the cyano groups of *26* to form the polymer *12*.

Polymeric Fe-phthalocyanine-coated carbon powders were prepared by the reaction of 1,2,4,5-tetracyanobenzene, FeCl$_2$ in the presence of carbon at 300°–500 °C [62, 63,122]. Afterwards the catalyst was mixed with poly(tetrafluoroethylene) emulsion.

### 3.2.3 B3: Plasma Polymerization

Smooth films of polymers are obtained by exposing monomers to plasma discharge. The rate of surface deposition of the plasma polymerized film depends on the geometry of the chamber, the radio-frequency power and the substrate temperature [123,124]. The details of the structure of the polymers prepared are unknown due to cross-linking during film formation.

Plasma polymerization of vinylferrocene on Pt or Ti/TiO$_2$ has been described [125–129]. The composition of the films obtained depends on the reaction conditions employed: $C_{9.7}H_{10.6}O_{1.7}Fe$, $C_{11}H_{12.6}O_2Fe$, $C_{10}H_{11}O_{1.7}Fe$ [ideal composition $[C_{10}H_{12}Fe]$]. In general, the degradation of ferrocene sites in the film is minimized by increasing the rate of deposition [126]. The thickness of the films is ~0.1 μm.

Coating with polymeric phthalocyanines is carried out by low-temperature plasma-induced polymerization of low molecular weight Cu-phthalocyanine (*6*) (thickness 3–0.6 μm) [130]. The electronic spectra of the insoluble solid films show the presence of phthalocyanine units.

Some other examples of films that immobilize redox pairs are those starting from vinylpyridine [131] or acrylic acid [131,132].

### 3.2.4 B4: Glow Discharge Polymerization

Polymeric electroactive films are prepared on carrier electrodes by cathodic glow discharge polymerization of monomeric vapours at an elevated temperature [133−135]. Highly cross-linked films of thickness <1 μm starting from metal acetylacetonates and acrylonitrile were obtained on carbon, Pt and other metals.

## 3.3 Multilayer-Coated Electrodes

A bilayer-coated electrode has two electroactive films, each having different reduction potentials. The inner layer is in direct contact with the carrier electrode surface and acts as a mediator to the outer layer which is mainly in contact with the solution. Provided that the redox levels in the two layers are appropriate, the interface between the two polymer films acts analogously to a semiconductor junction as a charge rectifying junction. The method of preparing first the inner layer on the electrode and then, in a second step, the outer layer may in principle be deduced from the conventional methods already described.

**Examples:**

a) Pt/[Ru(4-methyl-4′-vinyl-2,2′-bipyridine)$_3$]$^{2+}$ (electropolymerization, $\sim 10^{-9}$ mol $\times$ cm$^{-2}$)/polyvinylferrocene or derivatives of poly(2,2′-bipyridine) or poly-(4,4′-bipyridinium) (by droplet evaporation or electropolymerization, $\sim 10^{-8}$ to $10^{-9}$ mol cm$^{-2}$). [136]

b) Pt/poly[Os(bpy)$_3$ (vpy)$_2$]$^{2+}$/poly[Ru(vbpy)$_3$]$^{2+}$ (both layers by electropolymerization, each $\sim 10^{-8}$–$10^{-9}$ mol cm$^{-2}$). [10]

c) Pt/poly[Ru(bpy)$_2$ (vpy)$_2$]$^{2+}$ (electropolymerization, $\sim 10^{-8}$–$10^{-9}$ mol cm$^{-2}$)/electrodeposition of Cu, Si, Co or Ni [137]

d) C/polymer-bound Ru(bpy)$_3^{2+}$ (coating from DMF solution, thickness $\sim 0.45$ μm)/polymer-bound viologen (coating from CH$_3$OH solution, thickness $\sim 0.17$ μm) [138].

e) C/Prussian Blue (electropolymerization, thickness $\sim 1.3$ μm)/polymer-bound Ru(bpy)$_3^{2+}$ (_1_) (casting from DMF solution, thickness $\sim 0.16$ μm) [139].

# 4 Electron Processes at Polymer-Coated Electrodes

An interesting feature of polymer-coated electrodes is the electron processes that occur in the polymer membranes. They include, for instance, electron transport, redox reactions, catalysis, and doping-undoping of charges. In this section, electron transport, electrochemistry, redox reactions, and catalysis at polymer-coated electrodes will be described. Other types of electrochemical behaviours of polymer coatings, such as doping-undoping and electric properties, are described in Sect. 5.

## 4.1 Electron Transport and Electrochemistry

Electron transport in a film is achieved either by electronic conduction through a conductive material or by electron exchange between redox couples contained in the film.

Electronic conduction is observed for polymers containing conjugated aromatic rings such as polypyrrole (14, PP), polythiophene (17, PT) or poly(p-phenylene) (24, PPP), and for polymers containing aromatic rings connected with polarizable atoms, such as polyaniline (20, PAn) or poly(p-phenylenesulphide) (27, PPS). They constitute an important class among polymeric materials that can be made highly conductive upon appropriate electrochemical and chemical treatment (a so-called "doping" process). The remarkable increase of the conductivity achieved by oxidative doping with electron acceptors such as $I_2$ or $AsF_5$, or by reductive doping with electron donors such as alkali metals, in polyacetylene (28, PA) is well known[67,140-142]. In this review the conductivity of such materials will be mentioned only briefly.

27 (PPS)

28 (cis-PA)          28 (trans-PA)

The band gaps in the ground state of undoped aromatic structures of some polymers are as follows: PPP, calculated, 3.37 eV (experimental, 3.43 eV); PP, calculated, 3.9 eV (experimental, 3.43 eV); PT, 2.2 eV [67 d, 141 a]. Therefore the electron affinity is small and n-type doping should be difficult, whereas p-type doping should readily occur even with weak electron acceptors or at low oxidation potentials. In the case of PT, electrochemical reduction with cation doping at $-1.5$ V vs. Ag—AgCl was also described [143].

The intrinsic conductivities of the undoped materials are mainly of the order of $\sigma \leq 10^{-10}$ S cm$^{-1}$. PP, PT, PAn, and PPP are obtained by oxidative electrochemical polymerization (method B 1) in a highly conductive doped state with intercalation of anions. The oxidized dark-colored polymers can be discharged to a low conductivity state and charged again reversibly. The number of the intercalated anions corresponding to the positive charges in the polymer is usually of the order of 0.1—0.5 mole/mole monomer unit of the polymer. This number depends on the kind of the polymer, polymerization medium, and supporting electrolyte (e.g. see Ref. [78] for PP). Table 4 contains data for the conductivity of some polymers (see also Ref. [144]). Composites of saturated polymers as host with oxidized conductive polymers (PP or PT) as guest have the advantage of combining the properties of both the materials. The composites can exhibit high conductivities [151-153]: poly THF/PT $\leq 1$ S cm$^{-1}$, Nafion-impregnated polytetrafluoroethylene membranes/PP $< 50$ S cm$^{-1}$.

In order to preserve the conductivity and therewith also other properties, good

**Table 4.** Maximum conductivities of some polymers at room temperature

| Polymer | $\sigma$ (S cm$^{-1}$) | Refs. |
|---|---|---|
| Polypyrrole (PP, *14*) | 600[a] | 67 , 71 b, 73, 77, 80, 140, 141, 145 |
| Poly(1-methylpyrrole) | 0.001[a] | 67, 71 b, 77 |
| Poly(3,4-dimethylpyrrole) | 10[a] | 67, 141 |
| Polyindole (*16*) | 0.01[a] | 71 b |
| Polycarbazole (*15*) | 0.001[a] | 71 b |
| Polythiophene (PT; *17*) | 100[a] | 67, 71 b, 101, 140, 145, 146 |
| Poly(3-methylthiophene) | 100[a] | 71 b, 147 |
| Poly(3,4-dimethylthiophene) | 50[a] | 71 b |
| Poly(2,2'-bithiophene) | 0.1[a] | 71 b, 102 |
| Poly(dithienylalkylenes) (*18*) | 0.5[a] | 105 |
| Poly(dithienylalkylenes) (*18*) | 0.5[a] | |
| Polyaniline (PAn, *20*) | 5[a] | 145, 148 |
| Poly(p-phenylene) (PPP, *24*) | 400[a] | 140, 145, 149 |
| Poly(p-phenylenesulfide) | 1[a] | 140, 145 |
| Polyacetylene (PA, *28*) *trans* | 350[a] | 67, 140, 141 |
| *cis* | 10$^3$ [a] | 67, 140, 141 |
| Poly(azulene) | 1[a] | 71 b |
| Poly(fluorene) | 10$^{-4}$ [a] | 71 b |
| Poly(fluoranthene) | 0.001[a] | 71 b |
| Poly(triphenylene) | 10$^{-4}$ [a] | 71 b |
| Poly(pyrene) | 1[a] | 71 b |
| Octacyanophthalocyanine (*26*) | 0.1[b] | 62, 121 b, c |
| Polyphthalocyanine (*12*) | 0.001[b] | 62, 121 b, c |

[a] Conductivity in the oxidized state; kind and amount of counterion see literature;
[b] Intrinsic conductivity

long time stability of the polymers is necessary. Neutral (undoped) PA exhibits a conductivity of around $10^{-8}$ S cm$^{-1}$. After storage in air for one day the conductivity increases to $10^{-5}$ S cm$^{-1}$ (doping with dioxygen) (Fig. 10) [145]. Afterwards the conductivity decreases continuously to $10^{-10}$ S cm$^{-1}$. This is connected with the addition of dioxygen to PA (15%–30% weight gain after around 50 days). The films became more powdery. Also iodine-doped PA with 20 S cm$^{-1}$ shows a decrease in conductivity in air to 0.1 S cm$^{-1}$ (Fig. 9). Storage under argon gave a tenfold decrease of the conductivity after 200 days. In contrast, PP, PT and PAn exhibit a much

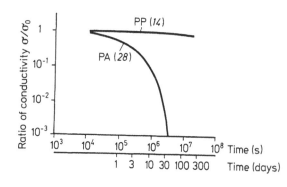

Fig. 9. Change of conductivity with time for oxidized polypyrrole (PP, *14*) (counterion phenylsulfonate) and polyacetylene (PA, *28*) (counterion iodine-) during storage under air

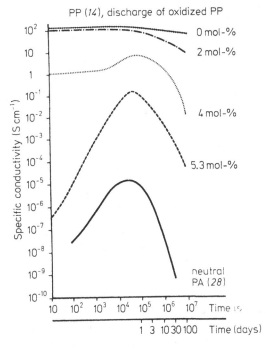

Fig. 10. Change of conductivity with time for polypyrrole (PP, *14*) (counterion $ClO_4^-$), in different charged states, and uncharged polyacetylene (PA, *28*) during storage under air

higher stability in air, as shown for oxidized PP in Fig. 10 [145]. Under argon no decrease in conductivity is observed. The stability of the conductivity of oxidized PP decreases with increasing temperature and smaller thickness of the films. For various applications, for example as electrodes in batteries, the long term stability in different charged states is important. Figure 10 shows that the stability in air becomes worse with greater loading. But the stability can be improved by using different kinds of anions for charge and by the pretreatment of the film with NaOH [145]. Graphite and PPP (*24*) have higher oxidation potentials and more negative reduction potentials than the previously discussed polymers (Table 3). Therefore these materials show a higher reactivity, for example with dioxygen, and an easier discharge. PT or composites of polyTHF/PT are stable in air after heating for 900 min [151].

It can be seen from Table 4 that the conductivity of polyPcs *12* and their monomer octacyanophthalocyanine exhibit high intrinsic (undoped) conductivities of $\leq 0.1$ S cm$^{-1}$.

The mechanism of charge carrier transport in oxidized conjugated aromatic polymers is not yet really clear [67, 71b, 145]. The electrical conduction of oxidized PP films was found to be of metallic type by thermo-electric power measurements and voltage shorted compaction measurements [78]. Four-probe conductivity measurements usually gave semiconductor-type temperature dependence of the conductivity (Fig. 11). However, it is explained that in this case the conductivity seems to be masked by intergranular contacts. A linear relation of $\log(\sigma \cdot T^{1/2})$ vs $T^{-1/4}$ was often found, indicating range hoping in disordered solids.

Electronic conductivity of poly[tris(5,5'-bis[(3-acryl-1-propoxy)carbonyl]-2,2'-bipyridine)ruthenium(0)] was studied without any concomitant redox-exchange conduction by locking the polymer complex into a single oxidation state [149].

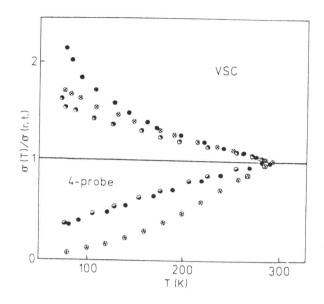

**Fig. 11.** Temperature dependence of conductivity of oxidized polypyrrole (*14*) by voltage shorted compaction (VSC) and by four-probe measurements. Counterions: ●, $BF_4^-$; ◑, $ClO_4^-$; ⊗, $HSO_4^-$

At polymer-coated electrodes where redox species are confined in the polymeric film, electron transport involves a "heterogeneous" electron transfer reaction between the electrode and the electroactive species confined to the polymer film as well as the "homogeneous" electron transport within the polymer film [59]. The "homogeneous" electron transport in the film is thought to occur either by electronic conduction by electron exchange between redox couples (Fig. 12a), or by physical diffusion of electroactive species itself (Fig. 12b).

An electron exchange mechanism has been proposed for polymer-confined redox species [154]. The rate-determining step for the electron transport in a film is not

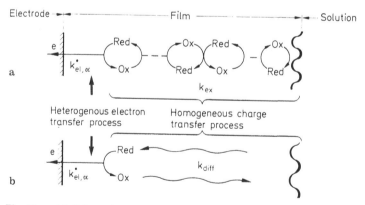

**Fig. 12a and b.** Schematic representation of electron transfer through polymer films by electron exchange between redox couples incorporated in the film. Ox and Red represent oxidized and reduced species of the redox couple, respectively. $k_{el}^0$: standard rate constant for the heterogeneous electron transfer between electrode and the redox site; $k_{ex}$: rate constant of the electron self-exchange; $k_{diff}$: rate constant of the physical diffusion of electroactive species

always the intrinsic electron transfer between the neighbouring redox groups; in some cases the motion of the charge-compensating counterion can be the rate-limiting process.

The electron transport by electron self-exchange can be regarded as a kind of diffusion process, and therefore represented by the diffusion coefficient. The apparent diffusion coefficient, $D_{app}$, for the charge transport in the polymer film was obtained from the slope of the cathodic limiting current in the normal pulse voltammogram ($i_{lim}$) vs $\tau^{-1/2}$ ($\tau$: the sampling time) plots by using the Cottrell equation [155] (Eq. (10)),

$$(i_d)^c_{Cott} = nFAc^0 \sqrt{D_{app}/\pi\tau} \qquad (10)$$

where $(i_d)^c_{Cott}$ denotes the cathodic limiting diffusion current, n the number of the electrons involved in the reaction, F the Faraday constant, A the electrode area, and $c^0$ the bulk concentration of the redox species in the film.

For the "heterogeneous" electron transfer between the electrode and the redox sites confined in the film, the standard rate constant, $k^0_{el}$, is derived from the current-potential relationship in the normal pulse voltammograms by using Eq. (11) [35, 156],

$$E = E^r_{1/2} + \frac{RT}{\alpha nF} \ln\left(\frac{4}{\sqrt{3}} \frac{k^0_{el}\sqrt{\tau}}{\sqrt{D}}\right)$$

$$-\frac{RT}{\alpha nF} \ln\left[x\left(\frac{1.75 + x^2[1 + \exp(\zeta)]^2}{1 - x[1 + \exp(\zeta)]}\right)^{1/2}\right] \qquad (11)$$

where E is the electrode potential, $E^r_{1/2}$ the reversible half-wave potential, $\alpha$ the cathodic transfer coefficient, R the gas constant, T the absolute temperature, $\zeta$ the dimensionless parameter $[(nF/RT)(E - E^r_{1/2})]$, and D the diffusion coefficient

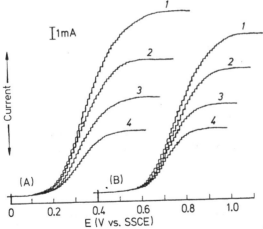

**Fig. 13.** Typical normal pulse voltammograms for the oxidation of (A) $W(CN)_8^{4-}$ and (B) $IrCl_6^{3-}$ incorporated into coatings of the protonated PVP film on BPG electrode dipped in a 0.2 M $CF_3COONa$—$CF_3COOH$ (pH 1.5) solution. The concentration of PVP coated on the electrode is $5.6 \times 10^{-7}$ unit mol of PVP $cm^{-2}$. The concentrations of $W(CN)_8^{4-}$ and $IrCl_6^{3-}$ are $3.1 \times 10^{-8}$ and $3.0 \times 10^{-8}$ mol $cm^{-2}$, respectively. Sampling time ($\tau$): (1) 1, (2) 2, (3) 4, (4) 10 ms

$[D_{app}$ (for cathodic process)] $[D_{app}$ (for anodic process)]$^{1-\alpha}$. Here x is the ratio of the current at potential E to the cathodic limiting diffusion current $(i_d)^c_{Cott}$.

Normal pulse voltammograms for the oxidation of (A) $W(CN)_8^{4-}$ and (B) $IrCl_6^{3-}$ incorporated into coatings of the poly(4-vinylpyridine) (PVP) film are shown in Fig. 13 [5]. Plots of the anodic limiting currents $(i_{lim})$ of these normal pulse voltammograms vs. $\tau^{-1/2}$ gave linear relations, as expected from Eq. (10). The $D_{app}$ values for this system were obtained as $10^{-7}$–$10^{-8}$ cm$^2$ s$^{-1}$ and depend on the concentration of the incorporated redox species. Figure 14 shows modified log plots of normal pulse voltammograms for the oxidation of $W(CN)_8^{4-}$ and $IrCl_6^{3-}$ in the film. The slope of the straight lines and the intersection of these lines with ther zero line give $\alpha$ and $k^0_{el}$, respectively. The $k^0_{el}$ is about $10^{-3}$ cm s$^{-1}$, depending on the concentration of the redox species. For both the complexes, $\alpha = 0.30$, which does not depend on their concentrations in the film.

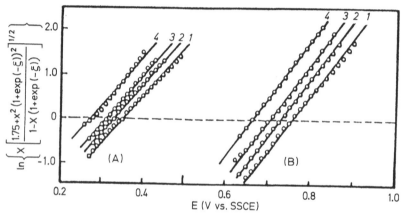

**Fig. 14.** Modified plots of normal pulse voltammograms for the oxidation of (A) $W(CN)_8^{4-}$ and (B) $IrCl_6^{3-}$ incorporated into a protonated PVP film. Sampling time $(\tau)$: (1) 1, (2) 2, (3) 4, (4) 10 ms. Other conditions are the same as in Fig. 13

The diffusion coefficient of methylviologen $(MV^{2+}$; 3) incorporated into films of porous agarose impregnated with Nafion has been reported [157]. In Nafion the diffusion coefficient decreases with increasing $MV^{2+}$ concentration, while in Nafion/agarose the opposite dependence is observed.

Electrochemical behaviour of methylviologen $MV^{2+}$ incorporated into a montmorillonite-modified electrode was studied [158]. The first reduction to the cation radical leads to the formation of a dimer that is not oxidizable within the clay film at potentials near the $MV^{2+}/MV^+$ wave. The dimeric viologen could be reoxidized via a mediation of charge from the electrode surface by $Fe(CN)_6^{3-}$.

The voltammetric behavior of polymer-coated stationary and rotating microelectrodes with critical dimensions of 0.25–25 μm were examined, and electron transfer rates at polymer/solution interface were determined [159].

## 4.2 Redox Reactions

Electrochemical reactions mediated by polymer coatings have been modelled [160,161]. Five limiting processes were considered for the polymer-mediated electrochemical redox reactions [7]: (1) the mass transfer of the substrate in the solution, (2) the rate of the cross exchange reaction between the film moieties and the substrate, (3) the rate of mass transfer of the substrate across the film-solution interface, (4) the diffusion of the substrate from the film-solution interface toward the electrode-film interface, and (5) the diffusion-like charge propagation from the electrode surface toward the solution. The representative profiles of the concentrations of the species concerned are shown in Fig. 15. The analysis of this model provided evidence for finite interfacial mass transfer rates in step (3). The resistance to interfacial mass transfer was greater for polymerized films such as poly(vinyl ferrocene) (*11*) and poly[Ru(Vbpy)$_3$]$^{2+}$ (only the Ru groups in *1*) than for the swollen polymers which incorporate redox-active species such as Mo(CN)$_8^{3-}$ [7]. Electron transfer kinetics were reported for the system containing poly[Ru(Vbpy)$_3$]$^{2+}$/solution interface [9]. The oxidized state of the complex attached to the polymer mediated oxidation of other metal poly-pyridine complexes.

Oxidation state trapping was achieved by utilizing interfacial electron transfer at a bilayer of the polymer coating [162]. The inner (next to the electrode) redox polymer film is chosen to have two stable redox couples, whose potentials (more positive a and more negative b) are located between that of the outer film (c). The redox sites in the outer film can become oxidized via inner film redox level a, and reduced via level b, but not reduced via level a or oxidized via level b. It is possible

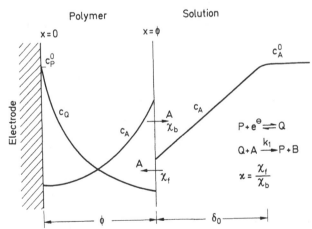

**Fig. 15.** Representative concentration profiles for the polymer-coated electrode. Within the film thickness (φ) the steady-state concentration profiles for the active form of the mediator in the film (Q) and the substrate from the solution (A) are shown. The electrode is held at a potential such that all the P (the inactive form of the mediator) which reaches the electrode surface is immediately reduced to Q. The concentration of Q at the electrode surface is therefore maintained at the maximum value of $c_p^0$. Within the solution the concentration of A is approximately linear. $\delta_0$ is the thickness of the diffusion layer. $\chi_f$ and $\chi_b$ are the rate constants for the transport of A into and out of the film, respectively

to trap positive charges at level c in the outer film by reaction with inner film level a (Eq. (12)), (see Fig. 16).

$$\text{Electrode/Ox(a)/Red(c)} \xrightarrow[\text{from c to a}]{\text{Interfacial electron transfer}} \text{Electrode/Red(a)/Ox(c)}$$

$$(12)$$

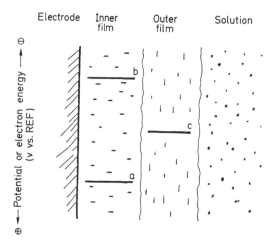

Fig. 16. Schematic electron energy diagram for a bilayer coated electrode

The untrapping of the trapped positive charges is carried out by the reduced state of b in the inner film (Eq. (13)).

$$\text{Electrode/Red(b)/Ox(c)} \xrightarrow[\text{from a to c}]{\text{Interfacial electron transfer}} \text{Electrode/Ox(b)/Red(c)}$$

$$(13)$$

Charge trapping and untrapping of a bilayer system $\text{Pt/poly[Ru(Vbpy)}_3]^{3+}/\text{poly[Fe(Vbpy)}_3]^{2+}$ was studied. It was shown that electron diffusion in the inner bilayer film is an important rate-limiting factor.

Even when the coated redox-active moieties are redox-active only near the surface of electrode, the presence of redox reagents dissolved in the solution phase can activate other coated species by helping propagation of the redox process through the film. A carbon electrode modified with polymeric anthraquinone (6) gave a dramatically enhanced anodic current due to the oxidation of the bound units when ferrocene carboxylic acid was present in the solution [163].

The electroactive species incorporated in polyionic films such as Nafion usually do not show a significant shift of the redox potential. However, the redox couples comprising neutral and positively charged forms exhibited substantial negative shift of the redox potential due to the larger difference in the interaction of the reduced and the oxidized forms with the polymer [164]. The ferrocene/ferricinium and $\text{MV}^+/\text{MV}^{2+}$ (3) couples incorporated in Nafion films showed fairly large negative shift of the redox potential.

The direct electrochemistry of redox proteins such as ferredoxin [165] and blue copper protein [166] was studied. The additional effect of poly(L-lysine) on the redox behaviour of horse heart cyt c at functional electrodes has been reported [167].

Electropolymerized films such as PAn undergo redox reactions producing a colour change. This is described in Sect. 5.2. The anodic oxidation of poly(N-vinylcarbazole) films was shown to involve initially the cross-linking of the polymer chains by oxidation of the carbazole moieties and dimerization of the resulting pendant carbazole cation radicals [168]. The resulting dimeric carbazole unit is more easily oxidized than the monomer and undergoes a further (reversible) two-electron oxidation.

## 4.3 Catalysis

Electrocatalysis is in principle a very important electrochemical process of great industrial interest. With the help of electrode coatings containing active surface states, oxidation or reduction of substrates in solution can be carried out. The aim is to drive reactions selectively and/or efficiently at a modest potential. In comparison to electrodes of inorganic materials (metals, metal compounds), those of organic compounds are at present at a fundamental stage of research. Electrodes modified with organic compounds may have the disadvantage of less stability, but they possess the advantage of a broad range of application due to great variation of the active species confined on the electrode. Studies of catalysis by polymer-modified electrodes have mainly focused on the reduction of dioxygen, aiming at fuel cells. It is suggested that the scope of research be enlarged to include other fields as well. Moreover, mechanisms have to be studied in detail.

An electroactive compound confined in a coated polymer film can act to relay electrons (in a reduction) from the electrode to the solution in two-step processes. First the electron is transferred from the electrode to the catalyst. Then follows the outer-sphere or inner-sphere electron transfer from the catalyst to the substrate. Another mechanism includes the interaction of the catalyst (which is not capable of mediating electrons) with the incoming substrate to alter its chemical reactions or electrochemical reactivities.

A Nafion coating incorporating $Os(bpy)_3^{2+}$ (with Os in *2*) catalyzed the electrochemical oxidation of ascorbate present in the solution [169]. The reaction between the catalyst and the substrate proceeded only within the outermost layer of the coating. In order to overcome this problem a polymeric coating with high porosity (copolymer gels of acrylamide/vinylpyridine) was used as an electrode film, and ferricyanide-mediated oxidation of ascorbate was studied [170]. A carbon paste electrode composed of polyvinylpyridine (PVP) and a graphite powder/Nujor oil matrix bound $Fe(CN)_6^{4-}$ under acidic conditions, which then catalyzed the oxidation of ascorbic acid [171].

Reactions in the polyelectrolyte medium of the coating are usually slower than in solution. Collisional restrictions may be responsible for the lower rates when the diffusion-controlled rate constant is much smaller within the polymeric phase. Mediated reactions were studied under conditions where the collisional restrictions on the rates were minimal. The $IrCl_6^{2-}$ incorporated into a cationic polyelectrolyte exhibited high diffusional rates. Oxidation of catechol or L-dopa (3-(3,4-dihydroxy-phenyl)alanine) by this catalyst system was studied. Collisional factors do not limit the reaction rates [172]. Oxidations of ascorbic acid, dihydroxyphenylacetic acid, and dopamine were studied at PP-coated and bare carbon electrodes [173]. The currents are mass-transport limited and not limited by permeation into or through the PP film.

Electrodes coated with poly(4-vinyl-4',4''-dibromotriphenylamine) mediated the oxidation of carboxylate anions [174].

Polymer-modified electrodes were used for alkyl dibromide reduction (Eq. (14)) [86-88].

$$C_6H_5-\overset{\overset{\displaystyle Br}{|}}{\underset{\underset{\displaystyle Br}{|}}{C}}H\overset{}{C}H-C_6H_5 + 2\ e^- \rightarrow C_6H_5-CH=CH-C_6H_5 + 2\ Br^-$$

$$(14)$$

This reaction is chemically irreversible and kinetically slow. On bare Pt electrodes the reduction of *meso*-1,2-dibromo-1,2-diphenylethane in acetonitrile/$n$-Bu$_4$NClO$_4$ occurs at $-2.1$ V vs Ag/Ag$^+$. Pt was covered by anodic electropolymerization (method B 1) with the polymer from viologen-substituted pyrrole (*13*) as electron transfer mediator [86]. The reduction of the alkyl bromide by the coated electrode was observed at $-1.4$ V vs Ag/Ag$^+$ and yielded selectively *trans*-stilbene. The reduction occurred catalytically over the doubly reduced form of the viologen. The disadvantage is that with a surface concentration of viologen species of $10^{-8}$ mol cm$^{-2}$, only $2.4 \times 10^{-5}$ mols of the substrate could be reduced (turnover $\sim 700$). It is assumed that the decrease of the activity is due to a loss of film permeability. The function of polymer backbone is not clear because the polymer works in the non-conducting state. The Pt electrodes covered with the non-conducting poly(4-nitrostyrene) obtained by dip coating also reduced the alkyl dibromide (turnover $10^4$) [88]. The reduction occurs at $-1.6$ V vs Ag/Ag$^+$, which is the potential of charge and discharge of the polymer film itself.

Electroreduction of hydrogen peroxide was carried out by montmorillonite clay incorporating Ru(NH$_3$)$_6^{3+}$ [175]. Electrocatalysis by polypyridine Os and Ru complexes confined to SnO$_2$ electrodes as a Langmuir-Blodgett (LB) monomolecular layer was studied [176] by electron exchange kinetics.

Catalytic multi-electron reaction is an important process in oxygen reduction and hydrogen evolution. Four-electron reduction of oxygen is attracting much attention in conjunction with the development of efficient and economic cathodes for fuel cells instead of the present expensive Pt-dispersed electrode systems. For the development of active catalysis for fuel cells — cathodic reduction of O$_2$ and anodic oxidation of, for example, H$_2$, N$_2$H$_4$, CH$_3$OH — some conditions must be met: high open circuit cell voltage, reduced polarization in the working mode of the cell and long-term stability. In the case of O$_2$ reduction, beside noble metals such as Pt and inorganic catalysts [177, 178], metal chelates have also been studied [62, 179, 180]. Metal chelates as catalysts are inserted on active carbons or graphite in the presence of polytetrafluoroethylene (polymer binder) as pressed electrodes.

The O$_2$ reduction may be carried out in either alkaline or acidic solution. Depending on the electrode surface the reduction proceeds either predominantly by HO$_2^-$ or by both the 4e$^-$ and peroxide pathways. These mechanisms are discussed elsewhere [178].

It is important to mention that the chelate contains a metal to bind $O_2$ and that the carbon-chelate mixture exhibits high electronic conductivity.

As low molecular weight chelates, iron- or cobalt-containing phthalocyanines (7 Pc), tetraazaannulenes (29 TAA), and tetraphenylporphyrins (30, R = —H, TPP) were investigated [62, 178, 179, 180].

29                                                    30

Heat treatment of the macrocycles such as Co tetramethoxy-TPP (30, R = —OCH$_3$) and TAA at temperatures of 450°–900 °C results in higher catalytic activity for $O_2$ reduction and good long time stability [177, 178]. Structure 30 (R = OCH$_3$) with Mt = Co or Fe suspended on high surface area carbon and heat-treated at 800 °C under argon shows in acidic electrolytes a significantly better performance than that of Pt/carbon electrodes [180b].

Another method uses polymeric Pc 12 as catalyst [62, 63, 180a]. Various methods of preparing the chelates on carbon were used: impregnation from solution (method A 6, mainly conc. $H_2SO_4$), painting of a mixture of catalyst and carbon on an electrode with a suspension of a catalyst (method A 1), evaporation of the catalyst and direct synthesis of the catalyst on carbon from its starting materials (method B 2). Besides electrochemical measurements, additional decomposition of $H_2O_2$ was investigated to determine the activity of the catalysts for the different reactions of $O_2$ reduction. Reviews of the literature up to 1983 are given elsewhere [62, 178–181].

Very recently a critical review of the work on heat-treated macrocyclic complexes on carbon was published [180c]. It showed that the results for gas diffusion electrodes obtained can hardly be compared with each other. The activity and pyrolysis behaviour of carbon-supported transition metal chelates is determined by various factors such as the chelate, the texture of the carbon, the dispersion of the catalyst on the support, the kinetics of the pyrolysis reactions, etc. The processes which occur during the metal loading on the carbon and the pyrolysis were discussed. The authors developed a new carbon modified rotating ring disc electrode which can be used for quick comparisons of both the activity and selectivity of carbon-supported catalysts [180c, d].

For low molecular weight chelates as Fe- or Co-Pc 7 and Co-TAA the influence of the method of preparation, catalyst handling and temperature treatment on the electrocatalytic activity were reported [181–183]. Polymeric Fe-Pc 12 is a better catalyst for the $O_2$ reduction than the low molecular weight FePc 7 [63, 181]. Recently the influence on activity and stability of three synthesis methods for 12 on carbon were examined in alkaline solutions [63]. The direct synthesis of 12 was carried out by reacting tetracyanobenzene with $FeCl_2$ either in bulk at 300°–500 °C or in ethylene glycol at 190 °C. both for different reaction times (Eq. (9)). The third method used the impregnation from concentrated $H_2SO_4$. The optimum coverage of 12 for the activity of melt synthesis electrodes was $7.9 \times 10^{-4}$ g m$^{-2}$ which is nearly equal to the calculated value of a monolayer of the polymer. Optimum conditions for the synthesis seem to be 350°–500 °C for more then 20 h. Figures 17 and 18 compare the polarization characteristics and stabilities of different electrodes. The mass synthesis electrode exhibits the lowest polarization in the range 2–100 mA cm$^{-2}$ and the smallest drop over 200 h during a 10 mA discharge test. Therefore this electrode shows excellent electrochemical activity and stability for $O_2$ reduction. The disadvantage of all the polymer coated electrodes is that the structural uniformity of the polymers is not really considered, as described elsewhere [184].

Face-to-face Co(II)porphyrins have been studied as catalysts for $O_2$ reduction [185]. An iron porphyrin complex was incorporated into the coating polymer layer to attain four-electron reduction of $O_2$. The copolymer of 1-vinyl-2-pyrrolidone with meso-[tri(phenyl)mono(p-methacrylamidophenyl)]porphin (31) was metallated and coated on a graphite electrode [186]. The four-electron reduction of $O_2$ by this modified electrode was achieved at sufficiently negative potentials.

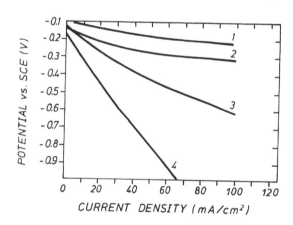

Fig. 17. Polarization curves for air electrodes in KOH solution. 1, polymeric Fe-Pc (12) on carbon from melt synthesis; 2, 12 on carbon from liquid-phase synthesis; 3, 12 on carbon by impregnation from $H_2SO_4$; 4, carbon support without catalyst

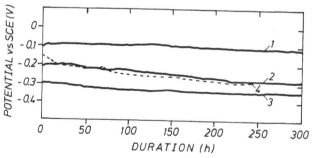

Fig. 18. Time dependence of potential drop at 10 mA cm$^{-2}$ for air electrodes in KOH solution. 1, 12 on carbon from melt synthesis; 2, 12 on carbon from liquid-phase synthesis; 3, 12 on carbon by gas-phase synthesis; 4, 12 on carbon by impregnation method

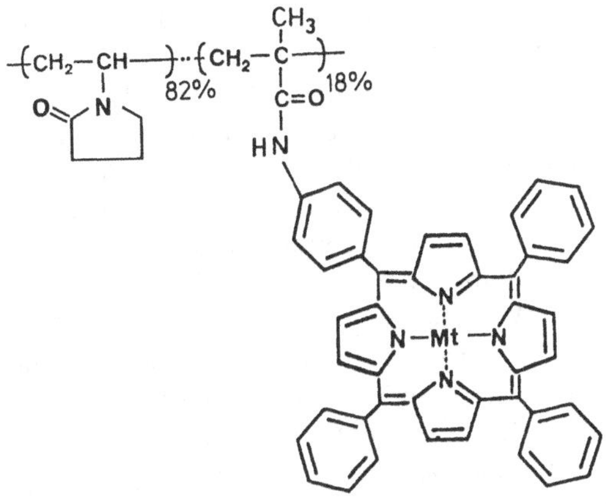

*31*

In the Nafion coating that incorporates both Co-TPP and $Ru(NH_3)_6^{2+}$, the former worked as a catalyst for $O_2$ reduction while the latter as an electron mediator between the electrode and the Co-TPP catalyst [187]. When Fe-TPP contained amino groups, the rate of $O_2$ reduction was larger for the Fe-TPP with more amino groups [188].

The cathodic reduction of $O_2$ in an alkaline solution has been studied on Mn(III) (TPP) X with different axial ligands (X = $F^-$, $Cl^-$, $Br^-$, $I^-$), so that the effects of X could be studied [189]. The order of the catalytic activity was $F^- > Cl^- > Br^- > I^-$. Only Mn(III) (TPP) F can reduce $O_2$ to $OH^-$, which has been attributed to the *d*-electron configuration with intermediate spin that facilitates a four-electron pathway through the side-on coordination of $O_2$ to the metal. The effects of axial ligands for Fe(III) (TPP) X were also studied [190].

Planar dinuclear Co chelate bis(3,5-di-2-pyridyl-1,2,4-triazole)dicobalt dichloride (*32*) adsorbed onto pyrolytic graphite catalyzed $O_2$ reduction to water in alkaline solution, whereas in acidic solutions only hydrogen peroxide was formed [191].

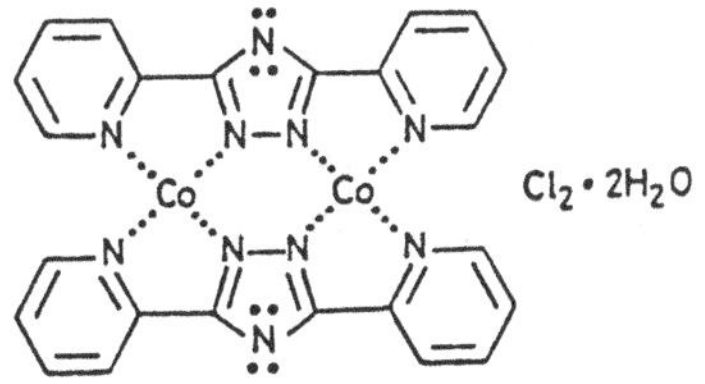

*32*

The mechanism of $O_2$ reduction by Co- and Fe-tetrasulfonated Pc coated as a mixture on a graphite electrode was discussed [192]. The mechanism of the electrocatalysis of $O_2$ reduction was studied on various catalysts taking into account the role of superoxide radicals, hydrogen peroxide and adsorbed $O_2$ [193]. The influence of electrode preparation was discussed for $O_2$ reduction by PP-coated electrodes incorporating Co-tetrasulfonated Pc [194].

PP can incorporate anionically charged metal porphyrin and MtPc electrochemically.

The incorporated film can catalyze $O_2$ reduction [195, 196]. Metal complexes of *meso*-tetrakis(4-sulfonatophenyl)porphyrin (*30* R = $-SO_3H$) were doped in a PP film and their catalyses of $O_2$ reduction were studied [196]. The PP film incorporating Fe-tetrasulfonated Pc exhibited significant catalytic activity at potentials between 250 and 800 mV less negative than at a bare glassy carbon electrode.

In acidic media and alkaline solutions, stable electrodes of polynaphthoquinone (*23*) prepared on Pt, Au and graphite by electrochemical oxidation (method B 1) of the corresponding naphthoquinone were investigated for their electrocatalytic effect on the $O_2$ reduction [111]. While on graphite the catalytic effect was quite low, the $O_2$ reduction of *23* on Au occurred at the reduction potential of the quinone group with high intensity. The *23* on Pt showed an additional shift of the half-wave potential of 190 mV more positive than the formal potential of the quinone/hydroquinone couple ($E° = 0.16$ V vs SCE). Immobilized laccase was used as a catalyst for cathodic molecular oxygen reduction [197]. Oxygen reduction by polyaniline films was studied at stationary and rotating film-coated electrodes [198].

A secondary fuel cell was constructed using the covalently bound polymeric Co-Pc (*33*) obtained by a dipping process

$$M = Fe^{III}, Co^{II}, Ni^{II}, \text{ or } Cu^{II}$$
$$m = 0.50, p = 0.48 - 0.39, q = 0.02 - 0.11$$

*33*

(method A 3) on a Pt electrode [199]. During charging (500 μA for 30 min at 1 cm² Pt sheets) $O_2$ is evolved from the surface of the working electrode (anode) and $H_2$ from the counter electrode (cathode) in 30% KOH aqueous solution. During discharging with 100 μA a quite stable plateau at $-0.24$ V vs SCE was observed at the Pc electrode (cathode) due to $O_2$ reduction. The binding of $O_2$ (and its storage) at the Co-Pc is responsible for the reversibility of this system. The discharge capacity of 833 Ah kg$^{-1}$ of the metal complex is quite high, but, considering also the amount of complex in the polymer (only 1.5%), the practical activity is much lower.

Hydrogen evolution by $H^+$ reduction is a two-electron process. Co-TPP and its derivatives coated on a carbon electrode by various coating techniques (e.g. methods A1 and A8) exhibited catalytic activity for $H_2$ reduction [200]. The ITO electrode modified with an $N,N'$-dialkyl-4,4'-bipyridinium-based polymer derived from trimethoxysilane derivatives of $MV^{2+}$ showed good catalytic activity for electrochemical evolution of $H_2$ in the presence of Pd deposited onto the outer surface of the polymer [201]. The same coating system was used for the electrochemical reduction of bicarbonate to formate [202]. Pt and Ag aggregates were electrochemically included in poly(3-methyl thiophene) films, and the system showed high catalytic activity for $H^+$ reduction [203]. Poly(p-phenylene) (24) catalyzed hydrogen production from water in the presence of a sacrificial donor such as triethanolamine or diethylamine [204]. A viologen-modified electrode (method B1) brought about redox conversion of peroxidase, and the kinetics were studied [205]. A clay (montmorillonite) coating catalyzed the electroreduction of $H_2O_2$ [206].

Coatings of polymeric Pc 12 (prepared by method B2) on different metals (mainly titanium) were investigated for their Faradaic activity.[120, 121, 207]. The coatings were obtained by an in situ gas phase synthesis of the metal substrates with 1,2,4,5-tetracyanobenzene (Eq. (9)). The anodic and cathodic peak currents and the peak separation $E_p$ for the $K_3Fe(CN)_6/K_4Fe(CN)_6$ redox couple demonstrate the high activity and reversibility of such electrodes, whose properties approach those of a Pt electrode (Fig. 19). The properties of coatings of 12 (Mt = 2II) on Ti were optimized by careful variation of the reaction conditions (temperature, time and amounts of tetracyanobenzene) for their preparation [121]. The oxidation of $I^-$ and the reduction of p-benzoquinone in comparison with bare Pt electrodes were also investigated [207]. Additionally, the electrodes modified with 12 exhibit an anodic photocurrent for the oxidation of various substrates [207]. The action spectrum corresponds to the absorption of the 12. The anodic photocurrents are in agreement with the n-conducting behaviour of the layers. For all the electrochemical reactions the modified electrodes show long-term stability. The activity may be due to the large surface area, high electronic conductivity ($10^{-3}$ S cm$^{-1}$), and redox behaviour of 12.

Gold electrodes coated with PP or poly(1-methylpyrrole) were used for the oxidation of hydroquinone and the reduction of p-benzoquinone in aqueous acidic solutions as well as in a water/organic solvent mixture [208] (and references cited therein). While with uncoated Au electrodes the anodic and cathodic peak separations are larger than 200 mV, the PP coating reduced the separation to 60 mV. This corresponds

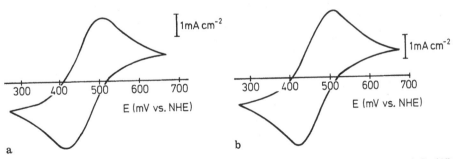

**Fig. 19.** Cyclic voltammograms at (a) Pt electrode and (b) Ti electrode coated with polymeric Pc (12) dipped in $2 \times 10^{-2}$ M Fe(CN)$_6^{4-/3-}$ containing 0.5 M $K_2SO_4$. Scan rate 20 mV s$^{-1}$

to a two-electron reaction with activation polarization. The mechanism for the reduction or oxidation of the species in solution is the same as that for the Pt electrode. The function of the polymer electrode is still unclear. Moreover, repeated cyclic scanning of the potentials results in a decrease of the current at the redox waves (aging effect). The polymer electrodes became increasingly blocked by immobilized redox compounds.

Low molecular weight Pc such as *7* and polymeric Pc *12* mixed with acetylene black (prepared according to method B2) were investigated as cathodes in secondary lithium cells [209, 210]. The Pcs exhibit electrocatalytic activity in the decomposition of the electrolyte propylene carbonate (Eq. (15)). In the case of *7* the specific discharge capacity at 1 mA discharge is 1008 Ah kg$^{-1}$. The cells were not optimized for maximum activity.

$$CH_3-CH-CH_2 + 2\,e^- \rightarrow CH_3-CH=CH_2 + CO_3^{2-} \qquad (15)$$

Electrocatalytic reduction of $CO_2$ at a PP-coated film containing $Re(bpy)\,(CO_3)\,Cl$ was reported [211]. Reduction of CO, formic acid and formaldehyde was mediated by thin films of Co(I)-4,4',4'',4'''-tetracarboxyphthalocyanine [212].

# 5 Charge Storage and Electric Functions at Polymer-Coated Electrodes

In addition to the reactions described in Sect. 4, polymer-coated electrodes exhibit various electric properties which can be used for electronic devices. The most important of them are charge storage, electrochromism, transport of ions or molecules, and control of electron processes. Their applications are described in this section.

## 5.1 Charge Storage

As described in Sect. 3.2, charges can be stored in polymer films containing redox groups such as metal ions, metal complexes, and redox reagents. Not only in such polymer films but also in conjugated polymers like polyacetylene (*28* PA), polypyrrole (*14* PP), polythienylene (*17* PT) and polyaniline (*20* PAn) are charges stored and released reversibly via doping and undoping. In this section charge storage by doping and undoping of polymer films as well as by redox reactions of film components are described (see also Table 3).

These electroactive polymers are being investigated intensively in some industrial companies as electrode materials in rechargeable batteries. The following polymers and their derivatives are interesting: PA [213-225], PAn [226-231], PP [80, 232-234], PT [100, 102, 235-238], and polyphenylene (*24*, PPP) [225, 239].

The reason why conducting polymers are being investigated as promising materials is that their specific weight is lower than that of ordinary inorganic materials. Therefore

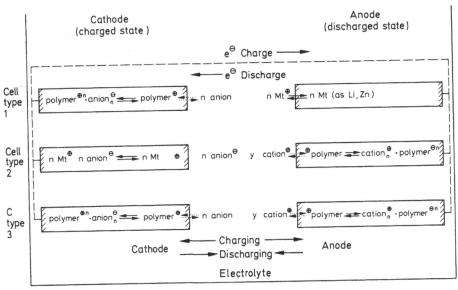

**Fi«. 20.** Principle and types of polymer batteries based on doping and undoping

cells of high energy density can be expected. Most authors have reported cells which utilize Li (the lightest metal element) as the anode and the polymer as the cathode (Fig. 20, cell type 1). In this cell the polymer is oxidized by charging under intercalation of anions. In cells of type 2 a metal such as Li can work as the cathode. In the anode the polymer is reduced under intercalation of cations. In cells with Li an aprotic organic solvent containing a salt is used as the electrolyte solution. Besides the many difficulties of using an aprotic solvent, the surface of Li is gradually covered by passive films such as Li-carbonate during repeated charging and discharging cycles. Additional electrocatalytic decomposition of the electrolyte at the electrode materials may occur. The problem of safeguarding against pollution also remains, which is a problem always present for Li. Therefore, some cells of types 1 and 2 that use no Li but have materials working in aqueous electrolytes have been described. Cells of type 3 (Fig. 20) use polymers as both cathode and anode. The electrode materials were obtained in the cases of PAn, PP and PT by oxidative polymerization (method B 1), but chemically synthesized polymers like PP [233], PA, PPP and polyPc (*12*) can also be used. Studies have addressed the problem of air stability, reactions with electrolyte components, mechanical strength, ease of processing, and control over ionization potential. To improve the properties the conducting polymers were mixed with a second component [216].

Since an enormous enhancement of the electronic conductivity of PA by doping from an almost insulator to a metal over a range of 11 orders of magnitude has been reported (Sect. 4.1) [238a, 239b, 240], the doping of conjugated polymers has been attracting more and more attention. This reversible process brings about not only a change in the conductivity of the polymer, but also charge storage. It also causes colour changes, which will be described in the next section.

PA batteries function in aprotic solvents. The electrochemical potential of the oxidized film (p-type; anions such as $ClO_4$, $AsF_6$, Br) is 3—4 V vs $Li^+/Li$ electrode. The potential of the reduced film (n-type; cations such as Li, Na) shows about 0.5–1.0 V vs $Li^+/Li$ electrode. Oxidation and reduction can occur with high Coulombic efficiency for doping levels of n 0.10 per CH unit of the polymer. The stability of PA is low (see, Ref. [145] in [221], and Sect. 4.1). Pure PA is an extremely sensitive electrode material which must be handled carefully during cell assembly. Highly purified electrolytes are needed and cycling must be carried out within a strict potential limit. Another disadvantage is the low volumetric energy density due to the low specific weight of the polymer. At present PA appears perhaps to be more an intriguing electrochemical curiosity than a material with immediate technological applications in high energy density cells.

Cells of type 1 (Fig. 20) use mainly Li as the anode [213, 222, 224] e.g.,

$$Li/LiClO_4 \text{ (propylenecarbonate; PC)}/-(CH)_x- .$$

The cells show an initial $V_{oc}$ (open circuit voltage) around 3.7 V and $J_{sc}$ (short circuit current) of around 50 mA cm$^{-2}$. The diffusion of the anions proceeds slowly. The cells can release only around 40 % of their charge at discharge rates up to 10 mA cm$^{-2}$ and less than 0.5 V decrease in the cell potential [222].

A large-scale PA battery comprising a 20–54 cm$^2$ area electrode was reported [214]. The battery,

$$Li-Al/LiClO_4 (PC + EC)/-(CH)_x- ,$$

where EC is ethylene carbonate, exhibited little difference in the charge-discharge characteristics from a small-scale battery. The energy density was 114 Wh kg$^{-1}$ with 5 mA discharge current, which is much higher than that of the conventional lead-battery ($\sim$40 Wh kg$^{-1}$). Scaling up of organic batteries is therefore not the main problem.

A PA solid-state battery was constructed using a superionic conductor as the solid electrolyte [215]. Thus, batteries such as

$$Ag/RbAg_4I_5/-(CHI_y)_x- ,$$
$$Na/-Al_2O_3/-(CHI_y)_x- ,$$

and

$$-(CHNa_y)_x-/-Al_2O_3/-(CHI_y)_x-$$

were tested. These batteries exhibit a long lifetime but so far only a low current and energy density, which is due to the solid electrolyte.

Reduced PA (n-types) was also investigated [216, 223, 225]):

$$Li/LiClO_4 , \quad \text{borate salts (PC, THF, or other ethers)}/-(CH)_x- .$$

The potential of -(Li-CH)- is 0.5–1.0 V more positive than that of Li. Compared with the Li anode the -(CH)$_x$- anodes show lower gravimetric and volumetric energy densities and a higher electrode potential. In order to stabilize reduced PA against reaction with electrolytes, reduced PA was modified by reaction with ethyleneoxide or with cyclic sulfonate [216]. Repeated cycling of the reduced form of the modified PA electrodes vs Li counterelectrodes at 0.5–1.0 mA cm$^{-1}$ displayed 96%–99% Coulombic efficiency. The PA battery of type 3 (Fig. 20) showed initial $V_{oc}$ = 2.5 V, and $J_{sc}$ = 22 mA cm$^{-2}$ [213, 221]. $Bu_4NClO_4$ was used in PC as the electrolyte.

Electropolymerized films such as PAn, PP, and PT are stable when used in polymer batteries (Sect. 4.1). They are promising candidates for practical use.

A PAn battery of type 1, Li/LiClO$_4$ (PC)/PAn, was constructed which had $V_{oc}$ = 3.3 V and a high energy density of 352 Wh kg$^{-1}$ with discharge capacity 106 Ah kg$^{-1}$ [226].

Two other batteries working with aqueous electrolytes have the following compositions [227, 228]:

Type 1    $Zn/ZnSO_4$ $(H_2O)/PAn$ ,

Type 2    $PAn/H_2SO_4$ $(H_2O)/PbO_2$ .

The first cell exhibited a maximum capacity of 108 Ah kg$^{-1}$ and an energy density of 111 Wh kg$^{-1}$ ($V_{oc}$ around 1 V). The Coulombic efficiency was close to 100% over at least 2000 cycles (cycling between 1.35 and 0.75 V) at a constant current of 1 mA cm$^{-2}$. The second cell also shows a high efficiency (more than 95% after 4000 complete cycles) [228].

The variation of chemical charge storage and electrical conductivity of electrolyte-wetted PP films has been studied. The charging and discharging rates of PP films were studied in AlCl$_3$:1-methyl-(3-ethyl)-imidazolium chloride molten salts and in CH$_3$CN [232]. The rates are limited by ion migration in the polymer, and for potential steps in the double layer charging region PP behaves as a porous electrode material. Chemically synthesized PP is used as an electrode material in battery applications [233]:

Li-Al/LiClO$_4$ (PC)/PP .

The PP can stand more than 100 charging-discharging cycles. The material shows a high capacity of 90–110 Ah kg$^{-1}$ (theoretical 88 Ah kg$^{-1}$), which is partly due to the large capacitance of the electrode. A cell of type 3 was also studied [234]:

PP/Et$_4$ClO$_4$, Et$_4$NBF$_4$ (H$_2$O or CH$_3$CN)/PP .

The mechanism of the reduction of PP working as an anode is not clear. The cell shows large Coulombic efficiencies but poor charge retention properties.

PT is also an excellent stable electrode material. The characteristics of PT as a stable polymer cathode material for organic batteries were studied [100, 102]:

Li/LiClO$_4$ (PC), Et$_4$BF$_4$ (CH$_3$CN)/PT .

Electrochemical doping and undoping of PT film using tetramethyl ammonium cations were reported [237]. The applicability of polydithienothiophene (19) as a cathode-active material was tested and it was concluded that it can be used as a stable material for a battery, although the fast self-discharge is a disadvantage which has to be overcome [236, 238b]. Electroactivity of polythiophenes obtained from various thiophene derivatives were reported: these films allowed current densities of 2 mA cm$^{-2}$ at an open-circuit voltage of 2.3 V [102]. Structure *19* as the anode in ZnBr$_2$/water shows a storage charge capacity of 90 Ah kg$^{-1}$, which is higher than the value of 54 Ah kg$^{-1}$ for *19* in Et$_4$ClO$_4$/CH$_2$Cl$_2$ or LiClO$_4$/PC [236].

Poly(*p*-phenylene) (PPP) was also used as an electroactive material for a battery [239]. The cell of type 1,

$$Li/LiAsF_6 (PC)/-(C_6H_4)_x-(AsF_6^-)_y ,$$

gave $V_{oc} = 4.4$ V and current density of 40 mA cm$^{-2}$. The storage capacities are smaller than those of PA using PPP as an anode. Other polymers without a conjugated structure can also be doped electrochemically and used as a battery. A poly(*N*-vinylcarbazole) (PVCz) battery,

$$Li/LiClO_4 (PC)/PVCz ,$$

showed $V_{oc} = 3.86-4.15$ V and $J_{sc} = 2.2-6.0$ mA cm$^{-2}$. The Coulombic efficiency of the cell was 90% at a current density of 4.4 μA cm$^{-2}$. Charging and discharging 100 times produced no appreciable change in the characteristics [241].

Iodine adducts of nylon-6 formed a battery with Li and lithium iodide solid electrolyte [242]. The current efficiency at 500 kΩ load was about 50% based on the iodine added. Activated carbon fibre (ACF) can also be doped to form a battery [237, 243] with Li:

$$Li/LiClO_4 (PC)/ACF(ClO_4^-)_y .$$

The performance was $V_{oc} = 3.9$ V, $J_{sc} = 203$ mA, and energy density of 248 Wh kg$^{-1}$ (ACF) with 200 charge-discharge cycles.

The cyanide-bridged polynuclear complex Prussian Blue (PB), whose unit cell may be written as $Fe_4^{3+}[Fe^{II}(CN)_6]_3^{4-}$, undergoes two-step reversible redox reactions

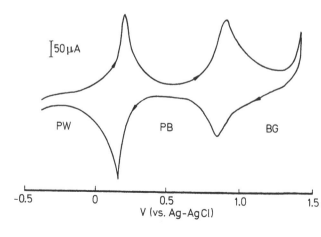

Fig. 21. Cyclic voltammogram of Prussian Blue (PB) coated on BPG dipped in 0.5 M K$_2$SO$_4$ aq. solution. Scan rate, 20 mV s$^{-1}$. (BG: Berlin Green; PW: Prussian White)

(Fig. 21). This complex can be deposited electrochemically on an electrode as a film (method B 1). These facts allow the construction of a simple battery of the form

PB/aqueous electrolyte/PB

by using electrodes coated with PB films [244]. On charging, one PB film is oxidized to form Berlin Green (BG; $Fe^{3+}$-$Fe^{3+}$ complex) and the other reduced to form Prussian White (PW; $Fe^{2+}$-$Fe^{2+}$ complex). On discharging, both the films recover their original PB states, thus making repeated charge-discharge cycles possible. PB films were able to be coated on both surfaces of a Nafion film, which then worked as a secondary battery to give $V_{oc}$ of 0.7 V [245].

Thin film electrodes of phthalocyanineoctacarbonitrile (26) on gold carriers were investigated for their electrochemical charge and discharge behaviour [246, 247].

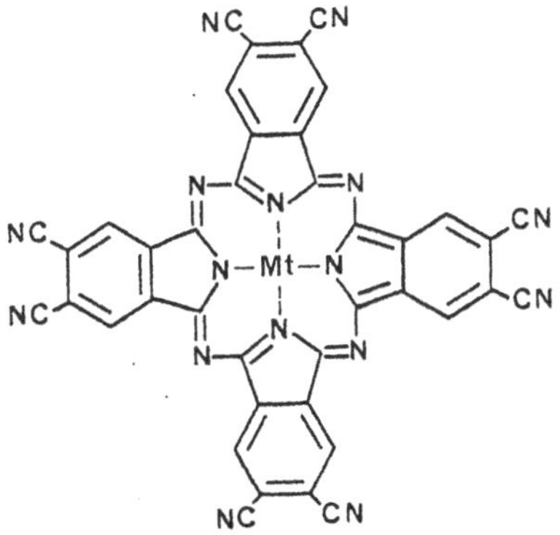

26

The films of pure 26 dispersed in the interior of a polymer matrix [poly(N-vinyl-carbazole), polyimide] were prepared by casting from solution (method A 1). The film thickness varied between 10 and 2600 nm. In contact with a weak acidic electrolyte pure films of 26 are remarkably stable and show reversible electrochromic reduction and oxidation (Eq. (16)) at ~200 mV vs NHE ($\Delta E_p$ = 19–38 mV, Fig. 22) [246]. Reduction takes place via transport of electrons (about 3 electrons per molecule of 26, Eq. (16)) from the gold carrier and intercalation of protons

$$MtPc + nH^+ + ne^- \rightleftharpoons MtPcH_n \quad (n = 1, 2, 3) \tag{16}$$

from the solution (reverse during reoxidation) up to the investigated scan rates of $2 \times 10^5$ mV s$^{-1}$.

Electrodes coated with 26 dispersed in the interior of poly(N-vinylcarbazole) exhibit very similar stable cyclic voltammograms (Fig. 22b, c) [247]. The 26 forms conducting pathways in the polymer and is not molecularly dispersed. Additionally, films of polymeric Pcs 12 (prepared by method B 2) exhibit a similar electrochemical behaviour [121b, c, 248]. The peak potentials shift in the cathodic direction (Fig. 22). The outstanding electrochemical properties of the films of 12 and 26 appear to be due to a combination of the high conductivity, also in the neutral state (Table 4), and the electron-accepting

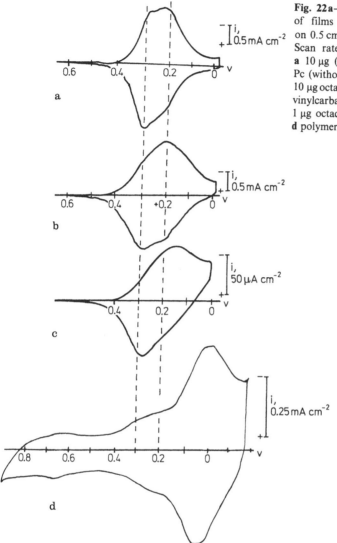

**Fig. 22a—d.** Cyclic voltammograms of films of phthalocyanines (Pcs) on 0.5 cm² gold carrier in 1 M HCl. Scan rate, 50 mV s⁻¹; V vs NHE. **a** 10 μg (1.4 × 10⁻⁸ mol) octacyano-Pc (without Mt in *26*); **b** mixture of 10 μg octacyano-Pc and 10 μg poly(N-vinylcarbazole) (PVCz); **c** mixture of 1 μg octacyano-Pc and 10 μg PVCz; **d** polymeric Pc film *(12)*

properties of the cyano groups (in *12* end groups of the polymer). Low molecular weight Pcs show no charge-discharge behaviour under the conditions used. In comparison to films of PP, PT and PAn, the films of *12* and *26* show — beside the IR drop — the anodic and cathodic peak waves at the same potentials up to scan rates of 500 mV/s. The activity of the films can be treated after Anson and Laviron in analogy to adsorbed mono- or multilayers of an electroactive component showing a reversible electrode reaction (Eq. (17)) [246] ($I_p$: peak current, A: surface area, $\Gamma$: surface coverage, v: scan rate, n: exchanged charge). For these electrodes, applications in batteries, redox active membranes and memory devices can be envisaged.

$$I_p = \frac{n^2 F^2 A \Gamma v}{4RT}.$$

$$(17)$$

In addition to solid/liquid cells, all solid state polyether (PEO) based batteries as Li/PEO/TiS$_2$ were investigated [249]

## 5.2 Display Devices

Polymer-coated electrodes often show electrochromic properties based on reactions of metal ions or metal complexes incorporated into the coating polymer layer, or based on doping-undoping of electrodeposited polymers such as PAn, PP and PT. This electrochromism can be used for display devices.

Prussian Blue (PB) film ($\lambda_{max} = 700$ nm) undergoes a colour change when reduced to Prussian White (PW) [116, 117, 250]. This phenomenon allows its use as an electrochromic device. Polymeric viologen (4, 5) films go blue ($\lambda_{max}$ 610 nm) on reduction, which makes their use in display devices possible. Electrochromism of heptylviologen incorporated in a solid polymer electrolyte cell has been reported [251]. Electrochromism of lutetium diphthalocyanine was studied by subliming it in vacuo an SnO$_2$ glass to form a film [252, 253]. Various colours, such as yellow, orange and red when oxidized and blue and purple when reduced, were reported. Polymeric solid-state electrochromic cells were fabricated by a polyelectrolyte to which cationic organic compounds such as methylene blue and MV$^{2+}$ were electrostatically bound [254].

Multicolour electrochromism was reported for the complex formed from tris(5,5'-dicarbo(3-acrylatoprop-1-oxy)-2,2'-bipyridine) (34) and ruthenium(II) [255]. The polymer showed a fairly rapid electrochemical response (as short as 250 ms for the total conversion through seven oxidation states).

34

It is well known that electropolymerized PAn, PP, PT, etc., show electrochromism on doping and undoping at different potentials. PAn, when polymerized under anodic acidic conditions, shows remarkable electrochromism, while no such property is observed when it is polymerized under neutral and alkaline conditions. PAn was coated on a Pt plate by polymerizing it in 2 M HCl aqueous solution (method B1) and the colour change between −0.15 V vs SCE (yellow) and 0.4 V (green) was studied continuously [256]. The colour change was complete within 100 ms, and the reflectance (at 740 nm) of the film changed little with potential step cycles of $1.36 \times 10^6$, indicating that the film is stable. Two kinds of redox reactions were involved in the electrochromism of PAn [257]. One of them is a proton addition/elimination reaction (Eq. (18)). The other reaction is related to the insertion/elimination of the electrolyte anions in the PAn film (Eq. (19)).

$$\left[\begin{array}{c} +\text{C}_6\text{H}_4-\overset{X^{\ominus}}{\underset{\text{H}_2}{\overset{\oplus}{\text{N}}}}-\text{C}_6\text{H}_4-\overset{X^{\ominus}}{\underset{\text{H}_2}{\overset{\oplus}{\text{N}}}}+ \end{array}\right]_n \underset{+e^{\ominus},\ +\text{H}^{\oplus}}{\overset{-e^{\ominus},\ -\text{H}^{\oplus}}{\rightleftharpoons}} \left[\begin{array}{c} +\text{C}_6\text{H}_4-\overset{X^{\ominus}}{\underset{\text{H}}{\overset{\oplus}{\text{N}}}}=\text{C}_6\text{H}_4=\overset{X^{\ominus}}{\underset{\text{H}}{\overset{\oplus}{\text{N}}}}+ \end{array}\right]_n$$

at 0.15 V (vs. SCE)

(18)

at 0.3–0.6 V (vs. SCE)

(19)

Three forms of PAn were reported, that is, the closed valence shell oxidized $+\text{C}_6\text{H}_4-\text{N}=\text{C}_6\text{H}_4=\text{N}+$ and reduced $+\text{C}_6\text{H}_4-\text{NH}+$ forms, and the radical cation intermediate form $+\text{C}_6\text{H}_4-\text{NH}+^{4+}$ [258], from the study of the absorption spectra and the conductivity [256, 259]. PAn exhibits multicolour electrochromism as

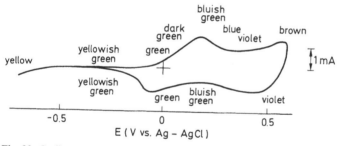

**Fig. 23.** Cyclic voltammogram and colour changes of polyaniline (PAn *20*) film coated on Pt and dipped in 10 mM HCl containing 0.2 M LiCl. Scan rate, 10 mV s$^{-1}$

**Table 5.** Electrochromic data for some polymers

| Polymer | Anion | Oxidized form λ (nm) | Reduced form λ (nm) | Frequency response time (ms) |
|---------|-------|----------------------|---------------------|------------------------------|
| Polypyrrole (PP, *14*) | $ClO_4^-$ | 660 | 420 | 20 |
| Polythiophene (PT, *17*) | $ClO_4^-$, $BF_4^-$ | 730 | 470 | 45 |
| Poly(3-methylthiophene) | $BF_4^-$ | 750 | 480 | 12 |
| Poly(3,4-dimethyl-thiophene) | $ClO_4^-$ | 750 | 620 | 60 |
| Poly(2,2'-bithiophene) | $CF_3SO_3^-$ | 680 | 460 | 40 |
| Octacyanophthalo-cyanine (*26*) | $H^+$ | 580 | 650 | <1 [246] |

shown in Fig. 23 [259]. See also the recent papers reporting on the redox mechanism of PAn films [260].

Polymers electropolymerized from thiophene derivatives, e.g. thiophene, 3-methyl-thiophene, 3,4-dimethylthiophene, and 2,2'-bithiophene, showed electrochromism with various colours and stabilities depending on the derivatives (Table 5) [261]. The oxidized state of poly(3-methylthiophene) was blue-green, and the reduced state was red. This polymer retains more than 80 % of its activity after $1.2 \times 10^5$ cycles of electro-chromism between −0.2 and 0.8 V vs SCE. High stability as regards doping-undoping cycles was also reported for PT [262].

Photoelectrochromic properties of PP-coated Si electrodes in propylene carbonate (PC) solution have been studied [263]. Optical image formation and its storage have been studied.

Although the electrochromism of polymer films contains problems in its relatively slow response and higher electricity consumption due to the redox or doping-undoping reactions in comparison with conventional liquid crystal devices, the rich variations of the coatings, the multi-coloured displays, and the easy construction of large-area devices with polymer films should lead to interesting applications.

## 5.3 Permeation and Transportation of Ions and Molecules

Since the charged state of electrode-coated polymer films can be changed easily by applied potentials, as described in Sect. 4.1, the permeability and transportation of ions and molecules in the films must be changed by changing the applied potential. This should lead to electrochemical control of permeation and transportation of ions and molecules through films.

When a polymer film coated on an electrode contains both anionic groups and redox-active groups, the latter being able to be oxidized to the cationic state electrochemically, the cation exchange capacity of the film can be controlled by applied potentials as shown in Fig. 24. When the redox-active groups on the polymer chain are in their reduced state, the counter cations (Cat$^+$) to anionic groups also attached to the poly-mer chain are present in the polymer film. When the redox-active groups are oxidized to become cations, the interaction of the formed cations with the anionic groups expels the Cat$^+$ into the solution.

Electropolymerization of N,N-dialkyl substituted aniline derivatives gave polymer films containing quaternary ammonium groups as permanently charged positive sites

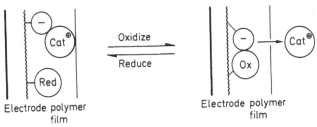

**Fig. 24.** Control of cation exchange capacity of polymer film by applied potential. Cat$^+$: cation; Red and Ox: reduced and oxidized groups, respectively

in the polymer backbone (21) [264]. These films work as anion-exchangers depositing multiply-charged redox species such as $Fe(CN)_6^{3-}$ and $Ru(CN)_6^{4-}$. They adsorb anions from solutions as dilute as $10^{-8}$ M and therefore possess potential utility as preconcentration materials for the anions prior to their electroanalysis.

Doping and undoping of PP films were utilized for ion sieving [265]. A PP film which was electrochemically synthesized in the presence of NaCl showed permeability in electrochemical doping of $Cl^-$ and anions of comparable diameter, and prevented anions of larger diameter such as $EtSO_3^-$ from penetrating the matrix.

The electroactive ion-exchange polymer containing both sulfonate and ferrocene groups (35) was coated on a glassy carbon electrode [266]. The ion-exchange characteristics of this polymer can be modulated electrochemically. When the ferrocene groups are reduced, this polymer is a cation exchanger, and can incorporate cations. When ferrocene is oxidized, the formed ferricenium cation becomes the counterion for the $-SO_3^-$ groups, and the $Cat^+$ present in the film is expelled from the film. The permeability of the ions is dependent on the charges on the film. The permeation rate of ferrocene through the cationic films polymerized from [Ru(4-methyl-4'-vinyl-2,2'-bipyridine)$_3$]$^{2+}$ and [Os(bpy)$_2$(4-vinylpyridine)$_2$]$^{2+}$ was about three times larger than that of the ferricenium cation [267]. Permeability of similar polycationic films has been measured for ferrocene, ferricenium, TCNQ (tetracyanoquinodimethane) and TCNQ$^-$ [268]. Ion exchange properties have been studied on poly(vinylpyridinium) films containing $Fe(CN)_6^{3-/4-}$ [269].

35

A Nafion film-coated micro-electrode is very advantageous for in vivo detection of neurotransmitter-related species such as dopamine and catecholamine [270]. The partition and diffusion of the substance in the coating film are generally important factors that influence the response characteristics of the coated electrode. A quinonoid polymer electrode, prepared by the electrochemical oxidation of mercaptohydroquinone and modified with mercaptide, functioned as a selective ion sensor for heavy metal ions such as $Ag^+$, $Hg^{2+}$, $Cd^{2+}$, and $Cu^{2+}$ in solution [271]. A PVC membrane electrode containing macrocyclic polyethers with sulfur atoms (36) exhibited appreciable selectivity for $Ca^{2+}$ relative to $Mg^{2+}$, alkali metal ions and $H^+$ [272].

n = 2, 3

X = S(O)Ph,  S(O)C$_8$H$_{17-n}$

36

Ionic permeability of polypyrrole (PP) coated on a minigrid was controlled electrochemically by changing the PP oxidation state [273]. The anionic permeability of oxidized PP is very high whereas that of the reduced PP is a factor of 1000 lower.

Electrochemical permeability control of organic molecules by a bilayer-immobilized film containing redox sites has been reported [274, 275]. Bilayer-forming amphiphiles (*37*) and sodium poly(styrene sulfonate) (PSSNa) were mixed in water, and the resulting precipitates were dissolved in $CHCl_3$ and cast on a Pt minigrid. The amphiphiles form extended lamellae parallel to the film plane in polyion complexes with polystyrene sulfonate (PSS⁻). The *37*-PSS⁻ and the reduced *37*-PSS⁻ film showed phase transition temperatures $T_c$ at 24° and 38 °C, respectively. When the *37* group

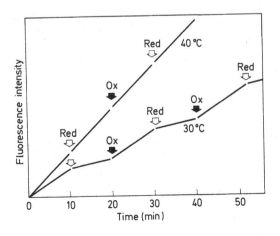

Fig. 25. Permeation control of probe *38* across the viologen-containing film on a Pt grid by electrochemical redox reactions. The potentials of −0.5 V and 0 V vs SCE were applied to Pt grid/film at Red and Ox, respectively

is reduced to the cation radical, the permeability of the film is decreased by a factor of 3.3 at 30 °C. The permeation of the film is not affected at temperatures higher than $T_c$ of the reduced 37-PSS$^-$ (38 °C). These phenomena were studied by using a fluorescent probe (38) as a permeant, and the results are shown in Fig. 25. Reversible permeability control was also achieved by using a polypeptide membrane containing pendant ferrocene groups [276].

Controlled loading and release of drug materials at PP films were reported [277], which might lead to new drug-delivery devices. PVC membranes containing valinomycine [278], —OH, —COOH [279], etc. were reported for K$^+$ ion-selective transport.

## 5.4 Polymer-Based Electronics

Electrochemical doping and undoping of polymer films bring about changes of not only the colours and charged states of the films but also the conductivity. Most conjugated polymers increase the electronic conductivity on doping by more than 8 orders of magnitude, as described in Sect. 4.1. This phenomenon can be used to construct polymer-based electronic devices such as transistors and diodes.

New developments can be envisaged from the interesting results on derivatized microelectrodes that can work as microelectronic devices when immersed in an electrolyte solution. The operation of these "chemiresistors" is due to different physical properties in the oxidized or reduced states of the polymers compared with the neutral state. The general route described below should be applicable to a wide variety of metal/polymer/metal as well as metal/polymer 1/polymer 2/metal interfaces.

The derivatized microelectronics were prepared on small gold electrodes (e.g. 2 µm wide × 100 µm long × 0.1 µm thick, with nominal separation between the adjacent Au electrodes 1–2 µm) on insulating carriers. These Au microelectrodes were functionalized with polypyrrole (14, PP) [77, 280], poly(1-methylpyrrole) [77], poly(3-methylthiophene) [147], and polyaniline (20, PAn) [148] by electropolymerization from the corresponding monomers (method B1). The thickness of the coating is about 1–10 µm. A cross-sectional view of a device either in contact with an electrolyte solution containing reference and counter electrodes or in contact with an electrolyte solution containing redox reagent is shown in Fig. 26. The devices can be addressed either electrochemically or chemically. Macroscopic electrodes derivatized with a redox polymer and coated with a porous metal contact and working as diodes and triodes are the fundamental concept for the operation of the microelectronic devices [281].

The arrangement with a counter electrode (Fig. 26, upper figure) is a triode system (transistor): the drain and source are Au microelectrodes coated with a polymer, and a counter electrode in an aqueous or organic medium electrolyte is the third terminal, called as gate.

The fundamental effects that cause such high capacity devices to function are the different conductivities in the low conductivity reduced state (neutral) and the high conductivity oxidized state. The change in conductivity between the source and the drain, which depends on the potential of the reference electrode (the so-called gate potential $V_G$) is shown in Fig. 27 for two polymer coatings. In the case of poly(3-methylthiophene) the resistance between the two connected microelectrodes decreases reversibly by 8 orders of magnitude upon a change in potential from 0.0 to 0.8 V vs

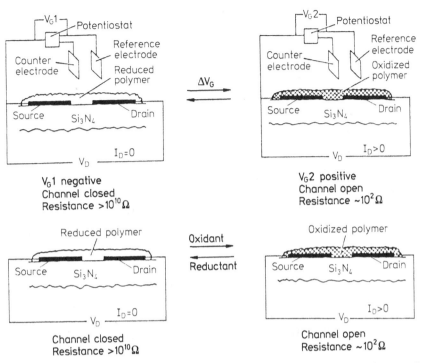

**Fig. 26.** Configuration of microelectronic devices for the amplification of electrical signals (upper figures) and chemical signals (lower ones)

SCE. Polyaniline is less conductive at both more negative and more positive potentials than 0.4 V vs SCE. More positive potentials lead to irreversible behaviour.

The characteristics of the polymer-based microelectrochemical transistors are as follows. The current between source and drain, $I_D$, is a function of the potential between source and drain, $V_D$, at various fixed gate potentials, $V_G$. When the $V_G$ using poly(3-methylthiophene) or PP is held at negative potentials where the polymer

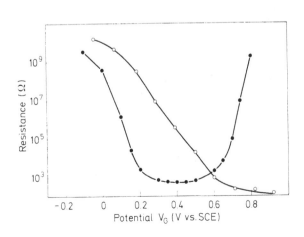

**Fig. 27.** Plots of polymer resistance between the two connected adjacent microelectrodes as a function of the gate potential $V_G$. ● polyaniline; ○ poly(3-methylthiophene)

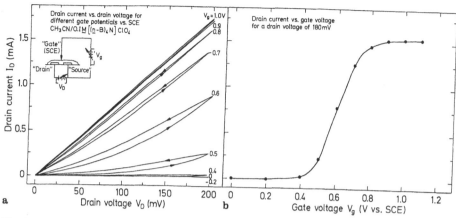

**Fig. 28a and b.** Transistor characteristics of polymer devices. **a** $I_D$ vs $V_D$ plots at fixed $V_G$ using poly(3-methylthiophene). **b** $I_D$ vs $V_G$ with $V_D = 0.18$ V for poly(3-methylthiophene)

is insulating, the device is "off" and $I_D$ is small. At potentials $V_G$ more positive than the oxidation potential of the polymer, the device turns "on" and a significant steady state value of $I_D$ can be observed. This is shown in Fig. 28. PAn-based devices can be turned "on" by either a positive or a negative shift of the electrochemical potential. This is due to the fact that the minimum resistance is in the vicinity of about 0.4 V vs SCE (Fig. 27). The charge necessary to turn the device "on" is $10^{-6}$–$10^{-8}$ C. The devices can show transconductance values only one order of magnitude lower than good Si MOSFET ones. In addition, excellent rectifying behaviour is shown by the diode characteristics of two-electrode systems. The devices are stable for thousands of on-off cycles. In the case of poly(3-methylthiophene) the quickest turn-on and turn-off are less than 20 ms. These devices show larger maximum values of $I_D$ ($\sim 1$ mA at $V_D = 0.2$ V) compared with the other coatings. The devices based on the polythiophene derivatives exhibit the best characteristics and stabilities of all the polymer coatings.

Some differences from solid-state MOSFETs must be noted. For the "wet" system (chemiresistors), the rate depends on the chemical reaction (intercalation of anions); potentials are referred to a reference electrode; channel thickness of the solid-state devices is smaller than the thickness of the polymer coating; the I–V characteristics are expressed by $V_D$ vs $I_D$.

Polymer-based devices show a response to chemical signals [147, 148]. This corresponds to the presence of a redox reagent that can equilibrate with the polymer (Fig. 26) to change the value of $I_D$ at a given value of $V_D$. Therefore the microsensors can turn "on" or "off". In the case of poly(3-methylthiophene), which may be used over a wide pH range (1 to 9), the device is turned "on" by an oxidant such as $IrCl_6^{2-}$ $[E^0(IrCl_6^{2-/3-}) = 0.68$ V vs SCE] and switched "off" by a reductant such as $Fe(CN)_6^{4-}$ $[E^0(Fe(CN)_6^{3-/4-}) = 0.12$ V vs SCE] [147]. In addition, amplification of the signal was tested. Also, the first reports of gas detected in the couples of $H_2O/H_2$ and $O_2/H_2O$ are included [148].

A microelectrochemical diode with an "open-face" sandwich arrangement of two closely spaced microelectrodes using dissimilar redox polymers has been described

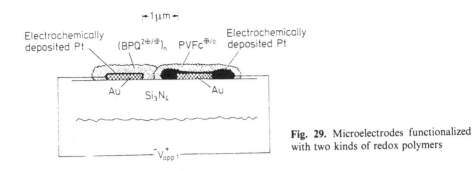

**Fig. 29.** Microelectrodes functionalized with two kinds of redox polymers

(Fig. 29) [13]. Two adjacent microelectrodes were derivatized by stepwise electrochemical polymerization. First a polymer viologen film $(BPQ^{2+})_n$ (related to *4*) on one electrode and then a polyvinylferrocene (PVFc) film (related to *11*) on the other electrode were obtained from the corresponding monomers by reductive and oxidative depositions, respectively. Small spacing between the two microelectrodes is crucial because for these materials the maximum conductivity is much lower than that of polymers such as *14*, *17*, and *20* The redox levels are as follows: $E^0(BPQ^{2+}/^+)$ $= -0.55\,V$, $E^0(PVFc^{+/0}) = 0.4\,V$ vs SCE. The redox reaction at the interface immersed in an aqueous electrolyte containing $LiClO_4$ occurs at a good rate only in one direction because the reaction in Eq. (20) is thermodynamically downhill.

$$(BPQ^+)_n + PVFc^+ \rightleftharpoons (BPQ^{2+})_n + PVFc^0 \tag{20}$$

The reverse reaction is then uphill. When the $(BPQ^{2+})_n$ is reduced a current passes between the electrodes (Eq. (20)). This means that a steady-state current passes when the negative lead is on the $(BPQ^{2+})_n$ electrode to effect its reduction and the positive lead is on the $PVFc^0$ electrode to effect its oxidation. Regeneration occurs because the the two polymers are in contact with each other and react.

If the $PVFc^0$ electrode is made more negative than the $(BPQ^{2+})_n$ electrode, no current passes because $PVFc^0$ is thermodynamically incapable of reducing $(BPQ^{2+})_n$. The switching time of the diode is of the order of seconds. Real diode current/voltage curves were obtained. In addition, the behaviour of the microdevices coated with only one polymer in contact with a redox couple present in a solution was tested.

A film of electroactive polymer $[Os(bpy)_2(vpy)_2]^{2+}$, sandwiched between Pt and Au electrodes, becomes conductive when both the electrode potentials have appropriate values such that the mixed valent state, Os(III)/(II), is generated. Such arrangements are used in two-terminal and three-terminal diodes [281]. Solid-state organic heterojunctions utilizing two conducting polymers [PA and poly(*N*-methylpyrrole) junctions] were reported [282]. The electronic properties can be modified chemically. A field-effect transistor was fabricated utilizing chemically prepared poly(*N*-methylpyrrole) [283].

A couple of a PP-coated electrode and a Pt counter electrode acted as pH modulator. Under an applied potential, the pH changes around the Pt electrode due to the reaction there, but the pH around the PP-coated electrode does not change because of the doping or undoping process at the PP film. The whole solution pH could then be changed by the applied potential [284]. A pH sensor was made from poly(mercaptohydro-quinone) which showed a Nernstian response to $H^+$ ions over pH 2–9 [285].

A synapse model was devised by using a microtip of a PP film (50 µm in diameter) [286]. The microtip incorporated a neurotransmitter such as glutamic acid on oxidation, and released it on the reductive pulse.

PP can be provided with various electronic properties by doping it with various anions. The PP film incorporating iron bathophenanthroline disulfonic acid showed electrochromism due to the redox reaction of the complex [287]. When pyrrole was electropolymerized in the presence of polymeric anions, the latter was incorporated in the formed PP film to give a new composite material [288]. Potassium polyvinylsulfate and sodium polystyrenesulfonate were incorporated into PP. Since the incorporated polymeric anion cannot move out of the film when the PP is reduced, cations enter the film from the solution to compensate the charges. The cation exchange ability of the polymeric anion is thus controlled by the applied potential.

New composite materials were produced by electropolymerizing pyrrole on an insulating polymer coating such as poly(vinyl chloride) (PVC) [289, 290]. Highly conductive PP-PVC alloy films were prepared [290]. By applying such techniques, patterning of conductive PP in the insulating polymer film was achieved [291]. Patterning was also carried out by focusing radiation onto the surface of a non-conductive silver-nitrate-doped polymer film to form silver [292]. Stable composite films of PP-Nafion, PP-clay [293], PP-Nafion-impregnated Gore-tex [294], and PT/poly-THF [295] have also been reported. Conducting PP polymers were formed on semiconductors by photo-electrochemical polymerization of pyrrole [296]. A scanning He—Ne laser beam on a n-Si wafers caused photodecomposition of PP [297]. To study an electrode surface in solution, scanning tunneling microscopy was developed [298]. A test with a Pt-coated structure yielded a spatial resolution of about 30 nm.

# 6 Electron Pumping by Photophysical Processes

As described in Sect. 2, electron pumping powered by solar radiation is one of the important electron processes found in the electron cycles of nature. Artificial electron pumping is achieved by two main types of processes: photophysical and photochemical ones. In this section electron pumping by photophysical processes is described.

A photophysical process utilizes a process called photovoltaic conversion that occur mainly in semiconductors or semiconductor-like materials. The process converts light energy into electricity or other forms of energy. Polymer-coated electrodes are receiving much attention as attempts are being made to construct photoenergy conversion systems based on photophysical processes. The conversion systems utilizing polymers are shown in Table 6 [299–313]. They are classified into systems composed of semiconducting polymers, liquid-junction inorganic semiconductors/polymer membranes, and dye-polymer membranes. They are described in Sects. 6.1, 6.2 and 6.3, respectively.

## 6.1 Semiconducting Polymer Systems

Semiconducting polymers can also be used to construct photovoltaic conversion systems instead of inorganic semiconducting materials. The discovery of high-

**Table 6.** Examples of photovoltaic conversion systems utilizing polymer membranes

| Systems | Type of devices | | Examples | Ref. |
|---|---|---|---|---|
| Semiconducting polymers | Solid-state devices | pn junction | $-(CH-)_x$/inorg. semiconductor | 299 |
| | | | $-(CH-)_x$/n-CdS | 300 |
| | | Schottky junction | Al-doped $-(CH-)_x$/undoped $-(CH-)_x$ | 301 |
| | | | Phthalocyanine in polymer/Al | 302 |
| | Liquid-junction devices | | $-(CH-)_x$ | 302, 303 |
| | | | Polypyrrole (PP) | 79, 304 |
| | | | Polyaniline (PAn) | 305 |
| | | | Polythienylene (PT) | 306 |
| Liquid-junction inorganic semiconductors/polymer membranes | Stabilization by polymer membranes | | n-Si/PP | 307 |
| | | | n-CdS/polymer $Ru(bpy)_3^{2+}$ | 308 |
| | | | n-CdS/PT | 309 |
| | Sensitization by dye membranes | | n-TiO$_2$/Ru(bpy(COO$^-$))$_3^{4-}$ | 310 |
| | Semiconductor microcatalyst | | (CdS + Pt)/Nafion | 311 |
| Dye membranes | Solid-state devices | | Al/AlPcF/Au | 312 |
| | Solution devices | | Pt/Chl a-Poly(vinyl acetate) | 313 |

conductivity doped polyacetylene (PA) films evoked great hopes for the development of a polymer film-based solar battery. The photovoltaic devices based on polymer films are classified into solid-state devices and liquid-junction devices.

### 6.1.1 Solid-State Devices

Solid-state photovoltaic cells composed of semiconducting polymer films were constructed based on p-n and Schottky junctions, etc. Since a homogeneous pn-junction of semiconducting polymer films is unstable because of the self-migration of the dopant, hetero pn-junctions composed of semiconducting polymer films have been well studied.

The heterojunction pn device, p-$(CH)_x$-/n-CdS prepared by direct polymerization of polyacetylene (28, PA) on CdS gave an open-circuit voltage ($V_{oc}$) of 0.8 V [299]. The device, p-$(CH)_x$-(n-CdS/Nesa glass, gave a short circuit current ($J_{sc}$) of 1.5 mA cm$^{-2}$ [300]. The similar p-$(CH)_x$-/n-CdS exhibited 0.5% conversion efficiency at air-mass (AM) 1 conditions (100 mW cm$^{-2}$ irradiation) as shown in Fig. 30 [314]. A Schottky junction is formed between a semiconductor and a metal whose work function is relatively small. The Schottky junction device Al/p-$(CH)_x$- gave $V_{oc}$ = 0.3 V, $J_{sc}$ = 6 µA cm$^{-2}$ under 1/20 AM 2 (3.5 mW cm$^{-2}$ irradiation) conditions with a conversion efficiency of 0.30% [315]. The device Al-doped -$(CH)_x$-/undoped -$(CH)_x$- had $V_{oc}$ = 0.65 V, $J_{sc}$ = 200 µA cm$^{-2}$, fill factor (FF) = 0.3 and conversion efficiency = 0.1%. The instability of the devices, shown for example as dopant migration, is one of the limiting factors for their practical use. A stable pn-junction was formed in a high-density p-PA film (200 µm thick) by injecting dopant by sodium ion implantation [316].

Electropolymerized semiconducting polymer films also form various solid-state junctions. An ITO/polythienylene (17, PT)/Al Schottky junction type cell gave $V_{oc}$ = 1.07 V and $J_{sc}$ = 1.35 µA cm$^{-2}$ [317]. The poly(p-phenylene) (24, PPP)/n-Si junction showed a conversion efficiency of 3.2% at a light intensity of 7 mW cm$^{-2}$ [318]. It gave an efficiency of ~1% even under AM 1 illumination. The n-Si/PP (14) cell gave $V_{oc}$ = 0.29 V, $J_{sc}$ = 9 mA cm$^{-2}$, FF = 0.46 and a conversion efficiency of 1.24% under AM 1 [319].

Various other polymer films can form solid-state cells. Al/poly(p-phenylenesulfide) (27, PPS)/Cu cell exhibited $V_{oc}$ = 0.5 V and $J_{sc}$ = 1.8 nA cm$^{-2}$ under monochromatic light of 300 nm with a conversion efficiency of 0.21% [320]. Electrochemically doped poly(N-vinylcarbazole) showed a photovoltaic effect when sandwiched between Al and gold (Au) electrodes [321]; $V_{oc}$ = 1.01 V, $J_{sc}$ = 182 nA cm$^{-2}$, and FF = 0.237 were reported. The power conversion efficiency was $2.8 \times 10^{-2}$% under 1.08 mW cm$^{-2}$ monochromatic light. Complexes of poly(ethylene oxide) with sodium polyiodide formed a photovoltaic cell when sandwiched between ITO and Pt [322].

39 (P)

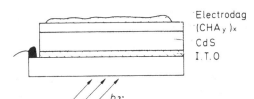

Fig. 30. Semiconducting polymer-based solid-state device

Photovoltaic cells using Schottky junctions composed of thin films of organic dye and metal have been reported [20]. Phthalocyanines (7, Pc), porphyrins (39, P), and other organic dyes such as merocyanines (40, Mc), triphenylmethanes, etc., have been shown to form thin film photovoltaic cells. The adsorption of oxygen on a Ni—Pc film increased the conversion quantum efficiency [323]. The irreversibly adsorbed oxygen causes an order of magnitude increase in the photovoltage, which is attributed to an increased efficiency of minority carrier injection.

40 (Mc)

When these dyes are used dispersed in polymer films with high dielectric constants, the photovoltaic output efficiency increases [31, 324]. Metal-free phthalocyanine (7, $H_2Pc$) was transformed by ball milling into the x-$H_2Pc$ form (diameter of particles, 80 nm). A mixture of 2.3% x-$H_2Pc$ and 1.5% polymer (polycarbonate, polyvinylacetate, polyvinylcarbazole) in methylenechloride was cast onto Nesa glass (method A1). Then a metal film was evaporated onto the organic film. Table 7 describes the composi-

Table 7. Performance of some photovoltaic cells (Nesa glass/$H_2Pc$ (7)-polymer/metal←hv) at room temperature. Content of x-$H_2Pc$ in the organic film 60%. Thickness of the metal film ~900 Å Irradiation with visible light of 80 mW/cm$^2$

| Cell | Thickness of org. film (μm) | Metal films | $I_T$ (mW cm$^{-1}$)[a] | $\bar{V}_{oc}$ (V)[b] | $J_{sc}$ (μA cm$^{-2}$)[c] | $\eta'$ (%)[d] |
|---|---|---|---|---|---|---|
| 1 | 0.5 | In | 1 | 0.40 | 179 | 2.4 |
| 2 | 0.9 | In | 1 | 0.40 | 110 | 1.42 |
| 3 | 1.8 | In | 1 | 0.38 | 48 | 0.63 |
| 4 | 1.8[e] | In | 1.1 | 0.40 | 130 | 1.58 |
| 5 | 1.8[e] | In | 0.135 | 0.35 | 27 | 2.30 |
| 6 | 1.8 | In | 0.135 | 0.39 | 7 | 0.64 |
| 7 | 1.8 | Al | 1 | 0.97 | 6 | 0.19 |
| 8 | 1.8 | Sn | 1 | 0.34 | 26 | 0.29 |
| 9 | 1.8 | Pb | 1 | 0.53 | 7.2 | 0.13 |

[a] $I_T$: Light energy after traversing the metal film;
[b] $\bar{V}_{oc}$: Open circuit voltage;
[c] $J_{sc}$: Short circuit current;
[d] Conversion efficiency $\eta' = (J_{sc} \cdot \bar{V}_{oc}/I_T) \cdot FF$ (FF = fill factor);
[e] Organic film doped with 14% trinitrofluorenone

tion of some cells. The cells exhibit good rectifying behaviour and diode parameters. Current flows with a negative potential at the metal (e.g. indium) electrode. The conversion efficiencies under irradiation with visible light in the cell Nesa glass/ $H_2$Pc-polymer/metal $\leftarrow$ hv can reach $\sim 3\%$ (Table 7). Disadvantages are the very low transmission of the metal film and the decreasing quantum efficiency with increasing light intensity. Comparing cells 1–3 (Table 7), it is seen that thinner mixed layers show more promising properties. Thicker layers favour recombination. The addition of acceptors such as trinitrofluorenone also leads to higher conversion due to a higher $J_{sc}$ (cells 4, 5). It is well known that doping of Pc with acceptors increases the conductivity (more charge carriers). Higher temperatures increase the efficiency by $4\%$–$5\%$ $K^{-1}$ also due to a higher concentration of charge carriers. The way the cell works is as follows. By contacting the p-conductor x-$H_2$Pc with indium (metal with low work function of 4 eV), the negative charge carriers generated by irradiation flow into the metal and the positive charge carriers into the Nesa glass (Fig. 31).

In the cells composed of Pc particles dispersed in a binder polymer, the effect of the polymer was studied. Polymers such as poly(vinylidene fluoride), poly(vinyl fluoride), polyacrylonitrile, poly(4-vinylpyridine), poly(p-vinyl phenol) and poly-(vinylidene chloride) gave a higher photocurrent and conversion efficiency [33]. This is attributed to the enhancement of exciton dissociation due to a large electric field formed by the presence of polar groups in these polymers. The effects of recrystallization of $H_2$Pc were studied in the cell Nesa glass/$H_2$Pc in oly(vinylcarbazole)/Au. For the lower-resistance $H_2$Pc, carriers were generated in $H_2$Pc, while for the higher-resistance $H_2$Pc, carriers were generated in the polymer [325]. The MIS (metal/insulator/semiconductor)-type cell Nesa glass/CuPc/polyethylene/Al showed higher efficiency than the corresponding MS(metal/insulator)-type cell [326].

Surfactant AlPc was electrodeposited to give the cell Al/AlPc/Ag [327]. It gave $0.7\%$ power conversion efficiency at 638 nm, This technique could easily produce large-scale depositions of pigments. A new type of cell was fabricated by contacting a chloro-

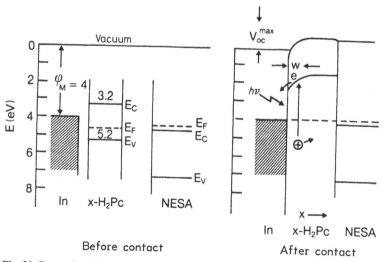

**Fig. 31.** Energy level diagram of a photovoltaic cell, Nesa glass/$H_2$Pc (7)/In before and after contact

phyll a (Chl a) layer with wet PVA to give a Au/Chl a/PVA/Au cell [328]. It was suggested that a junction formed between the Chl and PVA films. The power conversion efficiency was of the order of $10^{-2}\%$ with FF = 0.39.

Besides the macrocyclic metal complexes, many dyes have been used to fabricate organic photovoltaic cells. Correlations between the cell performance and the molecular structure of the dyes have been studied [329], and the desirable properties of photovoltaic dyes discussed [330].

A flexible photovoltaic cell was produced (Al/Mc/ITO) by vapour deposition under vacuum of Mc dye on a transparent polyester film coated with ITO [331]. The efficiency under AM 1 sunlight was 0.1%. Amorphous films of poly(2-vinylpyridine)-$I_2$ complex deposited on a ITO layer gave a photovoltaic response with a conversion efficiency of $5 \times 10^{-2}\%$ under white light illumination [332].

### 6.1.2 Liquid-Junction Devices

Films of organic semiconducting polymers and organic dyes form liquid junctions (a kind of Schottky junction) when dipped into an electrolyte solution. This is a simple method of constructing photovoltaic cells.

A photoelectrochemical photovoltaic cell was fabricated by dipping PA into an aqueous sodium polysulfide [302]. The cell gave $V_{oc}$ = 0.3 V and $J_{sc}$ = 40 μA cm$^{-2}$ under AM 1 illumination. A similar photoelectrochemical cell was fabricated with trans-PA which was dipped into an aqueous methylviologen (MV$^{2+}$) solution [303].

Since electrochemical doping occurs in semiconducting polymer films dipped in solution, the liquid-junction device composed of semiconducting polymer films can be doped photochemically (Eqs. (21), (22)).

$$\text{Polymer} + e^-(h\nu) + M^+ \rightarrow \text{Polymer-}M^+ \tag{21}$$

$$\text{Polymer} + h^+(h\nu) + X^- \rightarrow \text{Polymer-}X^- \tag{22}$$

Photodoping of PA films by dipping them into a $CH_3CN$ solution of $Et_4NClO_4$ has been reported [333]. The PA doped $Et_4N^+$ in the potential range above $-1.2$ V upon irradiation with visible light. Photochemical doping of PP was carried out in methylene dichloride containing diphenyliodonium, which was hexafluoroarsenated upon irradiation with uv light [334].

A liquid-junction PP photoelectrochemical cell was fabricated by dipping a PP (preparation, method B1) film into an aqueous solution of redox reagents such as MV$^{2+}$, hydroquinone, oxygen, Cu$^{2+}$, I$^-$ and Br$^-$ [309]. A cathodic photocurrent was obtained at potentials more negative than 0 V vs Ag with redox reactions on the PP film. Photocurrent generation by a liquid-junction PP film coated on Nesa glass was studied [79]. The ITO/PP/Al device showed that electropolymerized PP is p-type. Annealing of the PP film at 100 °C markedly improved the adherence to the substrate. The liquid-junction cell ITO/PP/aqueous electrolyte gave a higher photocurrent (45 μA cm$^{-2}$) for the doped PP than for the undoped one. A higher concentration of the electrolyte (LiClO$_4$) produced a larger photocurrent [79].

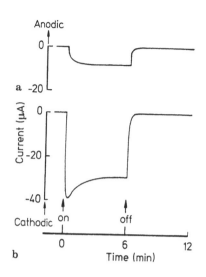

Fig. 32. Current changes induced by switching on and off radiation incident on an undoped polythienylene film (1.8 cm²) dipped in a 0.1 M LiClO₄ aq. solution at the applied potential of (a) 0.2 V and (b) —0.2 V vs Ag-AgCl

PT (preparation, method B1) was also fabricated as a liquid-junction photoelectro-chemical cell [306]. The current changes induced by irradiation of the ITO/PT/aqueous electrolyte are shown in Fig. 32. In the region more negative than the onset potential (~0.5 V), the device gave a cathodic photocurrent up to 55 $\mu A$ cm⁻². A poly($p$-phenylenesulfide) (27) film was used for liquid-junction photoelectrochemical cells by dipping it into the $CH_3CN$ electrolyte solution [335]. The presence of $MV^{2+}$ induced a more stable photocurrent.

Films of other sensitizers such as MtPc (7), P (39), cyanines and Ru complexes also form photovoltaic cells when dipped into an electrolyte solution. Thin films of chlorogallium Pc, vanadyl Pc, and other Pcs [336–341] prepared by vapour deposition gave a photoelectrochemical response when dipped into an aqueous electrolyte. The photochemical cell consisting of GaPc—Cl thin films backed by a conductive metal substrate (Mt) and in contact with an electrolyte solution develops a photopotential which is determined by the difference in the Fermi potential of Mt and the electrode in the redox couple. The cell Au/GaPc-Cl/Fe(CN)₆³⁻/⁴⁻/GaPc-1/Pt gave $V_{oc}' = 0.55$ V with power conversion efficiency of 0.03%–0.05% [339]. A thin film of CuPc coated on

$$O-\overset{\overset{\displaystyle O}{\|}}{\underset{\underset{\displaystyle OR_1}{|}}{P}}-OR_1 \cdot x\ H_2O$$

A: $R_1 = H$, $R_2 = C_{16}H_{33}$, $x = 1$
B: $R_1 = R_2 = C_{16}H_{33}$, $x = 0$

41

Au and dipped in an aqueous electrolyte responded to near-infrared light of wavelength ~1100 nm [342].

Electrodeposited surfactant AlPc (41) film also formed solid liquid-junction type cells with redox systems: $Fe(CN)_6^{3-/4-}$, benzoquinone/hydroquinone, $I_3^-/I^-$, $Fe^{3+/2+}$, $Sn^{4+/2+}$, and $Fe(EDTA)^{1-/2-}$ were present in the solution [343]. The maximum quantum efficiency was 4% at 630 nm and 0.26 mW cm$^{-2}$, and the power conversion efficiency was around 10$^{-2}$% under white light irradiation.

Various metal porphyrin complexes (MtP) have also been used to fabricate liquid-junction devices [344-347]. A photocurrent was generated by a liquid-junction Zn-tetraphenylporphyrin (ZnTPP) (30 R = —H) film [344]. The direction of the photocurrent is determined by the Schottky barrier formed at the ZnTPP/metal interface, whereas the charge carriers are produced at the ZnTPP/solution interface. Mn and CdTPP in contact with a $Fe(CN)_6^{3-/4-}$ aqueous solution gave $V_{oc} = 1$ V and a quantum yield of 0.2 under monochromatic light [348]. The photovoltage decreased in the order Pd > H$_2$ > Zn > Cu > Mg > Pb > Ni > Co.

The photoresponse of a VTPP-coated Pt depends on the pH of the solution with which the metal complex is in contact. A cathodic photocurrent is obtained at lower pH values whereas an anodic photocurrent is obtained at higher pH values [349]. This was explained as the combined effect of band-bending at the Pt/VTPP interface and the adsorption of both hydrogen and hydroxyl ions at the VTPP/solution interface.

When MgTPP film on ITO is used for an photoelectrochemical cell, it acts only as a photocathode. The ITO coated with ZnTPP, CuTPP or H$_2$TPP, which have oxidation potentials more positive than that of MgTPP, works both as photocathode and photoanode, depending on the electrode potential and the direction of the radiation [350]. The cathodic photocurrent is attributed to the downward band-bending in the space charge layer at the TPP/electrolyte contact, and the anodic one to that at the ITO/TPP contact.

A pn junction was formed between p-type Pc and n-type P. Thus the spectral cosensitization effect of the cell ITO/ZnPc/tetrapyridyl(TPy)P (30 with 4-pyridyl instead of —C$_6$H$_4$R)/electrolyte solution was studied [351]. The photocurrent at the Soret band for the ZnPc/TPyP electrode is 1 or 2 orders of magnitude greater than for the ITO/TPyP electrode. Such photocurrent enhancement at the ZnPc/TPyP interface may imply that the number of surface recombination centres at the pn junction is less than at metal/dye junctions. Organic photoelectrodes based on pp iso-type junctions such as ITO/ZnPc/ZnTPP also show a cosensitization effect [352].

The photoelectrochemical behaviour of Chl a dispersed in a poly(vinyl acetate) film (method E1) was studied. When the film cast from dioxane or DMSO was immersed in water, it showed a transformation from a Chl a-dioxane aggregate to the monomer or from a Chl a-DMSO aggregate to (Chl a · 2 H$_2$O)$_n$. The former induced

42

a photocathodic current, and the latter a photoanodic current [313]. Two photoactive electrodes were used for photoelectrochemical cells. A cell composed of a $SnO_2$/TPP photocathode and a $SnO_2$/Victorial Blue B photoanode gave $V_{oc} = 1$ V, $J_{sc} = 100$ µA and quantum efficiency greater than 1 % [353].

Mc dyes containing an octadecyl group (42) formed aggregates when coated on an electrode. Absorption by the aggregates contributes to the photocathodic current [354].

A carbon electrode coated with $[Ru(bpy)_2]_2L^{2+} \cdot 2$ $(PF_6^-)$ (method A1), where L represents the dianion of 1,5-dihydroxy anthraquinone, gave a cathodic photocurrent of the order of nanoamperes [355]. Multimolecular layers of $Ru(bpy)_3$ $(BPh_4)_2$ coated on a Pt electrode gave anodic photocurrents [356]. The reaction of a photogenerated Ru(III) complex with $OH^-$ ions was suggested. Photoelectrochemistry of a one-dimensional Ni-dithiolene complex (43) thin film coated on a Pt electrode (method A1) was studied [357]. Electrochemically grown dithio-oxamido Cu(II) (copper rubeanate) films on copper electrodes gave a cathodic photocurrent [358].

$$\left( \begin{array}{c} NC \\ NC \end{array} \underset{S}{\overset{S}{\diagdown}} Ni \underset{S}{\overset{S}{\diagup}} \begin{array}{c} CN \\ CN \end{array} \right)^{2\ominus}$$

43

# 6.2 Inorganic Semiconductor/Polymer Membrane Systems

Liquid-junction devices with semiconductor electrodes are attracting much attention, not only because they can be used in photoelectrochemical cells for light-to-electricity conversion, but also because the dipped semiconductor and the counter electrodes can bring about photochemical reactions on their surfaces. Such light-to-chemicals conversion systems have received attention in view of chemical conversion of solar energy [20]. Liquid-junction semiconductor systems have other advantages, such as being less sensitive to the quality of the semiconductors than pn junctions, and ease of fabrication of the systems. A liquid-junction n-$TiO_2$ semiconductor and Pt counter electrodes system caused water photolysis under uv irradiation to give simultaneously hydrogen at one electrode and oxygen at the other [359].

In order to utilize visible light, narrow band gap semiconductors (band gap $<3.1$ eV) must be used. The narrow band gap n-type semiconductors are, however, unstable in water under irradiation, so they must be stabilized against corrosion. Polymer coating is one way of achieving this objective. Another approach is to sensitize wide band gap semiconductors by coating them with sensitizers. In the following the liquid-junction semiconductors/polymer membrane systems are described.

## 6.2.1 Stabilization and Photoelectrochemical Catalysis by Polymer Coating

There are two reasons for the instability of liquid-junction narrow band gap n-type semiconductors in water under irradiation: (a) degradative corrosion (e.g., n-GaAs and n-CdS) and (b) inactive oxide film formation (e.g., n-Si) under photoanodic conditions. It was reported that liquid-junction n-GaAs, which is a typical narrow band gap semiconductor, can be stabilized against corrosion under irradiation by

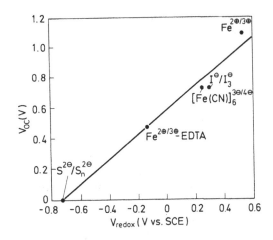

Fig. 33. Correlation between $V_{oc}$ and the redox potential $V_{redox}$ for n-GaAs photoanode coated with polymer-pendant $Ru(bpy)_3^{2+}$ (*1*) dipped in an aq. solution containing various redox reagents

coating it with polypyrrole (PP)[360]. The n-Si could also be stabilized by PP coating (method B1) so that it generated a stable photocurrent for 5 days with 5% conversion efficiency[307]. When an n-Si electrode was coated with a thin Pt film before coating with PP, it showed a longer-term stability with 5.5% efficiency in an iodide/triiodide electrolyte[361]. Polythienylene (PT)[309] and polyaniline (PAn)[362] were also effective for the stabilization of n-GaAs. In these systems so-called photoregenerative cells were constructed in which redox couples such as $I_3^-/I^-$, $Fe^{3+/2+}$, etc., are used in solution to transport charges between electrodes. The role of the polymer coating is not only to protect the semiconductor surface from contact with corrosive water molecules but also to transport holes from the semiconductor surface to the solution.

A thin film of polymer-pendant $Ru(bpy)_3^{2+}$ (*1*) was effective in stabilizing an n-GaAs photoanode used in water containing various redox couples[25]. The linear

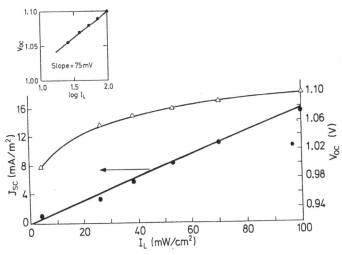

Fig. 34. Dependence of $J_{sc}$ and $V_{oc}$ on light intensity ($I_L$) for n-GaAs photoanode coated with *1* in contact with the $Fe^{2+/3+}$ redox couple in water

relation of the open-circuit photovoltage ($V_{oc}$) to the redox potential of the redox couple present in the solution implied the absence of Fermi level pinning at the interface (Fig. 33). The logarithmic dependence of $V_{oc}$ on light intensity is predicted by Eq. (23) [363],

$$V_{oc} \simeq \frac{nkT}{q} \ln \frac{J_{sc}}{J_o}$$ (23)

where n is the junction ideality factor, k is Boltzmann's constant, q is the electron charge and $J_o$ is the reverse saturation current density. When $J_{sc} \propto I_L$ (light intensity) and $J_{sc} \gg J_o$, a plot of $V_{oc}$ vs ln $I_L$ should yield a straight line whose slope gives the value of n. From Fig. 34 the value of n was estimated to be 1.3 for the *I*-coated n-GaAs. This is very close to the ideal value of 1, meaning that at the semiconductor/electrolyte interface the carrier recombination losses are negligible. The conversion efficiency for this stabilized n-GaAs dipped in $Fe^{2+/3+}$ aqueous electrolyte was ~12% under AM 1 illumination [25].

An n-Si electrode was stabilized by coating it with viologen-based polymer films (*4*) (method A1) [364]. Substantial improvements in photocurrent stability were obtained in the presence of electroactive anions, because they are adsorbed by the positively charged polymer film, so providing efficient scavenging of photogenerated holes at the semiconductor surface.

The liquid-junction semiconductor electrodes stabilized by polymer-coating can be used for photochemical conversion systems. The stabilized n-CdS coated with PP which incorporates $RuO_2$ as catalyst was used for visible-light-induced water cleavage [365]. Photochemical diodes were fabricated by coating CdS with PP and polystyrene films, the latter containing metal dispersions such as Pt, Rh and $RuO_2$ as a catalyst [366] (Fig. 35).

An n-CdS photoanode was also stabilized by coating with a polymer-pendant $Ru(bpy)_3^{2+}$ (*I*) (method A1) film which incorporated $RuO_2$ dispersions [308]. The modification brought about the disappearance of the redox waves in the cyclic voltammogram (CV) which are due to corrosion under irradiation. Figure 36a shows a cathodic wave at ~ −0.8 V and an anodic one at ~ −0.7 V in the CV of a bare CdS electrode irradiated in a KCl electrolyte. The first wave is due to reduction of $Cd^{2+}$ to $Cd^0$ and the second due to reoxidation of $Cd^0$ to $Cd^{2+}$. The CdS modified with a film of *I* and $RuO_2$ does not show such waves which are due to corrosion (Fig. 36b),

Catalyst (I)   Polymer support   Catalyst (II)

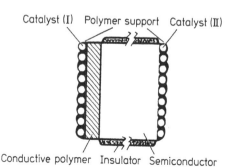

Conductive polymer   Insulator   Semiconductor

Fig. 35. Photochemical diode. Single crystal CdS coated on one side with polypyrrole (PP *14*) overlaid with $RuO_2$ powder immobilized in a polystyrene film and on the other face with Pt black immobilized in a polystyrene film

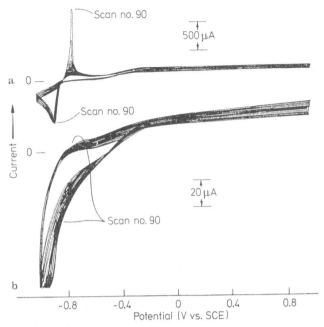

**Fig. 36a and b.** Typical contrast in the photoelectrochemical stability of (a) bare CdS and (b) CdS coated with *1* containing dispersed RuO₂, measured in 0.5 M KCl aqueous electrolyte. The I-V scans 1 through 90 were carried out while the electrode was illuminated with ~ 85 mW cm⁻²

indicating that the photoanode is stabilized against degradation. Both the pendant Ru complex and RuO₂ were needed for the stabilization. It is assumed that the Ru complex works as a mediator to transport holes from the CdS surface to RuO₂ where catalytic oxidation of Cl⁻ to Cl₂ takes place (Fig. 37). Visible-light photolysis of HCl was confirmed by analysis of the H₂ and Cl₂ formed at the Pt counter electrode and the modified CdS electrode, respectively.

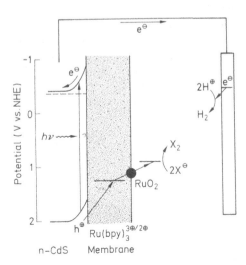

**Fig. 37.** Schematic diagram of the photocatalysis at polymer-modified CdS (pH 7)

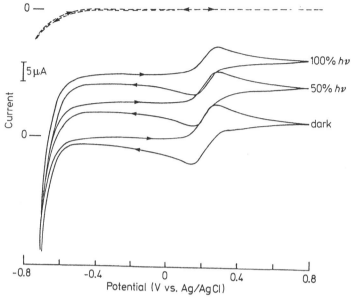

**Fig. 38.** Steady-state voltammograms (scan rate, 100 mV s$^{-1}$) for the *1*-coated CdS in the dark and under illumination dipped in 50 mM Fe(CN)$_6^{3-/4-}$/1 M KCl aq. electrolyte. A bare CdS electrode is shown for comparison (dashed curve, in the dark)

When the RuO$_2$ powder was first put on the surface of CdS and the polymer Ru(bpy)$_3^{2+}$ was coated over them, the surface of CdS/RuO$_2$ showed an ohmic character[367]. The coated cationic Ru complex film adsorbed Fe(CN)$_6^{3-/4-}$ by electrostatic binding, and the adsorbed Fe(CN)$_6^{3-/4-}$ gave an ohmic response at CdS superimposed on an anodic photocurrent under cyclic potential sweep conditions (Fig. 38).

### 6.2.2 Sensitization by Dye Coating

Sensitization of wide band gap semiconductor photoelectrodes by coating with dye molecules has been one of the major subjects of research in photoconversion systems. The typical sensitization mechanism can be represented by Fig. 39. Ru(bpy)$_3^{2+}$, phthalocyanines (Pc), porphin (P), organic dyes such as Rhodamine B and mero-

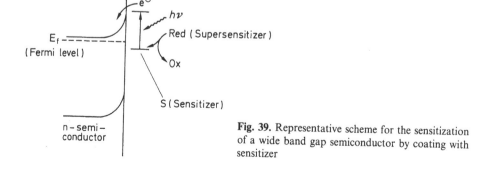

**Fig. 39.** Representative scheme for the sensitization of a wide band gap semiconductor by coating with sensitizer

cyanines (Mc) have been used as sensitizers. The sensitizer-coated systems can over-come the problem inherent to the solution systems for which the sensitizer present in the solution brings about a negative filter effect by absorbing the incident light before it reaches the electrode surface.

$Ru(bpy)_3^{2+}$ (2) was attached to n-SiO$_2$ via condensation of surface hydroxyl groups with Ru(4-trichlorosilylethyl)-4'-methyl-2,2'-bipyridine)bis(2,2'-bipyridine)bis(hexa-fluorophosphate), and the photoelectrochemical response of the coated SnO$_2$ was studied [368]. A polymer film containing $Ru(bpy)_3^{2+}$ pendant groups was formed on a SnO$_2$ electrode by polymerizing the mixture of tris(2,2'-bipyridine)ruthenium(II) cinnamate and poly(vinyl alcohol) cinnamate on the electrode to obtain a cross-linked polymer coating [369]. The coated SnO$_2$ gave a sensitized photocurrent when dipped in an aqueous electrolyte containing EDTA. An n-SnO$_2$ electrode coated with Nafion incorporating $Ru(bpy)_3^{2+}$ also gave a sensitized photocurrent (Fig. 40) [370]. The anodic photocurrent at the absorption maximum was larger for thin films (0.3–

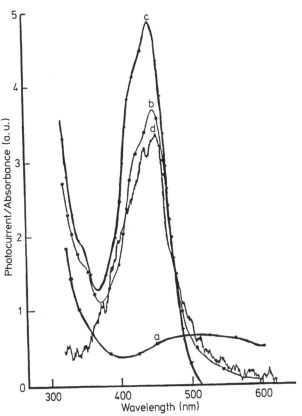

**Fig. 40.** Wavelength dependence of the Nafion-bound $Ru(bpy)_3^{2+}$-sensitized anodic photocurrent at n-SnO$_2$. a: Absorption spectrum of Nafion-coated SnO$_2$; b: absorption spectrum of Nafion-bound $Ru(bpy)_3^{2+}$; c: absorption spectrum of $Ru(bpy)_3^{2+}$ in solution; d: anodic photocurrent at SnO$_2$ coated with Nafion-bound $Ru(bpy)_3^{2+}$ in 0.1 N KCl containing 10 mM hydroquinone obtained at the applied potential of 0.2 V vs SCE; photocurrent normalized with respect to the lamp spectrum

0.5 μm) than for thicker films (1–10 μm). Monolayer assemblies composed of a Ru(bpy)$_3^{2+}$-containing surfactant molecule (44) and arachidic acid sensitized the anodic photocurrent obtained at the n-SnO$_2$ electrode [371].

$$\left[\begin{array}{c} (bpy)\,Ru\,(bpy) \\ \nearrow\quad\nwarrow \\ -N\qquad N- \\ \\ O=C\qquad C=O \\ | \qquad\quad | \\ O\qquad\quad O \\ | \qquad\quad | \\ H_{37}C_{18}\quad C_{18}H_{37} \end{array}\right]^{2\oplus}$$

**44**

The efficiencies of these systems are, however, not high, mainly because of the small quantum yield for charge injection. High efficiency was reported in the sensitization of n-TiO$_2$ using surface-adsorbed tris(2,2'-bipyridyl-4,4'-dicarboxylate)ruthenium(II)-dichloride as a sensitizer [310]. The photocurrent of 36 μA cm$^{-2}$ was obtained by monochromatic light flux of 0.22 mW cm$^{-2}$ at 460 nm, corresponding to a conversion efficiency of 44 %.

Chlorophyll (Chl) has been used for constructing photovoltaic cells [372–375]. Zn or Mn-tetrakis(4-sulfophenyl)porphyrin (30 R = —SO$_3$H) was incorporated into a film of poly(vinyl alcohol) (PVA) cinnamate coated on an In$_2$O$_3$ electrode [376]. An action spectrum showed the sensitization by the immobilized P complex (39). The presence of ascorbic acid in the solution enhanced the anodic photocurrent. Films of various MTPPs (30 R = —H) (Mt = Mg, Zn) coated on SnO$_2$ by dipping or vacuum sublimation were effective for the sensitization of the anodic photocurrent [377]. Co, Cu and NiTPP were ineffective for the sensitization. The study of film thickness showed that multilayer activity takes place. A film-type photovoltaic cell was fabricated by using Zn-tetramethylpyridiniumporphyrin (30 —C$_5$H$_4$N$^+$—CH$_3$ instead of —C$_6$H$_4$R) dispersed in a conventional polymer film such as polyacrylonitrile, celluloseacetate, PVP (method A1), etc. [378]. Thus, the cell ITO-coated polyester/polymer film containing sensitizer and supporting electrolyte/Au-coated polyester film was fabricated. It gave an action spectrum of the photocurrent corresponding to the absorption spectrum.

Films of CuPc (7) vacuum-deposited on a SnO$_2$ electrode produced a sensitized anodic photocurrent [379, 380]. They also gave cathodic photocurrents which are attributable to excitons generated and dissociated in the bulk of CuPc.

Mc and rhodamine B (Rh B) were incorporated into polyethyleneimine and PVA by using diethylphosphoryl cyanide or 1-methyl-2-chloropyridinium iodide as the coupling agent [381]. The SnO$_2$ coated with these films exhibited remarkably sensitized

**45**

photocurrents. Mc-coated $SnO_2$ showed an enhanced photocurrent due to the Mc aggregation [382]. The quantum efficiencies of the sensitized photocurrent obtained at ZnO were an order of magnitude higher for the surface-formed J aggregates of cyanine dyes (45) than for those adsorbed from solution [383]. Sensitized photocurrent spectra of ZnO electrodes show distinct peaks at the J-band of cyanide dyes adsorbed on the ZnO [384].

The quantum yields of the sensitized photocurrent were studied on $SnO_2$ coated with Rh B or rose bengal [385]. The quantum yield (q) is dependent on the donor density ($N_d$) of the semiconductor and shows an optimum point. The strong dependence of q on $N_d$ may be interpreted by considering the decay process of electrons injected from the excited dye through possible surface states. The increase in $N_d$ causes an increase in the electric field of the space charge region of the semiconductor and hence suppresses these decays up to the optimum $N_d$. The decrease in q for $N_d$ may be attributed to a leakage of the conduction electrons by tunneling through the thin space charge layer.

Arene (e.g. naphthalene, anthracene, pyrene, etc.)-derivatized electrodes gave sensitized photocurrents in the presence of Rh B and hydroquinone in solution [386]. The attached molecules were considered to function as energy or electron relays.

## 6.3 Miscellaneous

Colloidal or powdered semiconductors can be incorporated into polymer films, which then behave as photocatalysts. This should lead to some interesting applications as photocatalysts in the form of films. CdS particles were embedded in a polyurethane film and photolysis processes studied [387]. Nafion film was used to disperse CdS [388]. Nafion films were soaked in $Cd(NO_3)_2$ aqueous solution and then treated with $H_2S$ gas to give CdS-dispersed films [311]. A Pt catalyst was also incorporated into the Nafion film. Photochemical $H_2$ production was achieved by irradiating this CdS-Pt-dispersed film in water containing $Na_2S$ as a sacrificial reducing agent. Surface analyses were carried out on the CdS-incorporated Nafion films [389, 390].

Liquid-junction photovoltaic cells have advantages in their simplicity and ease of fabrication, as described before. Solid-state devices can also be constructed from liquid-junction cells when a solid polymer electrolyte is used. A tandem photovoltaic

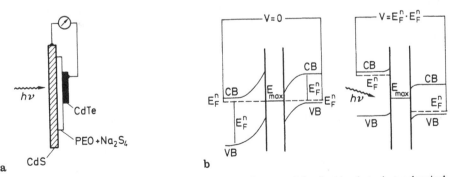

**Fig. 41a.** Schematic cell configuration. **b** Energy band diagram of the double photoelectrochemical cell

cell was constructed using a thin-film plastic electrolyte, n-CdS thin film and p-CdTe single crystal [391]. The solid electrolyte used was a thin film of poly(ethylene oxide) (PEO) with a polysulfide redox couple. The cell configuration and the energy band diagram are shown in Fig. 41. Values of $V_{oc} = 625$ mV and $J_{sc} = 35$ μA cm$^{-2}$ were obtained under AM 1 illumination conditions.

The absence of solvents in such solid-polymer-electrolyte photovoltaic cells presents the possibility of fabricating corrosion-free systems. The thin-film solid-state cells also allow fabrication of multispectral cells composed of more than one semiconductor in optical and electrical series. A solid-state photovoltaic cell, n-Si/Pt/PP/PEO(KI/I$_2$)/Pt/ITO, was studied [81]. The surface modifications of n-Si with PP can dramatically reduce the large activation energy barrier against efficient charge transfer between semiconductor and polymer-solid electrolyte. The efficiency of this cell is limited by a high surface recombination velocity associated with surface states of the n-Si. The cell had $V_{oc} = 225$ mV and $J_{sc} \sim 1.1$ mA cm$^{-2}$ at 100 mW cm$^{-2}$ illumination with junction ideality factor of 1.5. This implies the existence of deleterious surface states acting as recombination centres.

In a solid-state cell composed of n-CdS and a solid-polymer electrolyte (SPE), redox species such as FeCp$_2$ (Cp: cyclopentadienyl), Ru(bpy)$_3^{2+}$ (2), Fe(bpy)$_3^{2+}$, Fe(CN)$_6^{4-}$, and Fe(acac)$_3$ (Fe(III) acetylacetonate) were introduced into the SPE film [392]. The flatband potential of the cell can be modified by introducing redox species into the SPE. A photopotential of 500 mV was obtained. The solid-state cell p-InP/Nafion (porphin)/Nafion/Nafion (Ru(bpy)$_3^{2+}$)/n-CdS was fabricated by using two semiconductor electrodes [393]. A photopotential in excess of 1 V was obtained by this cell.

The cell ITO/CdS/SAlPc/Au was reported, where SAlPc is a surfactant aluminium Pc (41) [394]. It was made by sequential electrodeposition of CdS and SAlPc thin films onto ITO. The electrodeposited CdS acts as a blocking contact and all the band bendings occur in the SAlPc. A heterojunction cell was formed between n-CdS and p-type Pcs such as H$_2$Pc, ZnPc, MgPc, CuPc, MnPc, PbPc and VOPc [395]. The cell ITO/CdS/Pc/Au was fabricated by electrodeposition of a CdS thin film and sequential vacuum deposition of Pc and Au. $V_{oc} = 0.54$ V, $J_{sc} = 285$ μA cm$^{-2}$ and conversion efficiency of 0.10% were obtained for the ITO/CdS/ZnPc/Au cell.

Poly(N-vinylcarbazole) film made by casting and containing tetracyanoethylene was used for a photocell by dipping it into an aqueous electrolyte. A potential difference was induced between the solutions in contact with the film under irradiation [396]. Photopotentials of 200–300 mV were induced. An amorphous film of PVP-I$_2$ complex deposited on SnO$_2$ gave a photocurrent of $\sim 100$ μA cm$^{-2}$ and $V_{oc} = 65$ mV [332].

# 7 Electron Pumping by Photochemical Processes

In the previous section so-called photovoltaic cells were described which utilize photophysical processes of semiconducting materials. Photocells can also be constructed by utilizing photochemical reactions [24]. These types of photocells are called photogalvanic cells. Photochemical reactions used for photogalvanic cells are divided into two categories: (a) the so-called photoredox reaction for which the equilibrium of the system shifts under illumination because of the relatively slow back reaction and returns to the original state in the dark by a reversible back reaction; (b) the non-photo-redox reaction in which the illumination does not shift the equilibrium because of the rapid back reaction although electron transfer occurs between the excited state of the light-absorbing species and another component. In the following these two types of photochemical reactions are treated separately.

## 7.1 Photoredox Systems

The equilibrium between a redox couple present in a solution shifts upon illumination because of a redox reaction in the excited state, and returns to the original state in the dark. A typical example of a photoredox couple is thionine $(TH^+)$ (46)/$Fe^{2+}$ for which the photochemical reaction is given by Eq. (24).

$$TH^+ + Fe^{2+} \underset{dark}{\overset{h\nu}{\rightleftharpoons}} TH^+ + Fe^{3+} \qquad (24)$$
$$\text{(purple)} \qquad\qquad \text{(colorless)}$$

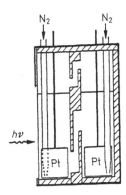

$(TH^\oplus)$

*46*

Induced by the shift of the equilibrium under illumination, the potential of the aqueous solution shifts to a lower value upon irradiating the mixture [397]. When the solution circulates between light and dark chambers [398] or when the solution is in a photo-galvanic cell composed of light and dark chambers (Fig. 42) [399] a photovoltage of ~150 mV is induced between the electrodes dipped in the solution.

The $TH^+/Fe^{2+}$ photoredox system was incorporated into polymeric gel films used to fabricate thin layer photogalvanic cells [400] represented by

$$SnO_2/TH^+ - Fe^{2+}/polymer\ gel/Pt\ .$$

Polymer gels such as poly(vinyl alcohol) (PVA) and polyethyleneimine were used as carriers for the redox couple $TH^+/Fe^{2+}$. The $TH^+$ polymer gel (47) was prepared from

**Fig. 42.** Photoelectrochemical cell composed of dark and light chambers for photocurrent generation by photoredox systems

poly(epichlorohydrin), $TH^+$ and triethylamine, and used as a film in combination with $Fe^{2+}$ ions.

$$----+CHCH_2O+_x----+CHCH_2O+_y----+CHCH_2O+_z---$$

47

If both the photochemical reaction products of a photoredox couple react at an electrode equally, the cell system cannot give a photocurrent because of the so-called short-circuit reaction occurring on the electrode surface. It is desirable therefore that the electrode discriminate between the two photochemical products in order to generate the photoeffect at the electrode. It has been reported that $TH^+$-coated $SnO_2$ and Pt electrodes are suitable for a photogalvanic cell in this respect [401,402]. $TH^+$ was coated on an electrode by holding the potential at 1.1–1.5 V for several minutes in a solution containing $TH^+$ (method A5). The $TH^+$-coated electrode showed selectivity for the reaction of photochemical reaction products of $TH^+$.

Polymer-pendant $TH^+$ was used for electrode coating. One example is poly(acrylamidomethylthionine-co-methylolacrylamide), which was coated on a Pt electrode by casting as a thick film of about 10 μm [403] (method A1). The cell composed of this polymer-coated Pt and counter electrodes dipped in $10^{-2}$ M ferrous sulfate aqueous solution gave $V_{oc} = 37$ mV and $J_{sc} = 2.8$ μA cm$^{-2}$. A recent study showed, however, that coating the electrode with $TH^+$ reduces the power conversion efficiency of the $TH^+$ cell by a factor of 19 in comparison with the conversion efficiency of $4 \times 10^{-4}$ % for the uncoated Au electrode [404].

A polypyrrole(PP)-coated electrode was selective for the $Fe^{2+/3+}$ couple but not for the $TH^+$/leucothionine couple, giving improved output in the $TH^+/Fe^{2+}$ photogalvanic cell [74]. The cell output was $V_{oc} = 138$ mV and $J_{sc} = 26.3$ μA cm$^{-2}$.

$TH^+$ attached to poly($N$-methylolacrylamide-co-acrylic acid), etc., was coated on a Pt electrode. Its cyclic voltammogram showed the formation of a complex between the coated dye and ferric/ferrous cyanide present in the solution. When the electrode is illuminated in a $Fe^{2+}$ aqueous solution, cathodic polarization at the coated electrode is observed, in contrast to the bare electrode dipped in a mixture of $TH^+$ and $Fe^{2+}$, which gave an anodic response at the electrode [405]. It was proposed that, for the polymer-coated electrode, the excited states of $TH^+$ and $Fe^{2+}$ form a complex, which is stabilized by the polymer network and accepts an electron from the electrode. A flash photolysis study showed the formation of such a complex [406].

Protonated polyvinylpyridine (PVP) coated on Nesa glass bound rose bengal electrostatically. This modified electrode, when dipped into an aqueous solution containing a redox reagent such as $Fe^{2+}$, induced an anodic photogalvanic effect [49]. A photovoltage of 135 mV and photocurrent of 1.24 μA cm$^{-2}$ were reported. Enhanced photogalvanic performance has been attributed to the presence of a high dye concentration at the electrode surface, to a longer-lived excited state of the dye in the polymer film, and to the electrostatic characteristics of the polymer in controlling access of charged redox couples to the illuminated electrode.

A photogalvanic cell was fabricated from ion-exchange resins. Cation- and anion-exchange membranes adsorb $TH^+$ and ascorbic acid, respectively. These membranes are confined and irradiated to produce a photopotential of about 30 mV [407]. The couple of thiazine or phenothiazine dye/aliphatic amines such as EDTA or triethanolamine gives a much higher photopotential, as exemplified by the high photopotential produced by the phenosafranine/EDTA couple [408]. Although such amines undergo decomposition during the photochemical reaction, they are regarded as model compounds for the composition of photogalvanic cells.

Thionine was incorporated into a clay-modified electrode and a photoelectrochemical investigation was reported [49, 409]. The modified electrode gave a photopotential in the presence of EDTA in the solution.

The irreversible photochemical reaction of the excited state of $(2,2'\text{-bipyridine})_2$-$Ru[poly(4\text{-vinylpyridine})_2]_2^{2+}$ coated on a Pt electrode with $Co(OX_3)^{3-}$ (where OX is oxalate, $C_2O_4^{2-}$) present in solution induced a cathodic photocurrent of the order of 45 $\mu A$ $cm^{-2}$ at the applied potential of 0.3 V vs SSCE [410]. A blue-green alga, *Anabena cylindrica*, immobilized on $SnO_2$ coated with $MV^{2+}$/polystyrene sulfonate/PVA showed an anodic photocurrent [411].

## 7.2 Non-Photoredox Systems

There are many photochemical redox reaction systems for which back electron transfer is so rapid that the equilibrium of the redox couple does not shift to the product side even under irradiation. In such systems, unlike the photoredox systems described in the last section, a photoresponse cannot be obtained at an electrode which is simply dipped into a mixed solution of the photoreaction couple. When the electrode is coated with one of the photoreaction couple, however, a photoresponse may be obtained at the coated electrode because the dynamics of the electron transfer process between the electrode and the photochemical reaction products may overcome the rapid back electron transfer reaction of the products.

The electron transfer between the excited state of $Ru(bpy)_3^{2+}$ (2) and $MV^{2+}$ (3) which has been studied intensively as a candidate for a water photolysis system [20] is a typical example of this rapid back electron transfer (Eq. (25)).

$$Ru(bpy)_3^{2+} + MV^{2+} \rightarrow Ru(bpy)_3^{2+*} + MV^{2+} \rightarrow Ru(bpy)_3^{3+} + MV^+$$

Rapid
(25)

A photoresponse cannot be obtained at an electrode which is dipped into the mixed solution of these photoreaction components. When the Ru complex is coated on an electrode as a polymer membrane, however, a photoresponse is obtained at the coated electrode in the presence of $MV^{2+}$ in the aqueous phase [412, 24] or in the second polymer layer coated on the top of the Ru complex layer [138]. As the Ru complex polymer coating, the $Ru(bpy)_3^{2+}$/polystyrene sulfonate adsorbed system, $Ru(bpy)_3^{2+}$/Nafion adsorbed system [26], or polymer-pendant $Ru(bpy)_3^{2+}$ (1) was used. The monolayer-coated system composed of a polymer membrane containing the Ru complex and

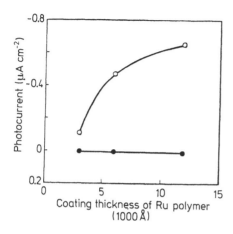

**Fig. 43.** Dependences of photocurrent on coating thickness of a bilayer (*1* and *5*) coated electrode dipped in 0.2 M CF₃COOH adjusted to pH 2. Applied potential of (○) −0.19 V and (●) 0.61 V vs Ag-AgCl, measured under argon. The molar ratio of Ru/MV²⁺ is taken constant (0.55)

$MV^{2+}$ present in the solution induced both cathodic and anodic photocurrents depending on the applied potential at the coated electrode [24, 412]. This is due to the electrode reaction of both the photoreaction products, i.e., $Ru(bpy)_3^{3+}$ and $MV^+$, the latter of which can be present also near the electrode surface because of adsorption on the polymer layer.

When the $MV^{2+}$ is coated as a second layer on top of the Ru complex layer forming the bilayer-coated system

$$\text{electrode/polymer } Ru(bpy)_3^{2+} \ (1)/\text{polymer } MV^{2+} \ (5),$$

the illumination induces a selective cathodic photocurrent because the photochemically formed $MV^+$ can hardly react at the electrode due to the intervening Ru polymer layer. The dependence of the cathodic and anodic photocurrents on the membrane thickness for this bilayer coated system is shown in Fig. 43 [138]. A selective cathodic photocurrent was induced which increases with increasing membrane thickness, while the small anodic photocurrent is almost independent of the thickness.

The fluxes of the electron transport processes of this bilayer-coated photoresponse system are estimated as in Fig. 44. The fluxes due to the photoreactions are very large. The electron transport in the polymer membrane occurs by electron exchange between

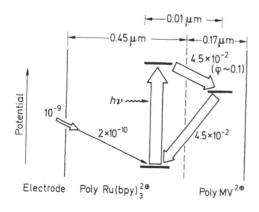

**Fig. 44.** Estimation of electron flux in photoelectrochemical processes at bilayer coated electrodes. Units of the values: mol cm⁻² s⁻¹

the redox groups as described in Sect. 4.1. This step is the slowest process in the photo-electrochemical processes concerned. The quantum efficiency must be higher when the membrane thickness is thinner. However, concerning the overall photocurrent value, the thicker membrane gave a higher photocurrent (cathodic) up to the thickness studied, as shown in Fig. 43. This means that the larger amount of the light-absorbing complex gives a higher engineering conversion efficiency under the conditions studied.

In the water photolysis system composed of an aqueous solution of colloidal Prussian Blue (PB, see Sect. 5.1) and $Ru(bpy)_3^{2+}$ [413, 414], photoinduced electron transfer was considered to occur with electrons being transferred from the excited Ru complex adsorbed on the PB colloidal particle to the PB. Such a photochemical reaction must also be applicable to the coated electrode system used for photogalvanic cells. This kind of coated electrode system for photoelectrochemical conversion must in addition give information on heterogeneous photochemical reactions. Thus a basal plane pyrolytic graphite (BPG) electrode was first coated with a polymer-pendant $Ru(bpy)_3^{2+}$ (1) membrane and then with a PB membrane by reductive electrodeposition from the aqueous mixture of $Fe(CN)_6^{3-}$ and $Fe^{3+}$. The bilayer-coated system

$$BPG/polymer\ Ru(bpy)_3^{2+}\ (1)/PB$$

gave a photoresponse at the electrode [415]. This device produces mainly cathodic photocurrent.

The action spectra for the photocurrent are shown in Fig. 45. In this figure the absorbances of the Ru complex ($\lambda_{max}$ 460 nm) and PB ($\lambda_{max}$ 700 nm) nearly correspond to the coating amounts. For the cathodic photocurrent the major process is electron transfer from the electrode to PB via the excited state of the Ru complex (Fig. 46a). Such a photoelectrochemical process supports the mechanism of water photolysis

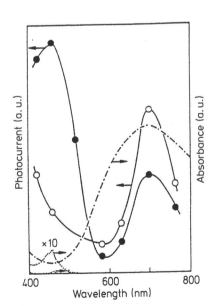

**Fig. 45.** Action spectra for the anodic (○) and cathodic (●) photocurrents obtained at BPG/polymer-pendant $Ru(bpy)_3^{2+}$ (1)/PB at the applied potentials of 1 V and −0.3 V vs Ag-AgCl, respectively). Absorbances of the Ru complex (· · · · · ·) and PB (—·—·—·—) which correspond to the coating amount are shown. The photocurrents are normalized by the number of incident photons

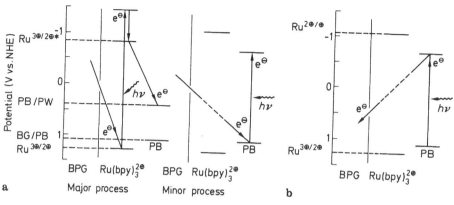

**Fig. 46a.** Mechanism for the cathodic photocurrent obtained in a BPG/*1*/PB bilayer coated system.
**b** Mechanism for the anodic photocurrent for the same device as **a**

described above. Direct electron transfer from the electrode to excited PB is the minor process for the cathodic photocurrent. This direct photoelectrochemical process between the electrode and PB was confirmed by a separate experiment using a PB-coated electrode [416]. For the anodic photocurrent, direct electron transfer occurs from the excited state of PB to the electrode (Fig. 46b). The bilayer coated system

$$BPG/PB/polymer\ Ru(bpy)_3^{2+}$$

was also reported [417].

A film of electropolymerized $Ru(bpy)_3^{2+}$ pendant polymer containing pyrrole groups (*48*) gave a cathodic photocurrent when irradiated in the presence of 4-methyl-benzenediazonium tetrafluoroborate ($4\text{-}CH_3C_6H_4N^+BF_4^-$) in acetonitrile [418]. The beneficial effect on the photocurrent intensity effected by increasing film thickness is rapidly cancelled by the resulting electron diffusion limitation.

*48*

*49*

*50*

Polymer pendant MtPcs ($M$; $Fe^{3+}$, $Co^{2+}$, $Ni^{2+}$, $Cu^{2+}$) were used as films to produce a photoresponse. $MV^{2+}$ present in the solution was not effective as an acceptor, but $Fe^{3+}$ EDTA worked as an acceptor to enhance the photocurrent about 10 times [419].

A polymer bilayer coated electrode consisting of an electron-mediator film in the inner layer (49) and a sensitizer film (50) in the outer layer produced a rectified photocurrent [420]. In the presence of a sacrificial reductant (triethanolamine) in the solution, the bilayer coated electrode gave exclusively anodic photocurrent independent of the applied potential between —0.2 and —0.3 V vs SCE.

# 8 Future Prospects

Polymer-coated electrodes belong to a new scientific field that combines modern aspects of electrochemistry, polymer chemistry, organic chemistry and physics. Research on this topic will be extended in the future with particular interest focused on the industrial applications of these promising new materials.

Although electrochemistry is an established field of science with many applications, it is still expected that new areas of important applications can be found. In order to reach this objective, the development of new functional materials is required. Polymer-coated electrodes are the most promising candidates, as can be seen from this review article.

The preparation, mechanisms and possible applications of polymer-coated electrodes have been described in this review. Processes based on the transportation, storage, activation, pumping, and utilization of electrons (Tables 1, 2, 6) which form important electron cycles in nature (Fig. 1) have been explained.

For the preparation of polymer-coated electrodes, mainly commercially available starting materials are used, and the electrochemically or electrocatalytically active coatings are obtained by simple one- or two-step procedures. In many cases more consideration has to be given to the reproducibility of the synthesis and the analytical characterization, e.g.

— structure,
— morphology,
— density and surface profiles.

The function of the coating on the molecular level and the consequences for transportation, storage, activation, and pumping of electrons are not yet understood. In order to enhance properties such as the stability and durability of known coatings and to develop new materials, these gaps in our knowledge must be filled.

The range of industrial applications seems to be unexpectedly large, but at present it is impossible to predict which polymer coating will take the place of known materials. Applications to be envisaged are:
— storage of energy: batteries
— storage of information by energy pulse: reversible and irreversible information storage, displays
— transformation of energy by catalysis: electrocatalysis, photocatalysis
— transformation of light energy: photovoltaic and photogalvanic cells, photodevices.

Polymer coating also seems to be an effective method for the fabrication of micro-electronic and molecule-based devices, such as diodes, transistors and conductors, that are attracting much attention at present. Considering the great advantages of the polymer-coated electrodes in the variety of the properties and of the materials of the polymer coatings, the authors believe that new fields of science and industry should be opened up by the coatings. This hope could be realized by organized collaboration between research institutes, universities and industry.

A considerabel numbers of papers on fundamental studies of polymer-coated electrodes were published in various journals (e.g., Synth. Met., J. Chem. Soc., J. Electroanal. Chem., Electrochim. Acta, Makromol. Chem., Macromol. Symp. *8* (1987)) in the year after this manuscript was completed. In several companies applied research works are on the way to realize application of these unconventional materials. Recently secondary batteries (polyaniline/Li, polypyrrole/Li) were commercialized. Developments for practical uses on electrochromic display and sensors are also underway.

# 9 References

1. Oyama N, Anson FC (1978) J. Electroanal. Chem. *87*: 289
2. Oyama N, Anson FC (1979) J. Am. Chem. Soc. *101*: 739
3. Oyama N, Anson FC (1980) Anal. Chem. *52*: 1192
4. Anson FC (1980) J. Phys. Chem. *84*: 3336
5. Oyama N, Ohsaka T, Ushiroguchi T (1984) J. Phys. Chem. *88*: 5274
6. Ohsaka T, Oyama N, Sato K, Matsuda H (1985) J. Electrochem. Soc. *132*: 1871
7. Leddy J, Bard AJ, Maloy JT, Savéant JM (1985) J. Electroanal. Chem. *187*: 205
8. Leddy J, Bard AJ (1985) ibid. *189*: 203
9. Feldman BJ, Ewing AG, Murray RW (1985) ibid. *194*: 63
10. Leidner CR, Murray RW (1985) J. Am. Chem. Soc. *107*: 551
11. Andrieux CP, Savéant JM (1984) J. Electroanal. Chem. *171*: 65
12. Harrison DJ, Daube KA, Wrighton MS (1984) ibid. *163*: 93
13. Kittlesen GP, White HS, Wrighton MS (1985) J. Am. Chem. Soc. *107*: 7373
14. Diaz AF, Kanazawa K, Gardini GP J. Chem. Soc., Chem. Commun. *1979*: 635
15. Tourillon G, Garnier F (1982) J. Electroanal. Chem. *135*: 173
16. Tourillon G, Garnier F (1984) ibid. *161*: 51
17. Kobayashi T, Yoneyama H, Tamura H (1984) ibid. *161*: 419
18. Hjertberg T, Salaneck WR, Lundström I, Samasiri NLD, MacDiarmid AG (1985) J. Polym. Sci., Polym. Lett. Ed. *23*: 503
19. Schiavello M (1984) (ed) Photoelectrochemistry, ·Photocatalysis and Photoreactors. Reidel Publishing Co., Dordrecht
20. Kaneko M, Yamada A (1984) Adv. Polym. Sci. *55*: 1
21. Murray RW (1984) Electroanal. Chem. *13*: 191
22. Albers MO, Coville NJ (1984) Coord. Chem. Rev. *53*: 227
23. Oyama N, Ohsaka T, Kaneko M, Sato K, Matsuda H (1983) J. Am. Chem. Soc. *105*: 6003
24. Kaneko M, Yamada A, Oyama N, Yamaguchi S (1982) Makromol. Chem., Rapid Commun. *3*: 769
25. Rajeshwar K, Kaneko M, Yamada A (1983) J. Electrochem. Soc. *130*: 38
26. Oyama N, Yamaguchi S, Kaneko M, Yamada A (1982) J. Electroanal. Chem. *139*: 215
27. Oyama N, Oki N, Ohno H, Ohnuki Y, Matsuda H, Tsuchida E (1983) J. Phys. Chem. *87*: 3642
28. Ohno H, Hosoda M, Tsuchida E (1983) Makromol. Chem. *184*: 1061
29. Ohsaka T, Yamamoto H, Kaneko M, Yamada A, Nakamura N, Nakamura S, Oyama N (1984) Bull. Chem. Soc. Jpn. *57*: 1844
30. Hong PM, Holdcroft S, Funt BL (1985) J. Electrochem. Soc. *132*: 2129
31. Loutfy RO, Sharp J (1979) J. Chem. Phys. *71*: 1211

32. Loutfy RO, McIntyre LF (1983) Can. J. Chem. *61*: 72
33. Minami N, Sasaki K, Tsuda K (1983) J. Appl. Phys. *44*: 6764
34. Oyama N, Shimomura T, Shigehara K, Anson FC (1980) J. Electroanal. Chem. *112*: 271
35. Sato K, Yamaguchi S, Matsuda H, Ohsaka T, Oyama N (1983) Bull. Chem. Soc. Jpn. *56*: 2004
36. Tsou YM, Anson FC (1984) J. Electrochem. Soc. *131*: 595
37. Krishnan M, Zhang X, Bard AJ (1984) J. Am. Chem. Soc. *106*: 7371
38. Oyama N, Ohsaka T, Sato K, Yamamoto H (1983) Anal. Chem. *55*: 1429
39. Henning TP, Bard AJ (1983) J. Electrochem. Soc. *130*: 613
40. Oyama N, Ohsaka T, Okajima T, Hirokawa T, Maruyama U, Ohnuki Y (1985) J. Electroanal. Chem. *187*: 79
41. Rubinstein I (1985) ibid. *195*: 431
42. Mazur S, Reich S (1986) J. Phys. Chem. *90*: 1365
43. Ghosh PK, Bard AJ (1983) J. Am. Chem. Soc. *105*: 5691
44. Ghosh PK, Mau AWH, Bard AJ (1984) J. Electroanal. Chem. *169*: 315
45. Itaya K, Bard AJ (1985) J. Phys. Chem. *89*: 5565
46. Rudzinski WE, Bard AJ (1986) J. Electroanal. Chem. *199*: 323
47. Scott NS, Oyama N, Anson FC (1980) ibid. *110*: 303
48. Rubinstein I, Bard AJ (1980) J. Am. Chem. Soc. *102*: 6642
49. Kamat PV, Fox MA (1984) J. Electrochem. Soc. *131*: 1032
50. Emsile AG, Bonner FT, Peck LG (1958) J. Appl. Phys. *29*: 858
51. Acrivos A, Shah MG, Peterson EE (1960) ibid *31*: 963
52. Meyerhofer D. (1978) ibid. *49*: 3993
53. Washo BD (1977) IBM J. Res. Dev. *21*: 190
54. Braun H, Storck W, Doblhofer K (1983) J. Electrochem. Soc. *130*: 807
55. Braun H, Decker F, Doblhofer K, Sotobayashi H (1984) Ber. Bunsenges. Phys. Chem. *88*: 345;
    Niwa K, Doblhofer, K (1986) Electrochim. Acta *31*: 549
56. Kaufman FB, Schroeder AH, Engler, EM, Kramer SR, Chambers JQ (1980) J. Am. Chem. Soc. *102*: 483
57. Inzelt G, Chambers JQ, Kinstle JF, Day RW (1984) ibid. *106*: 3396
58. Diaz A, Miller RD (1985) J. Electrochem. Soc. *132*: 834
59. Pearce PJ, Bard AJ (1980) J. Electroanal. Chem. *112*: 97; (1980) ibid. *108*: 121
60. Umana M, Rolison DR, Nowak R, Daum P, Murray RW (1980) Surf. Sci. *101*: 295
61. Rolison DR, Umana M, Burgmayer P, Murray RW (1981) Inorg. Chem. *20*: 2996
62. Wöhrle D (1983) Adv. Polym. Sci. *50*: 45
63. Hirai T, Yamaki J (1985) J. Electrochem. Soc. *132*: 2125
64. Rubinson JF, Behymer TD, Mark HB (1983) ibid. *130*: 121
65. Itaya K, Bard AJ (1978) Anal. Chem. *100*: 1487
66. Bookbinder DC, Wrighton MS (1983) J. Electrochem. Soc. *130*: 1080
67. Kuzmany H, Mehring M, Roth S (eds) Electronic Properties of Polymers and Related Compounds. Springer-Verlag, Berlin Heidelberg New York (1985);
    (a) Bloòr D, Hercliffe RD, Galiotis CG, Young RJ ibid. p. 179;
    (b) Yueqiang S, Carneiro K, Jacobsen C, Frelltoft T, Renyan Q, Xiantong B ibid. p. 187;
    (c) Lindenberger H, Roth S, Hanack M ibid. p. 194;
    (d) Bredas JL ibid p. 166
68. Ohsaka T, Ohnuki Y, Oyana N, Katagiri G, Kamisako K (1984) J. Electroanal. Chem. *161*: 399
69. Tanaka S, Sato M, Kaeriyama K (1985) Makromol. Chem. *186*: 1093; (1984) ibid. *135*: 1295
70. Oyama N, Ohsaka T, Shimizu T (1985) Anal. Chem. *57*: 1526
71 a. Waltman RJ, Diaz AF, Bargon J (1984) J. Phys. Chem. *88*: 4343
   b. Waltman RJ, Bargon J (1986) Can. J. Chem. *64*: 76
72 a. Heinze J, Hinkelmann K, Dietrich M, Mortensen J (1986) DECHEMA-Monogr. *102*: 209
   b. Heinze J, Hinkelmann K, Dietrich M, Mortensen, J (1985) Ber. Bunsenges. Phys. Chem. *89*: 1225
   c. Breitenbach JW, Olaj OF, Sommer F (1971) Adv. Polhm. Sci. *9*: 47
73 a. Diaz AF, Castillo JI J. Chem. Soc., Chem. Commun. *1980*: 397
   b. Diaz AF, Castillo JI (1981) J. Electroanal. Chem. *129*: 115

74. Murthy ASN, Reddy KS (1983) Electrochim. Acta *28*: 473
75. Asavapiriyanont S, Chandler GK, Gunawasdena GA, Pletcher D (1984) J. Electroanal. Chem. *177*: 229; 245
76. Pickup PG, Osteryoung RA (1984) J. Am. Chem. Soc. *106*: 2294
77. Kittlesen GP, White HS, Wrighton MS (1984) ibid. *106*: 7389
78. Bi X, Yao Y, Wan M, Wang P, Xiao K, Yang Q, Qian R (1985) Makromol. Chem. *186*: 1101
79. Kaneko M, Okuzumi K, Yamada A (1985) J. Electroanal. Chem. *183*: 407
80. Feldmann BJ, Burgmayer P, Murray RW (1985) J. Am. Chem. Soc. *107*: 872
81. Skotheim TA, Inganäs O (1985) J. Electrochem. Soc. *132*: 2116
82. Yoneyama H, Wakamoto K, Tamura H (1985) J. Electrochem. Soc. *132*: 2414
83. Diaz AF, Rock MC (1984) ibid. *131*:1802
84. Solama M, Aguilar M, Salmon M (1985) ibid. *132*: 2379
85. Inganäs O, Lundström I (1984) ibid. *131*: 1129
86. Coche L, Deronzier A, Moutet JC (1986) J. Electroanal. Chem. *198*: 187
87. Willmann KW, Murray RW (1983) ibid. *133*: 211
88. Kerr JB, Miller LL, Van de Mark MR (1980) J. Am. Chem.Soc. *102*: 3383
89. Skotheim T, Velazquez M, Linhous CA, J. Chem. Soc., Chem. Commun. *1985*: 612
90. Bull RA, Fan FR, Bard AJ (1984) J. Electrochem. Soc. *131*: 687
91. Noufi RN (1983) ibid. *130*: 2126
92. Chandler GK, Pletcher D (1986) J. Appl. Electrochem. *16*: 62
93. Niwa O, Tamamura T J. Chem, Soc., Chem. Commun. *1984*: 1015
94. DePaoli MA, Waltman RJ, Diaz AF, Barson J ibid. *1984*: 1015
95. Linsey SE, Street GB (1984) Synth. Metal *10*: 67
96. O'Brien RN, Sundaresan MS, Santhaman KSV (1984) J. Electrochem. Soc. *131*: 2028
97. Shirota Y, Noma N, Kanega H, Mikawa H J. Chem. Soc., Chem. Commun. *1984*: 470
98. Monvernay AD, Tsamantakis A, Lacaza PC, Dubois, JE (1986) J. Electroanal. Chem. *199*: 449
99. Tourillon G, Garnier F (1984) J. Polym. Sci., Polym. Phys. Ed. *22*: 33
100. Kaufmann JH, Chung TC, Heeger AJ, Wudl F (1984) J. Electrochem. Soc. *131*: 2092
101. Cao Y, Wang P, Qian R (1985) Makromol. Chem. *186*:1093
102. Waltman RJ, Diaz AF, Bargon J (1984) J. Electrochem. Soc. *131*: 740; 1452; (1983) J. Phys. Chem. *87*: 1459
103. Czerwinski A, Zimmer H, Van Pham C, Mark HB (1985) J. Electrochem. Soc. *132*: 2669
104. Horowitz G, Tourillon G, Garnier F (1984) J. ibid. *131*: 151; (1985) ibid. *132*: 635
105. Tanaka S, Sato M, Kaeriyama K (1985) Makromol. Chem. *186*: 1685
106. Biserni M, Marinangeli A, Mastrogostino M (1985) J. Electrochem. Soc. *132*: 1597
107. Oyama N, Ohnuki Y, Chiba K, Ohsaka T Chem. Lett. *1983*: 1759
108. Carlin CM, Kepley LJ, Bard AJ (1985) J. Electrochem. Soc. *132*: 353
109. Ohnuki Y, Matsuda H, Ohsaka T, Oyama N (1983) J. Electroanal. Chem. *158*: 55
110. Pham MC, Dubois JE, Lacaz PC (1983) J. Electrochem. Soc. *130*: 346
111. Pham MC, Dubois JE (1986) J. Electroanal. Chem. *199*: 153
113. Bargon J, Mohmand S, Waltman RJ (1983) Mol. Cryst. Liq. Cryst. *93*: 279
112. Rubinstein I (1983) J. Electrochem. Soc. *130*: 1506
114. Waltman RJ, Diaz AF (1985) ibid. *132*: 631
115. Neff VD (1978) ibid. *125*: 886
116. Itaya K, Ataka T, Toshima S (1982) J. Am. Chem. Soc. *104*: 4767
117. Itaya K, Shibayama K, Akahoshi H, Toshima S (1982) J. Appl. Phys. *53*: 804
118. Bailey CL, Bereman RD, Rillema DP, Nowak R (1986) Inorg. Chem. *25*: 933
119a. Leidner CR, Murray RW (1984) J. Am. Chem. Soc. *106*: 1606
    b. Denevich P (1983) Polym. Prepr., Am. Chem. Soc., Div. Polym. Chem. *23*: 330
    c. Kim OK (1982) J. Polym. Sci., Polym. Lett. Ed. *20*: 663
120. Bannehr R, Jaeger N, Meyer G, Wöhrle D (1981) Makromol. Chem. *182*: 2633; (1983) J. Mol. Catal. *21*: 255
121a. Wöhrle D, Bannehr R, Schumann B, Jaeger N (1983) Angew. Makromol. Chem. *117*: 103
    b. Wöhrle D, Schumann B, Schmidt V, Jaeger N (1987) Makromol. Chem. Macromol. Symp. *8*: 195
    c. Wöhrle D, Jaeger N, Schumann B, Schmidt V, Yamada A, Shigeharak Ber. Bunsenges. Phys. Chem. in press

122. Kreja L, Plewka A (1982) Electrochim. Acta 27: 251
123. Yasuda H, Hirotsu T (1978) J. Polym. Sci., Polym. Chem. Ed. 16: 743
124. Hiratsu H, Akovali G, Schen M, Bell AT (1978) J. Appl. Polym. Sci. 22: 917
125. Oyama N, Anson FC (1980) J. Eletrochem. Soc. 127: 640
126. Nowak RJ, Schlz FA, Umana M, Lam R, Murray RW (1980) Anal. Chem. 52: 315
127. Daum P, Murray RW (1979) J. Electroanal. Chem. 103: 289
128. Daum P, Lenhard JR, Rolison DR, Murray RW (1980) J. Am. Chem. Soc. 102: 4649
129. Rolinson DR, Murray RW (1984) J. Electrochem. Soc. 131: 337
130. Osada Y, Mizumoto A (1986) J. Appl. Phys. 59: 1776
131. Facci J, Murray RW (1982) Anal. Chem. 54: 772
132. Heider GH, Gelbert MB, Yacynych AM (1982) ibid. 54: 322
133. Doblhofer K, Dürr, W (1980) J. Electrochem. Soc. 127: 1041
134. Doblhofer K, Nölte D, Ulstrupp J (1978) Ber. Bunsenges. Phys. Chem. 82: 403
135. Brosset D, Ai B, Segui Y (1978) Appl. Phys. Lett. 33: 87
136. Denisevich P, Willman KW, Murray RW (1981) J. Am. Chem. Soc. 103: 4727
137. Pickup PG, Kuo KN, Murray RW (1983) J. Electrochem. Soc. 130: 2205
138. Kaneko M, Moriya S, Yamada A, Yamamoto H, Oyama N (1984) Electrochim. Acta 29: 115
139. Kaneko M, Yamada A (1986) ibid. 31: 273
140. Wegner G (1981) Angew. Chem. 93: 352
141a. Simon J, Andre JJ (1985) Molecular Semiconductors. Springer-Verlag, Berlin Heidelberg New York
    b. Chien JCW (1984) Polyacetylene-Chemistry-Physics-Material Science. Academic Press London
142. Takeno A, Miyata S, Masuda T, Higashimura T, Adachi C, Wada T, Iwaki M, Sasabe N (1986) Polym. Prepr. Jpn. 35: 597
143. Aizawa M, Watanabe S, Shinohara H, Shirakawa H J. Chem. Soc., Chem. Commun. 1985: 264
144. Several articles (1985) in: Mol. Cryst. Liq. Cryst. 118
145. Mair HJ, Roth S (eds) (1986) Elektrisch leitfähige Kunststoffe. Carl Hansen Verlag, München;
    a) Menke H, Roth S ibid. p 225;
    b) Münstedt H ibid. p 207
146. Kossmehl G, Chatzitheodorou G (1983) Makromol. Chem., Rapid Commun. 4: 639
147. Thackeray JW, White HS, Wrighton MS (1985) J. Phys. Chem. 89: 5133
148. Paul EW, Ricco AJ, Wrighton MS (1985) ibid. 89: 1441
149. Elliott CM, Redepenning JG, Balk EM (1985) J. Am. Chem. Soc. 107: 8302
150. Schumann B, Schmidt V, Wöhrle D, Yamada A in preparation
151. Druy MA (1986) J. Electrochem. Soc. 133: 353
152. Penner RM, Martin CR (1986) ibid. 133: 311
153. Niwa O, Hikita M, Tamamura T (1983) Makromol. Chem., Rapid Commun. 6: 375
154. Kaufman FB, Engler EM (1979) ibid. 101: 547
155. Cottrell FG (1903) Z. Phys. Chem. (Leipzig) 42: 385
156. Matsuda H (1980) Bull. Chem. Soc. Jpn. 53: 3439
157. Morman KD, Majda M (1986) J. Electroanal. Chem. 207: 73
158. White JR, Bard AJ (1986) ibid. 197: 233
159. Mallouk TE, Cammarata V, Crayston JA, Wrighton MS (1986) J. Phys. Chem. 90: 2150
160. Andrieux CP, Bouchiat JMD, Savéant JM (1982) J. Electroanal. Chem. 131: 1
161. Anson FC, Savéant JM, Shigehara K (1983) J. Phys. Chem. 87: 214
162. Pickup PG, Leidner CR, Denisevich P, Murray RW (1984) J. Electroanal. Chem. 164: 39
163. Degrand C, Roullier L, Miller LL, Zinger B (1984) ibid. 178: 101
164. Rubinstein I (1985) ibid. 188: 227
165. Armstrong FA, Hill HAO, Oliver BN, Walton NJ (1985) J. Am. Chem. Soc. 106: 921
166. Armstrong FA, Hill HAO, Oliver BN, Whitford D (1985) ibid. 107: 1473
167. Taniguchi I, Funatsu T, Umekita K, Yamaguchi H, Yasukouchi K (1986) J. Electroanal. Chem. 199: 455
168. Compton RG, Davis FJ, Grant SC (1986) J. Appl. Electrochem. 16: 239
169. Anson FC, Tsou YM, Savéant JM (1984) J. Electroanal. Chem. 178: 113
170. Koppenhagen JEV, Majda M (1985) ibid. 189: 379
171. Geno PW, Ravichandran K, Baldwin RP (1985) ibid. 183: 155
172. Sharp M, Montgomery DD, Anson FC (1985) ibid 194: 247

173. Saraceno RA, Pack JG, Ewing AG (1986) ibid. *197*: 265
174. Compton RG, Day MJ, Ledwith A, Abdour IIA J. Chem. Soc., Chem. Commun. *1986*: 328
175. Oyama N, Anson FC (1986) J. Electroanal. Chem. *199*: 467
176. Daifuku H, Yoshimura I, Hirata I, Aoki K, Tokuda K, Matsuda H (1986) ibid. *199*: 47
177. Kordesch K (1984) Brenstoffbatterien. Springer-Verlag, Berlin Heidelberg New York
178. Yeager E (1984) Electrochim. Acta *29*:1527 and literatures cited therein
179. Jahnke H (1980) Chimia *34*: 58
180a. Jahnke H, Schönborn M, Zimmermann G (1976) Top. Curr. Chem. *61*: 133
   b. Holze R, Vogel I, Vielstich W J. Electroanal. Chem. in press
   c. van der Putten A, Elzing A, Visscher W, Barendrecht E (1986) J. Electroanal. Chem. *205*: 233
   d. van der Putten A, Visscher W, Barendrecht E (1985) ibid. *195*: 63
181. Hirai T, Yamaki J, Yamaji A (1985) J. Appl. Electrochem. *15*: 441
182. Hirai T, Yamaki J, Yamaji A (1985) ibid. *15*: 77
183. Dhar HP, Darby R, Young VY, White RE (1983) Electrochim. Acta *30*: 423
184. Wöhrle D, Marose V, Knoop R (1985) Makromol. Chem. *186*: 2209
185a. Collman JP, Denisevich P, Konai Y, Marrocco M, Koval C, Anson FC (1980) J. Am. Chem. Soc. *102*: 6027
   b. Collman JP, Chong AO, Jameson GB, Oakley RT, Rose E, Schmittou ER, Ibers JA (1981) ibid. *103*: 516
186. Wan GX, Shigehara K, Tsuchida E, Anson FC (1984) J. Electroanal. Chem. *179*: 239
187. Buttry DA, Anson FC (1984) J. Am. Chem. Soc. *106*: 59
188. Kobayashi N, Nishiyama Y (1984) J. Electroanal. Chem. *181*: 107
189. Ikeda O, Kojima T, Tamura H (1986) ibid. *200*: 323
190. Ikeda O, Fukuda H, Kojima T, Tamura H (1985) J. Electrochem. Soc. *132*: 3069
191. van der Putten A, Elzing A, Visscher W, Barendrecht, E J. Chem. Soc., Chem. Commun. *1986*: 477
192. Zagal JH, Paez M, Sturn J, Zanartu SU (1984) J. Electroanal. Chem. *181*: 295
193. Yeager E (1984) Electrochim. Acta *29*: 1527
194. Elzing A, van der Putten A, Visscher W, Barendrecht E (1986) J. Electroanal. Chem. *200*: 313
195. Bedioui F, Bongars C, Devynck J, Charreton CB, Hinnen C (1986) ibid. *207*: 87
196. Ikeda O, Okabayashi K, Yoshida N, Tamura H (1985) ibid. *191*: 157
197. Tarasevich MR, Bogdanovskaya VA, Gavrilova EF, Orlov SE (1986) ibid. *206*: 217
198. Mongoli G, Nusiani MN, Zotti G, Valcher S (1986) ibid. *202*: 217
199. Hirabaru O, Hanabusa K, Shirai H, Takemoto K, Hojo N J. Chem. Soc., Dalton Trans. *1984*: 1485
200. Kellett RM, Spiro TG (1985) Inorg. Chem. *24*: 2378
201. Harrison DJ, Wrighton MS (1984) J. Phys. Chem. *88*: 3932
202a. Stalder CJ, Chao S, Wrighton MS (1984) J. Am. Chem. Soc. *106*: 3673
   b. Andre JF, Wrighton MS (1985) Inorg. Chem. *24*: 4288
203. Tourillon G, Garnier F (1984) J. Phys. Chem. *88*: 5281
204. Yanagida S, Kabumoto A, Mizumoto K, Pac C, Yoshino K J. Chem. Soc., Chem. Commun. *1985*: 474
205. Razumas VJ, Gudavicius AV, Kulys JJ (1986) J. Electroanal. Chem. *198*: 81
206. Oyama N, Anson FC (1986) ibid. *199*: 467
207. Wöhrle D, Bannehr R, Schumann B, Meyer G, Jaeger N (1983) J. Mol. Catal. *21*: 255
208. Jakobs RCM, Janssen LJJ, Barendrecht E (1985) Electrochim. Acta *30*: 1313
209. Wöhrle D, Kirschenmann M, Jaeger N (1985) J. Electrochem. Soc. *132*: 1150
210. Yamaki J, Yamaji J (1982) ibid. *129*: 1150
211. Cosnier S, Deronizier A, Moutet JC (1986) J. Electroanal. Chem. *207*: 315
212. Kusuda K, Ishihara R, Yamaguchi H, Izumi I (1986) Electrochim. Acta *31*: 657
213. MacInnes D Jr, Druy MA, Nigrey PJ, Nairns DP, MacDiarmid AG, Heeger AJ J. Chem. Soc., Chem. Commun. *1981*: 317
214. Nagatomo T, Kakehata H, Ichikawa C, Omoto O (1985) J. Electrochem. Soc. *132*: 1380
215. Beniere F, Boils D, Canepa H, Franco J, Le Corre A, Louboutin JP (1985) ibid. *132*: 2100
216. Maxfield M, Wolf JF, Miller GG, Frommer JE, Schacklette LW (1986) ibid. *133*: 117
217. Nigrey PJ, MacInnes D, Nairns DP, MacDiarmid AG, Heeger AJ (1981) ibid. *128*: 1651

218. Kaneto K, Maxfield M, Nairns DP, MacDiarmid AG, Heeger AJ (1982) J. Chem. Soc., Faraday Trans. 1, *78*: 3417
219. Kaner RB, MacDiarmid AG (1984) ibid. *80*: 2109
220. Caja J, Kaner RB, MacDiarmid AG (1984) J. Electrochem. Soc. *131*: 2744
221. Farrington GC, Scrosati B, Frydrych D, DeNuzzio J (1984) ibid. *131*: 7
222. Maxfield M, Mu SL, MacDiarmid AG (1985) ibid. *132*: 838
223. Hug R, Farrington GC (1985) ibid. *132*: 1432
224. Will FG (1985) ibid. *132*: 2351
225. Schacklette LW, Toth JE, Murthy NS, Baughman RH (1985) ibid. *132*: 1529
226. Kitani A, Kaya M, Hiromoto Y, Sasaki K (1985) Denki Kagaku *53*: 592
227. Kaya M, Sasaki K (1984) Denki Kagaku *52*: 847
228. Kitani A, Kaya M, Sasaki K (1986) J. Electrochem. Soc. *133*: 1069
229. DeSurville R, Josefowicz M, Yu LT, Perichon J, Buvet R (1968) Electrochim. Acta *13*: 1451
230. Genies EM, Syed AA, Tsintavis C (1985) Mol. Cryst. Liq. Cryst. *121*: 181
231. MacDiarmid AG, Mu SL, Somasiri NLD, Wu W (1985) ibid. *121*: 187
232. Pickup PG, Osteryoung RA (1985) Electroanal. Chem. *195*: 271
233. Mermilliod M, Tanguy J, Petiot F (1986) J. Electrochem. Soc. *133*: 1073
234. Mohammadi A, Inganäs O, Lundström I (1986) ibid. *133*: 947
235. Aizawa M, Watanabe S, Shinohara H, Shirakawa H J. Chem. Soc., Chem. Commun. *1985*: 264
236. Biserni M, Marinangeli A, Mastragostino M (1985) J. Electrochem. Soc. *132*: 1597; (1986) Electrochim. Acta *31*: 119
237. Nogami T, Nawa M, Mikawa H J. Chem. Soc., Chem. Commun. *1982*: 1158
238 a. Shirakawa H, Louis EJ, MacDiarmid AG, Chiang CK, Heeger AJ J. Chem. Soc., Chem. Commun. *1977*: 578
   b. Buttol P, Mastragostino M, Panero S, Scrosati B (1986) Electrochim. Acta *31*: 783
239 a. Schacklette LW, Elsenbaumer RL, Chance RR, Sowa JM, Ivory DM, Miller GG, Baughman RH J. Chem. Soc. Chem. Commun. *1982*: 361
   b. Chiang CK, Fincher CR Jr, Park YW, Heeger AJ, Shirakawa H, Louis EJ, Gau SC, MacDiarmid AG (1977) Phys. Rev. Lett. *39*: 1098
240. Nigrey PJ, MacDiarmid AG, Heeger AJ J. Chem. Soc., Chem. Commun. *1979*: 594
241. Kakuta T, Shirota Y, Mikawa H ibid. *1985*: 553
242. Yamamoto T, Hishinuma M, Yamamoto A (1985) J. Electroanal. Chem. *185*: 273
243. Nawa M, Nogami T, Mikawa H (1984) J. Electrochem. Soc. *131*: 1457
244. Kaneko M (1986) J. Polym. Sci., Polym. Lett. Ed. *24*: 435
245. Honda K, Hayashi H (1986) Prog. Batteries Solar Cells *6*: 255
246. Schumann B, Wöhrle D, Jaeger N (1985) J. Electrochem. Soc. *132*: 2144
247. Wöhrle D, Kaune H, Schumann B, Jaeger N (1986) Makromol. Chem. *187*: 2947
248. Schumann B, Wöhrle D, Jaeger N in preparation
249. Gauthier M, Fauteux D, Vassort G, Belanger A, Duval M, Ricoux P, Chabagno JM, Muller D, Rigaud P, Armand MB, Deroo D (1985) J. Electrochem. Soc. *132*: 1333
250. Itaya K, Uchida I, Ataka T, Toshima S (1982) Denki Kagaku *50*: 436
251. Sammells AF, Pujare NU (1986) J. Electrochem. Soc. *133*: 1270
252. Riou MT, Auregan M, Clasisse C (1985) J. Electroanal. Chem. *187*: 349
253. Sammells AF, Pujare NU (1986) J. Electrochem. Soc. *133*: 1065
254. Calvert JM, Manuccia TJ, Nowak RJ (1986) ibid. *133*: 951
255. Elliott CM, Redepenning JG (1986) J. Electroanal. Chem. *197*: 219
256. Kobayashi T, Yoneyama H, Tamura H (1984) ibid. *161*: 419
257. Kobayashi T, Yoneyama H, Tamura H (1984) ibid. *177*: 281
258. McManus PM, Yang SC, Cushman RJ J. Chem. Soc., Chem. Commun. *1985*: 1556
259. Kaneko M, Nakamura H (1987) Makromol. Chem., Rapid Commun. *8*: 179
260. Genies EM, Tsintavis C (1985) J. Electroanal. Chem. *195*: 109
261. Garnier F, Tourillon G, Gazard M, Dubois JC (1983) ibid. *148*: 299
262. Tourillon G, Garnier F (1984) ibid. *161*: 407
263. Yoneyama H, Wakamoto K, Tamura H (1985) J. Electrochem. Soc. *132*: 2414
264. Ohsaka T, Okajima T, Oyama N (1986) J. Electroanal. Chem. *200*: 159
265. Shinohara H, Aizawa M, Shirakawa H J. Chem. Soc., Chem. Commun. *1986*: 87
266. Espenscheid MW, Martin CR (1985) J. Electroanal. Chem. *188*: 73

267. Ewing AG, Feldman BJ, Murray RW (1984) ibid. *172*: 145
268. Ewing AG, Feldman BJ, Murray RW (1985) J. Phys. Chem. *89*: 1263
269. Niwa K, Doblehofer K (1986) Electrochim. Acta *31*: 549
270. Nagy G, Gerhardt GA, Oke AF, Rice ME, Adams RN (1985) J. Electroanal. Chem. *188*: 85
271. Arai G, Fujii A, Yasumori I Chem. Lett. *1985*: 1091
272. Sugihara H, Okada T, Hiratani K J. Chem. Soc., Chem. Commun. *1985*: 957
273. Burgmayer P, Murray RW (1984) J. Phys. Chem. *88*: 2515
274. Okahata Y, Enna G, Taguchi K, Seki T (1985) J. Am. Chem. Soc. *107*: 5300
275. Okahata Y, Taguchi K, Seki T J. Chem. Soc., Chem. Commun. *1985*: 1122
276. Okahata Y, Takenouchi K, Seki T ibid. *1986*: 558
277. Zinger B, Miller LL (1984) J. Am. Chem. Soc. *106*: 6861
278. Armstrong RD, Lockhart JC, Todo M (1986) Electrochim. Acta *31*: 591
279. Satchwill T, Harrison DJ (1986) J. Electroanal. Chem. *202*: 75
280. White HS, Kittlesen GP, Wrighton MS (1985) J. Am. Chem. Soc. *106*: 5375
281. Pickup PG, Murray RW (1984) J. Electrochem. Soc. *131*: 833
282. Koezuka H, Hyodo K, MacDiarmid AG (1985) J. Appl. Phys. *58*: 1279
283. Tsumura A, Koezuka H, Tsunoda S, Ando T Chem. Lett. *1986*: 863
284. Okano M, Fujishima A, Honda K (1985) J. Electroanal. Chem. *185*: 393
285. Arai G, Koiko T, Yasumori I Chem. Lett. *1986*: 867
286. Shinohara H, Aizawa M, Shirakawa H ibid *1985*: 179
287. Iyoda Y, Ohtani A, Shimidzu T (1984) Polym. Prepr. Jpn. *33*: 1747
288. Ohtani A, Iyoda T, Shimidzu T, Honda K (1985) ibid. *34*: 2829
289. Niwa O, Tamamura T J. Chem. Soc., Chem. Commun. *1984*: 817
290. Niwa O, Hikita M, Tamamura T (1985) Makromol. Chem., Rapid Commun. *6*: 375
291. Niwa O, Hikita M, Tamamura T (1985) Polym. Prepr. Jpn. *34*: 2821
292. Auerbach A (1985) J. Electrochem. Soc. *132*: 1437
293. Fan FRF, Bard AJ (1986) ibid. *133*: 301
294. Penner RM, Martin CR (1986) ibid. *133*: 310
295. Druy MA (1986) ibid. *133*: 353
296. Okano M, Itoh K, Fujishima A, Honda K Chem. Lett. *1986*: 469
297. Yoneyama H, Kitayama M ibid. *1986*: 657
298. Liu HY, Fan FRF, Lin CW, Bard AJ (1986) J. Am. Chem. Soc. *108*: 3838
299. Ozaki M, Peebles DL, Weinberger BR, Chiang CK, Gau SC, Heeger AJ, MacDiarmid AG (1979) Appl. Phys. Lett. *35*: 83
300. Abadie MJM (1985) Mol. Cryst. Liq. Cryst. *121*: 297
301. Kanicki J, Fedorko P (1984) J. Phys. D. *17*: 805
302. Chen SN, Heeger AJ, Kiss Z, MacDiarmid AG, Gau SC, Peebles DL (1980) Appl. Phys. Lett. *36*: 96
303. Yamase T, Harada H, Ikawa T, Ikeda S, Shirakawa H (1981) Bull. Chem. Soc. Jpn. *54*: 2817
304. Inoue T, Yamase T (1983) ibid. *56*: 985
305. Kaneko M, Nakamura H J. Chem. Soc., Chem. Commun. *1985*: 346
306. Kenmochi T, Tsuchida E, Kaneko M, Yamada A (1985) Electrochim. Acta *30*: 1405
307. Noufi RN, Frank AJ, Nozik AJ (1981) J. Am. Chem. Soc. *103*: 1849
308. Rajeshwar K, Kaneko M, Yamada A, Noufi RN (1985) J. Phys. Chem. *89*: 806
309. Horowitz G, Garnier F (1985) J. Electrochem. Soc. *132*: 634
310. Desilvestro J, Grätzel M, Kavan L, Moser J (1985) J. Am. Chem. Soc. *107*: 2988
311. Mau AWH, Huang CB, Kakuta N, Bard AJ, Campion A, Fox MA, White JM, Webber SE (1984) ibid. *106*: 6537
312. Shimura M, Toyoda A (1984) Jpn. J. Appl. Phys. *23*: 1462
313. Uehara K, Shibata K, Tanaka M Chem. Lett. *1985*: 897
314. Cadene M, Rolland M, Abadie MJM (1983) Rev. Phys. Appl. *18*: 691
315. Weinberger BR, Gau SC, Kiss Z (1981) Appl. Phys. Lett. *38*: 555
316. Wada T, Takeno A, Hiroyuki M, Sasabe H, Kobayashi Y J. Chem. Soc., Chem. Commun. *1985*: 1194
317. Kaneko M, Yamada A, Tsuchida E, Kenmochi T (1985) J. Polym. Sci., Polym. Lett. Ed. *23*: 629
318. Niwa O, Hikita M, Tamamura T (1984) Polym. Prepr. Jpn. *33*: 2515

319. Skotheim T, Inganäs O, Prejza J, Lundström I (1982) Mol. Cryst. Liq. Cryst. *83*: 329
320. Misoh K, Tasaka S, Miyata S, Sasabe H Nippon Kagaku Kaishi *1983*: 763
321. Shirota Y, Kakuta T, Kanega H, Mikawa H J. Chem. Soc., Chem. Commun. *1985*: 1201
322. Hardy LC, Shriver DF (1986) J. Am. Chem. Soc. *108*: 2887
323. Musser ME, Dahlberg SC (1980) Surf. Sci. *100*: 605
324. Loutfy RO, Sharp JH, Hsiao CK, Ho R (1981) J. Appl. Phys. *52*: 5218
325. Shimura M, Baba H (1982) Denki Kagaku *50*: 678
326. Soeda Y, Tasaka S, Miyata S, Yamada A, Sasabe H (1984) Polym. Prepr. Jpn. *33*: 475
327. Dodelet JP, Pommier HP, Ringuet M (1982) J. Appl. Phys. *53*: 4270
328. Uehara K, Yoshikawa T, Katoh H, Tanaka M, Isomatsu N, Hiraishi M Chem. Lett. *1984*: 1499
329. Morel DL, Stogryn EL, Ghosh AK, Feng T, Purwin PE, Shaw RF, Fishman C, Bird GR, Piechowski AP (1984) J. Phys. Chem. *88*: 923
330. Piechowski AP, Bird GR, Morel DL, Stogryn EL (1984) ibid. *88*: 934
331. Moriizumi T, Kudo K (1981) Appl. Phys. Lett. *38*: 85
332. Donckt EV, Noirhomme B, Kanicki J (1982) J. Appl. Polym. Sci. *27*: 1
333. Aizawa M, Watanabe S, Shinohara H, Shirakawa, H J. Chem. Soc., Chem. Commun. *1985*: 62
334. Pitchumani S, Willig F ibid. *1983*: 809
335. Kamat PV, Basheer RA (1984) Chem. Phys. Lett. *103*: 503
336. Linkous C, Klofta T, Armstrong NR (1983) J. Electrochem. Soc. *130*: 1050
337. Rieke PC, Armstrong NR (1984) J. Am. Chem. Soc. *106*: 47
338. Rieke PC, Linkous CL, Armstrong NR (1984) J. Phys. Chem. *88*: 1351
339. Buttner WJ, Rieke PC, Armstrong NR (1985) J. Am. Chem. Soc. *107*: 3738
340. Klofta T, Linkous C, Armstrong NR (1985) J. Electroanal. Chem. *185*: 73
341. Klofta TJ, Linkous CA, Buttner WJ, Nanthakumar A, Mewborn TD, Armstrong NR (1985) J. Electrochem. Soc. *132*: 2134
342. Minami N (1980) J. Chem. Phys. *72*: 6317
343. Belanger D, Dodelet JP, Dao LH, Lombos BA (1984) J. Phys. Chem. *88*, 4288
344. Kawai T, Tanimura K, Sakata T (1978) Chem. Phys. Lett. *56*: 541
345. Kampas FJ, Yamashita K, Fajer J (1980) Nature *284*: 40
346. Jimbo H, Yoneyama H, Tamura H (1980) Photochem. Photobiol. *32*: 319
347. Langford CH, Hollebone BR, Nadezhdin D (1981) Can. J. Chem. *59*: 652
348. Katsu T, Tamagake K, Fujita Y Chem. Lett. *1980*: 289
349. Basu J, Bhattacharya A, Das K., Chatterjee AB, Kundu KK, Mukherjee KKR (1983) Indian J. Chem. *22A*: 695
350. Yamashita K, Harima Y, Matsumura Y (1985) Bull. Chem. Soc. Jpn. *58*: 1761
351. Harima Y, Yamashita K (1985) J. Electroanal. Chem. *186*: 313
352. Harima Y, Yamashita K (1985) J. Phys. Chem. *89*: 5325
353. Tien HT, Higgins J (1982) Chem. Phys. Lett. *93*: 276
354. Mizutani F, Iijima S, Sasaki K, Yamashita Y Nippon Kagaku Kaishi *1983*: 933
355. Pool K, Buck RP (1979) J. Electroanal. Chem. *95*: 241
356. Yamamura T, Umezawa Y J. Chem. Soc., Dalton Trans. *1982*: 1977
357. Umezawa Y, Yamamura T, Kobayashi A (1982) J. Electrochem. Soc. *129*: 2378
358. Decker F, Decker MF, Zotti G, Mengoli G (1985) Electrochim. Acta *30*: 1147
359. Fujishima A, Honda K (1972) Nature *238*: 37
360. Noufi RN, Tench D, Warren LF (1980) J. Electrochem. Soc. *127*: 2310
361. Skotheim T, Petersson LG, Inganäs O, Lundström I (1982) ibid. *129*: 1737
362. Noufi RN, Frank AJ, White J, Warren LF (1982) ibid. *129*: 2261
363. Rajeshwar K, Thompson L, Singh P, Kainthla RC, Chopra KL (1981) ibid. *128*: 1744
364. Rosenblum MD, Lewis NS (1984) J. Phys. Chem. *88*: 3103
365. Frank AJ, Honda K (1982) ibid. *86*: 1933
366. Honda K, Frank AJ (1984) ibid. *88*: 5577
367. Rajeshwar K, Kaneko M (1985) ibid. *89*: 3587
368. Ghosh PK, Spiro TG (1980) J. Am. Chem. Soc. *102*: 5543
369. Kawai W, Yamamura S Nippon Kagoku Kaishi *1981*: 1217
370. Krishnan M, Zhang X, Bard AJ (1984) J. Am. Chem. Soc. *106*: 7371
371. Memming R, Schröppel F (1979) Chem. Phys. Lett. *62*: 207
372. Tien HT (1976) Photochem. Photobiol. *24*: 97

373. Miyasaka T, Watanabe T, Fujishima A, Honda K (1978) Nature *227*: 638
374. Jow T, Wagner JB (1978) J. Electrochem. Soc. *125*: 613
375. Tennekone K, Divigalpitiya WMR (1981) Jpn. J. Appl. Phys. *20*: 299
376. Yamamura S, Kawai W Nippon Kagaku Kaishi *1982*: 1287
377. Breddels PA, Blasse G (1984) J. Chem. Soc., Faraday Trans. 2, *80*: 1055
378. Shimidzu T, Iyoda T, Koide Y (1984) Polym. J. *16*: 919
379. Minami N, Watanabe T, Fujishima A, Honda K (1979) Ber. Bunsenges. Phys. Chem. *83*: 476
380. Leempoel P, Fan FRF, Bard AJ (1983) J. Phys. Chem. *87*: 2948
381. Horishima Y, Isono M, Itoh Y, Nozakura S Chem. Lett. *1981*: 1149
382. Iriyama K, Mizutani F, Yoshiura M ibid. *1980*: 1399
383. Natoli LM, Ryan MA, Spitler MT (1985) J. Phys. Chem. *89*: 1448
384. Hiroshi H, Yonezawa Y, Inabe H Chem. Lett. *1980*: 467
385. Nakao M, Itoh Y, Honda K (1984) J. Phys. Chem. *88*: 4906
386. Fox MA, Nobs FJ, Voynick TA (1980) J. Am. Chem. Soc. *102*: 4036
387. Meissner D, Memming R, Kastening B (1983) Chem. Phys. Lett. *96*: 34
388. Kuczynski JP, Milosavljevic BH, Thomas JK (1984) J. Phys. Chem. *88*: 980
389. Kakuta N, White JM, Campion A, Bard AJ, Fox MA, Webber, SE (1985) ibid. *89*: 48
390. Kakuta N, Park KH, Finlayson MF, Bard AJ, Campion A, Fox MA, Webber SE, White JM (1985) ibid. *89*: 5028
391. Skotheim T (1981) Appl. Phys. Lett. *38*: 712
392. Cook RL, Sammells AF (1985) J. Electrochem. Soc. *132*: 2429
393. Sammells AF, Schmidt SK (1985) ibid. *132*: 520
394. Lawrence MF, Dodelet JP (1985) J. Phys. Chem. *89*: 1395
395. Hor AM, Loutfy RO (1983) Can J. Chem. *61*: 901
396. Hattori M, Sasaki H, Toshima S (1980) Denki Kagaku *48*: 290
397. Rabinowitch E (1940) J. Chem. Phys. *8*: 551, 560
398. Eisenberg M, Silverman HP (1961) Electrochim. Acta *5*: 1
399. Kaneko M, Yamada A (1976) Rep. Inst. Phys. Chem. Res. *52*: 210
400. Shigehara K, Nishimura M, Tsuchida E (1977) Bull. Chem. Soc. Jpn. *50*: 3397
401. Albery WJ, Foulds AW, Hall KJ, Hillman AR, Egdell RG, Orchard AF (1979) Nature *282*: 793
402. Albrey WJ, Foulds AW, Hall KJ, Hillman AR (1980) J. Electrochem. Soc. *127*: 654
403. Tamilarasan R, Natarajan P (1981) Nature *292*: 224
404. Quickenden TI, Harrison IR (1985) J. Electrochem. Soc. *132*: 81
405. Tamilarasan R, Ramaraj R, Subramanian R, Natarajan P (1984) J. Chem. Soc., Faraday Trans. 1, *80*: 2405
406. Ramaraj R, Tamilarasan R, Natarajan P (1985) J. Chem. Soc., Faraday Trans. 1, *81*: 2763
407. Yoshida M, Oshida I (1964) Jpn. J. Appl. Phys. *33*: 34
408. Kaneko M, Yamada A (1977) J. Phys. Chem. *81*: 1213
409. Kamat PV (1984) J. Electroanal. Chem. *163*: 389
410. Westmoreland TD, Calvert JM, Murray RW, Meyer TJ J. Chem. Soc., Chem. Commun. *1983*: 65
411. Kobayashi K, Sagara T, Okada M, Niki K Chem. Lett. *1983*: 373
412. Kaneko M, Ochiai M, Yamada A (1982) Makromol. Chem., Rapid Commun. *3*: 299
413. Kaneko M, Takabayashi N, Yamada A Chem. Lett. *1982*: 1647
414. Kaneko M, Takabayashi N, Yamauchi Y, Yamada A (1984) Bull. Chem. Soc. Jpn. *57*: 156
415. Kaneko M (1987) J. Macromol. Sci., Chem. *A24*: 357
416. Kaneko M, Hara S, Yamada A (1985) J. Electroanal. Chem. *194*: 165
417. Kaneko M, Yamada A (1986) Electrochim. Acta *31*: 273
418. Cosnier S, Deronzier A, Moutet JC (1985) J. Phys. Chem. *89*: 4895
419. Kaneko M, Shimadzu M, Teratani S, Shirai H submitted
420. Morishima Y, Fukushima Y, Nozakura S J. Chem. Soc., Chem. Commun. *1985*: 912

Editor: H.-J. Cantow
Received March 27, 1987

# Author Index Volumes 1–84

*Allegra, G.* and *Bassi, I. W.:* Isomorphism in Synthetic Macromolecular Systems. Vol. 6, pp. 549–574.

*Andrade, J. D., Hlady, V.:* Protein Adsorption and Materials Biocompability: A. Tutorial Review and Suggested Hypothesis. Vol. 79, pp. 1–63.

*Andrews, E. H.:* Molecular Fracture in Polymers. Vol. 27, pp. 1–66.

*Anufrieva, E. V.* and *Gotlib, Yu. Ya.:* Investigation of Polymers in Solution by Polarized Luminescence. Vol. 40, pp. 1–68.

*Apicella, A.* and *Nicolais, L.:* Effect of Water on the Properties of Epoxy Matrix and Composite. Vol. 72, pp. 69–78.

*Apicella, A., Nicolais, L.* and *de Cataldis, C.:* Characterization of the Morphological Fine Structure of Commercial Thermosetting Resins Through Hygrothermal Experiments. Vol. 66, pp. 189–208.

*Argon, A. S., Cohen, R. E., Gebizlioglu, O. S.* and *Schwier, C.:* Crazing in Block Copolymers and Blends. Vol. 52/53, pp. 275–334.

*Aronhime, M. T., Gillham, J. K.:* Time-Temperature Transformation (TTT) Cure Diagram of Thermosetting Polymeric Systems. Vol. 78, pp. 81–112.

*Arridge, R. C.* and *Barham, P. J.:* Polymer Elasticity. Discrete and Continuum Models. Vol. 46, pp. 67–117.

*Aseeva, R. M., Zaikov, G. E.:* Flammability of Polymeric Materials. Vol. 70, pp. 171–230.

*Ayrey, G.:* The Use of Isotopes in Polymer Analysis. Vol. 6, pp. 128–148.


*Bässler, H.:* Photopolymerization of Diacetylenes. Vol. 63, pp. 1–48.

*Baldwin, R. L.:* Sedimentation of High Polymers. Vol. 1, pp. 451–511.

*Balta-Calleja, F. J.:* Microhardness Relating to Crystalline Polymers. Vol. 66, pp. 117–148.

*Barbé, P. C., Cecchin, G.* and *Noristi, L.:* The Catalytic System Ti-Complex/MgCl₂. Vol. 81, pp. 1–83.

*Barton, J. M.:* The Application of Differential Scanning Calorimetry (DSC) to the Study of Epoxy Resins Curing Reactions. Vol. 72, pp. 111–154.

*Basedow, A. M.* and *Ebert, K.:* Ultrasonic Degradation of Polymers in Solution. Vol. 22, pp. 83–148.

*Batz, H.-G.:* Polymeric Drugs. Vol. 23, pp. 25–53.

*Bell, J. P.* see *Schmidt, R. G.:* Vol. 75, pp. 33–72.

*Bekturov, E. A.* and *Bimendina, L. A.:* Interpolymer Complexes. Vol. 41, pp. 99–147.

*Bergsma, F.* and *Kruissink, Ch. A.:* Ion-Exchange Membranes. Vol. 2, pp. 307–362.

*Berlin, Al. Al., Volfson, S. A.,* and *Enikolopian, N. S.:* Kinetics of Polymerization Processes. Vol. 38, pp. 89–140.

*Berry, G. C.* and *Fox, T. G.:* The Viscosity of Polymers and Their Concentrated Solutions. Vol. 5, pp. 261–357.

*Bevington, J. C.:* Isotopic Methods in Polymer Chemistry. Vol. 2, pp. 1–17.

*Bhuiyan, A. L.:* Some Problems Encountered with Degradation Mechanisms of Addition Polymers. Vol. 47, pp. 1–65.

*Bird, R. B., Warner, Jr., H. R.,* and *Evans, D. C.:* Kinetik Theory and Rheology of Dumbbell Suspensions with Brownian Motion. Vol. 8, pp. 1–90.

*Biswas, M.* and *Maity, C.:* Molecular Sieves as Polymerization Catalysts. Vol. 31, pp. 47–88.

*Biswas, M., Packirisamy, S.:* Synthetic Ion-Exchange Resins. Vol. 70, pp. 71–118.

*Block, H.:* The Nature and Application of Electrical Phenomena in Polymers. Vol. 33, pp. 93–167.

*Bodor, G.:* X-ray Line Shape Analysis. A. Means for the Characterization of Crystalline Polymers. Vol. 67, pp. 165–194.

*Böhm, L. L., Chmeliř, M., Löhr, G., Schmitt, B. J.* and *Schulz, G. V.:* Zustände und Reaktionen des Carbanions bei der anionischen Polymerisation des Styrols. Vol. 9, pp. 1–45.

*Boué, F.:* Transient Relaxation Mechanisms in Elongated Melts and Rubbers Investigated by Small Angle Neutron Scattering. Vol. 82, pp. 47–103.

*Bovey, F. A.* and *Tiers, G. V. D.:* The High Resolution Nuclear Magnetic Resonance Spectroscopy of Polymers. Vol. 3, pp. 139–195.

*Braun, J.-M.* and *Guillet, J. E.:* Study of Polymers by Inverse Gas Chromatography. Vol. 21, pp. 107–145.

*Breitenbach, J. W., Olaj, O. F.* und *Sommer, F.:* Polymerisationsanregung durch Elektrolyse. Vol. 9, pp. 47–227.

*Bresler, S. E.* and *Kazbekov, E. N.:* Macroradical Reactivity Studied by Electron Spin Resonance. Vol. 3, pp. 688–711.

*Brosse, J.-C., Derouet, D., Epaillard, F., Soutif, J.-C., Legeay, G.* and *Dušek, K.:* Hydroxyl-Terminated Polymers Obtained by Free Radical Polymerization. Synthesis, Characterization, and Applications. Vol. 81, pp. 167–224.

*Bucknall, C. B.:* Fracture and Failure of Multiphase Polymers and Polymer Composites. Vol. 27, pp. 121–148.

*Burchard, W.:* Static and Dynamic Light Scattering from Branched Polymers and Biopolymers. Vol. 48, pp. 1–124.

*Bywater, S.:* Polymerization Initiated by Lithium and Its Compounds. Vol. 4, pp. 66–110.

*Bywater, S.:* Prepa: ition and Properties of Star-branched Polymers. Vol. 30, pp. 89–116.

*Candau, S., Bastide, J.* und *Delsanti, M.:* Structural. Elastic and Dynamic Properties of Swollen Polymer Networks. Vol. 44, pp. 27–72.

*Carrick, W. L.:* The Mechanism of Olefin Polymerization by Ziegler-Natta Catalysts. Vol. 12, pp. 65–86.

*Casale, A.* and *Porter, R. S.:* Mechanical Synthesis of Block and Graft Copolymers. Vol. 17, pp. 1–71.

*Cecchin, G.* see Barbé, P. C.: Vol. 81, pp. 1–83.

*Cerf, R.:* La dynamique des solutions de macromolecules dans un champ de vitresses. Vol. 1, pp. 382–450.

*Cesca, S., Priola, A.* and *Bruzzone, M.:* Synthesis and Modification of Polymers Containing a System of Conjugated Double Bonds. Vol. 32, pp. 1–67.

*Chiellini, E., Solaro, R., Galli, G.* and *Ledwith, A.:* Optically Active Synthetic Polymers Containing Pendant Carbazolyl Groups. Vol. 62, pp. 143–170.

*Cicchetti, O.:* Mechanisms of Oxidative Photodegradation and of UV Stabilization of Polyolefins. Vol. 7, pp. 70–112.

*Clark, A. H.* and *Ross-Murphy, S. B.:* Structural and Mechanical Properties of Biopolymer Gels. Vol. 83, pp. 57–193.

*Clark, D. T.:* ESCA Applied to Polymers. Vol. 24, pp. 125–188.

*Colemann, Jr., L. E.* and *Meinhardt, N. A.:* Polymerization Reactions of Vinyl Ketones. Vol. 1, pp. 159–179.

*Comper, W. D.* and *Preston, B. N.:* Rapid Polymer Transport in Concentrated Solutions. Vol. 55, pp. 105–152.

*Corner, T.:* Free Radical Polymerization — The Synthesis of Graft Copolymers. Vol. 62, pp. 95–142.

*Crescenzi, V.:* Some Recent Studies of Polyelectrolyte Solutions. Vol. 5, pp. 358–386.

*Crivello, J. V.:* Cationic Polymerization — Iodonium and Sulfonium Salt Photoinitiators, Vol. 62, pp. 1–48.

*Dave, R.* see Kardos, J. L.: Vol. 80, pp. 101–123.

*Davydov, B. E.* and *Krentsel, B. A.:* Progress in the Chemistry of Polyconjugated Systems. Vol. 25, pp. 1–46.

*Derouet, F. see Brosse, J.-C.*: Vol. 81, pp. 167–224.

*Dettenmaier, M.*: Intrinsic Crazes in Polycarbonate Phenomenology and Molecular Interpretation of a New Phenomenon. Vol. 52/53, pp. 57–104.

*Diaz, A. F., Rubinson, J. F.*, and *Mark, H. B., Jr.*: Electrochemistry and Electrode Applications of Electroactive / Conductive Polymers. Vol. 84, pp. 113–140.

*Dobb, M. G.* and *McIntyre, J. E.*: Properties and Applications of Liquid-Crystalline Main-Chain Polymers. Vol. 60/61, pp. 61–98.

*Döll, W.*: Optical Interference Measurements and Fracture Mechanics Analysis of Crack Tip Craze Zones. Vol. 52/53, pp. 105–168.

*Doi, Y. see Keii, T.*: Vol. 73/74, pp. 201–248.

*Dole, M.*: Calorimetric Studies of States and Transitions in Solid High Polymers. Vol. 2, pp. 221–274.

*Donnet, J. B., Vidal, A.*: Carbon Black-Surface Properties and Interactions with Elastomers. Vol. 76, pp. 103–128.

*Dorn, K., Hupfer, B.*, and *Ringsdorf, H.*: Polymeric Monolayers and Liposomes as Models for Biomembranes How to Bridge the Gap Between Polymer Science and Membrane Biology? Vol. 64, pp. 1–54.

*Dreyfuss, P.* and *Dreyfuss, M. P.*: Polytetrahydrofuran. Vol. 4, pp. 528–590.

*Drobnik, J.* and *Rypáček, F.*: Soluble Synthetic Polymers in Biological Systems. Vol. 57, pp. 1–50.

*Dröscher, M.*: Solid State Extrusion of Semicrystalline Copolymers. Vol. 47, pp. 120–138.

*Duduković, M. P. see Kardos, J. L.*: Vol. 80, pp. 101–123.

*Drzal, L. T.*: The Interphase in Epoxy Composites. Vol. 75, pp. 1–32.

*Dušek, K.*: Network Formation in Curing of Epoxy Resins. Vol. 78, pp. 1–58.

*Dušek, K.* and *Prins, W.*: Structure and Elasticity of Non-Crystalline Polymer Networks. Vol. 6, pp. 1–102.

*Dušek, K. see Brosse, J.-C.*: Vol. 81, pp. 167–224.

*Duncan, R.* and *Kopeček, J.*: Soluble Synthetic Polymers as Potential Drug Carriers. Vol. 57, pp. 51–101.


*Eastham, A. M.*: Some Aspects of the Polymerization of Cyclic Ethers. Vol. 2, pp. 18–50.

*Ehrlich, P.* and *Mortimer, G. A.*: Fundamentals of the Free-Radical Polymerization of Ethylene. Vol. 7, pp. 386–448.

*Eisenberg, A.*: Ionic Forces in Polymers. Vol. 5, pp. 59–112.

*Eiss, N. S. Jr. see Yorkgitis, E. M.* Vol. 72, pp. 79–110.

*Elias, H.-G., Bareiss, R. und Watterson, J. G.*: Mittelwerte des Molekulargewichts und anderer Eigenschaften. Vol. 11, pp. 111–204.

*Elsner, G., Riekel, Ch.* and *Zachmann, H. G.*: Synchrotron Radiation Physics. Vol. 67, pp. 1–58.

*Elyashevich, G. K.*: Thermodynamics and Kinetics of Orientational Crystallization of Flexible-Chain Polymers. Vol. 43, pp. 207–246.

*Enkelmann, V.*: Structural Aspects of the Topochemical Polymerization of Diacetylenes. Vol. 63, pp. 91–136.

*Entelis, S. G., Evreinov, V. V., Gorshkov, A. V.*: Functionally and Molecular Weight Distribution of Telchelic Polymers. Vol. 76, pp. 129–175.

*Epaillard, F. see Brosse, J.-C.*: Vol. 81, pp. 167–224.

*Evreinov, V. V. see Entelis, S. G.* Vol. 76, pp. 129–175.


*Ferruti, P.* and *Barbucci, R.*: Linear Amino Polymers: Synthesis, Protonation and Complex Formation. Vol. 58, pp. 55–92.

*Finkelmann, H.* and *Rehage, G.*: Liquid Crystal Side-Chain Polymers. Vol. 60/61, pp. 99–172.

*Fischer, H.*: Freie Radikale während der Polymerisation, nachgewiesen und identifiziert durch Elektronenspinresonanz. Vol. 5, pp. 463–530.

*Flory, P. J.*: Molecular Theory of Liquid Crystals. Vol. 59, pp. 1–36.

*Ford, W. T.* and *Tomoi, M.*: Polymer-Supported Phase Transfer Catalysts Reaction Mechanisms. Vol. 55, pp. 49–104.

*Fradet, A.* and *Maréchal, E.*: Kinetics and Mechanisms of Polyesterifications. I. Reactions of Diols with Diacids. Vol. 43, pp. 51–144.

*Franz, G.*: Polysaccharides in Pharmacy. Vol. 76, pp. 1–30.

*Friedrich, K.:* Crazes and Shear Bands in Semi-Crystalline Thermoplastics. Vol. 52/53, pp. 225–274.

*Fujita, H.:* Diffusion in Polymer-Diluent Systems. Vol. 3, pp. 1–47.

*Funke, W.:* Über die Strukturaufklärung vernetzter Makromoleküle, insbesondere vernetzter Polyesterharze, mit chemischen Methoden. Vol. 4, pp. 157–235.

*Furukawa, H.* see Kamon, T.: Vol. 80, pp. 173–202.


*Gal'braikh, L. S.* and *Rigovin, Z. A.:* Chemical Transformation of Cellulose. Vol. 14, pp. 87–130.

*Galli, G.* see Chiellini, E. Vol. 62, pp. 143–170.

*Gallot, B. R. M.:* Preparation and Study of Block Copolymers with Ordered Structures, Vol. 29, pp. 85–156.

*Gandini, A.:* The Behaviour of Furan Derivatives in Polymerization Reactions. Vol. 25, pp. 47–96.

*Gandini, A.* and *Cheradame, H.:* Cationic Polymerization. Initiation with Alkenyl Monomers. Vol. 34/35, pp. 1–289.

*Geckeler, K., Pillai, V. N. R.,* and *Mutter, M.:* Applications of Soluble Polymeric Supports. Vol. 39, pp. 65–94.

*Gerrens, H.:* Kinetik der Emulsionspolymerisation. Vol. 1, pp. 234–328.

*Ghiggino, K. P., Roberts, A. J.* and *Phillips, D.:* Time-Resolved Fluorescence Techniques in Polymer and Biopolymer Studies. Vol. 40, pp. 69–167.

*Gillham, J. K.* see Aronhime, M. T.: Vol. 78, pp. 81–112.

*Glöckner, G.:* Analysis of Compositional and Structural Heterogeneitis of Polymer by Non-Exclusion HPLC. Vol. 79, pp. 159–214.

*Godovsky, Y. K.:* Thermomechanics of Polymers. Vol. 76, pp. 31–102.

*Goethals, E. J.:* The Formation of Cyclic Oligomers in the Cationic Polymerization of Heterocycles. Vol. 23, pp. 103–130.

*Gorshkov, A. V.* see Entelis, S. G. Vol. 76, 129–175.

*Graessley, W. W.:* The Etanglement Concept in Polymer Rheology. Vol. 16, pp. 1–179.

*Graessley, W. W.:* Entagled Linear, Branched and Network Polymer Systems. Molecular Theories. Vol. 47, pp. 67–117.

*Grebowicz, J.* see Wunderlich, B. Vol. 60/61, pp. 1–60.

*Greschner, G. S.:* Phase Distribution Chromatography. Possibilities and Limitations. Vol. 73/74, pp. 1–62.


*Hagihara, N., Sonogashira, K.* and *Takahashi, S.:* Linear Polymers Containing Transition Metals in the Main Chain. Vol. 41, pp. 149–179.

*Hasegawa, M.:* Four-Center Photopolymerization in the Crystalline State. Vol. 42, pp. 1–49.

*Hatano, M.:* Induced Circular Dichroism in Biopolymer-Dye System. Vol. 77, pp. 1–121.

*Hay, A. S.:* Aromatic Polyethers. Vol. 4, pp. 496–527.

*Hayakawa, R.* and *Wada, Y.:* Piezoelectricity and Related Properties. of Polymer Films. Vol. 11, pp. 1–55.

*Heidemann, E.* and *Roth, W.:* Synthesis and Investigation of Collagen Model Peptides. Vol. 43, pp. 145–205.

*Heitz, W.:* Polymeric Reagents. Polymer Design, Scope, and Limitations. Vol. 23, pp. 1–23.

*Helfferich, F.:* Ionenaustausch. Vol. 1, pp. 329–381.

*Hendra, P. J.:* Laser-Raman Spectra of Polymers. Vol. 6, pp. 151–169.

*Hendrix, J.:* Position Sensitive "X-ray Detectors". Vol. 67, pp. 59–98.

*Henrici-Olivé, G.* and *Olivé, S.:* Oligomerization of Ethylene with Soluble Transition-Metal Catalysts. pp. 496–577.

*Henrici-Olivé, G.* und *Olivé, S.:* Koordinative Polymerisation an löslichen Übergangsmetall-Katalysatoren. Vol. 6, pp. 421–472.

*Henrici-Olivé, G.* and *Olivé, S.:* Oligomerization of Ethylene with Soluble Transition-Metal Catalysts. Vol. 15, pp. 1–30.

*Henrici-Olivé, G.* and *Olivé, S.:* Molecular Interactions and Macroscopic Properties of Polyacrylonitrile and Model Substances. Vol. 32, pp. 123–152.

*Henrici-Olivé, G.* and *Olivé, S.:* The Chemistry of Carbon Fiber Formation from Polyacrylonitrile. Vol. 51, pp. 1–60.

*Hermans, Jr., J., Lohr, D.* and *Ferro, D.*: Treatment of the Folding and Unfolding of Protein Molecules in Solution According to a Lattic Model. Vol. 9, pp. 229–283.

*Higashimura, T.* and *Sawamoto, M.*: Living Polymerization and Selective Dimerization: Two Extremes of the Polymer Synthesis by Cationic Polymerization. Vol. 62, pp. 49–94.

*Higashimura, T.* see *Masuda, T.*: Vol. 81, pp. 121–166.

*Hlady, V.* see *Andrade, J. D.*: Vol. 79, pp. 1–63.

*Hoffman, A. S.*: Ionizing Radiation and Gas Plasma (or Glow) Discharge Treatments for Preparation of Novel Polymeric Biomaterials. Vol. 57, pp. 141–157.

*Holzmüller, W.*: Molecular Mobility, Deformation and Relaxation Processes in Polymers. Vol. 26, pp. 1–62.

*Hori, Y.* see *Kashiwabara, H.*: Vol. 82, pp. 141–207.

*Hutchison, J.* and *Ledwith, A.*: Photoinitiation of Vinyl Polymerization by Aromatic Carbonyl Compounds. Vol. 14, pp. 49–86.

*Iizuka, E.*: Properties of Liquid Crystals of Polypeptides: with Stress on the Electromagnetic Orientation. Vol. 20, pp. 79–107.

*Ikada, Y.*: Characterization of Graft Copolymers. Vol. 29, pp. 47–84.

*Ikada, Y.*: Blood-Compatible Polymers. Vol. 57, pp. 103–140.

*Imanishi, Y.*: Synthese, Conformation, and Reactions of Cyclic Peptides. Vol. 20, pp. 1–77.

*Inagaki, H.*: Polymer Separation and Characterization by Thin-Layer Chromatography. Vol. 24, pp. 189–237.

*Inoue, S.*: Asymmetric Reactions of Synthetic Polypeptides. Vol. 21, pp. 77–106.

*Ise, N.*: Polymerizations under an Electric Field. Vol. 6, pp. 347–376.

*Ise, N.*: The Mean Activity Coefficient of Polyelectrolytes in Aqueous Solutions and Its Related Properties. Vol. 7, pp. 536–593.

*Isihara, A.*: Irreversible Processes in Solutions of Chain Polymers. Vol. 5, pp. 531–567.

*Isihara, A.*: Intramolecular Statistics of a Flexible Chain Molecule. Vol. 7, pp. 449–476.

*Isihara, A.* and *Guth, E.*: Theory of Dilute Macromolecular Solutions. Vol. 5, pp. 233–260.

*Iwatsuki, S.*: Polymerization of Quinodimethane Compounds. Vol. 58, pp. 93–120.

*Janeschitz-Kriegl, H.*: Flow Birefrigence of Elastico-Viscous Polymer Systems. Vol. 6, pp. 170–318.

*Jenkins, R.* and *Porter, R. S.*: Unpertubed Dimensions of Stereoregular Polymers. Vol. 36, pp. 1–20.

*Jenngins, B. R.*: Electro-Optic Methods for Characterizing Macromolecules in Dilute Solution. Vol. 22, pp. 61–81.

*Johnston, D. S.*: Macrozwitterion Polymerization. Vol. 42, pp. 51–106.

*Kamachi, M.*: Influence of Solvent on Free Radical Polymerization of Vinyl Compounds. Vol. 38, pp. 55–87.

*Kamachi, M.*: ESR Studies on Radical Polymerization. Vol. 82, pp. 207–277.

*Kamide, K.* and *Saito, M.*: Cellulose and Cellulose Derivatives: Recent Advances in Physical Chemistry. Vol. 83, pp. 1–57.

*Kamon, T., Furukawa, H.*: Curing Mechanisms and Mechanical Properties of Cured Epoxy Resins. Vol. 80, pp. 173–202.

*Kaneko, M.* and *Wöhrle, D.*: Polymer-Coated Electrodes: New Materials for Science and Industry. Vol. 84, pp, 141–228.

*Kaneko, M.* and *Yamada, A.*: Solar Energy Conversion by Functional Polymers. Vol. 55, pp. 1–48.

*Kardos, J. L., Dudukovic, M. P., Dave, R.*: Void Growth and Resin Transport During Processing of Thermosetting — Matrix Composites. Vol. 80, pp. 101–123.

*Kashiwabara, H., Shimada, S., Hori, Y.* and *Sakaguchi, M.*: ESR Application to Polymer Physics — Molecular Motion in Solid Matrix in which Free Radicals are Trapped. Vol. 82, pp. 141–207.

*Kawabata, S.* and *Kawai, H.*: Strain Energy Density Functions of Rubber Vulcanizates from Biaxial Extension. Vol. 24, pp. 89–124.

*Keii, T., Doi, Y.:* Synthesis of "Living" Polyolefins with Soluble Ziegler-Natta Catalysts and Application to Block Copolymerization. Vol. 73/74, pp. 201–248.

*Kelley, F. N.* see LeMay, J. D.: Vol. 78, pp. 113–148.

*Kennedy, J. P.* and *Chou, T.:* Poly(isobutylene-*co*-β-Pinene): A New Sulfur Vulcanizable, Ozone Resistant Elastomer by Cationic Isomerization Copolymerization. Vol. 21, pp. 1–39.

*Kennedy, J. P.* and *Delvaux, J. M.:* Synthesis, Characterization and Morphology of Poly(butadiene-g-Styrene). Vol. 38, pp. 141–163.

*Kennedy, J. P.* and *Gillham, J. K.:* Cationic Polymerization of Olefins with Alkylaluminium Initiators. Vol. 10, pp. 1–33.

*Kennedy, J. P.* and *Johnston, J. E.:* The Cationic Isomerization Polymerization of 3-Methyl-1-butene and 4-Methyl-1-pentene. Vol. 19, pp. 57–95.

*Kennedy, J. P.* and *Langer, Jr., A. W.:* Recent Advances in Cationic Polymerization. Vol. 3, pp. 508–580.

*Kennedy, J. P.* and *Otsu, T.:* Polymerization with Isomerization of Monomer Preceding Propagation. Vol. 7, pp. 369–385.

*Kennedy, J. P.* and *Rengachary, S.:* Correlation Between Cationic Model and Polymerization Reactions of Olefins. Vol. 14, pp. 1–48.

*Kennedy, J. P.* and *Trivedi, P. D.:* Cationic Olefin Polymerization Using Alkyl Halide — Alkyl-Aluminium Initiator Systems. I. Reactivity Studies. II. Molecular Weight Studies. Vol. 28, pp. 83–151.

*Kennedy, J. P., Chang, V. S. C.* and *Guyot, A.:* Carbocationic Synthesis and Characterization of Polyolefins with Si–H and Si–Cl Head Groups. Vol. 43, pp. 1–50.

*Khoklov, A. R.* and *Grosberg, A. Yu.:* Statistical Theory of Polymeric Lyotropic Liquid Crystals. Vol. 41, pp. 53–97.

*Kinloch, A. J.:* Mechanics and Mechanisms of Fracture of Thermosetting Epoxy Polymers. Vol. 72, pp. 45–68.

*Kissin, Yu. V.:* Structures of Copolymers of High Olefins. Vol. 15, pp. 91–155.

*Kitagawa, T.* and *Miyazawa, T.:* Neutron Scattering and Normal Vibrations of Polymers. Vol. 9, pp. 335–414.

*Kitamaru, R.* and *Horii, F.:* NMR Approach to the Phase Structure of Linear Polyethylene. Vol. 26, pp. 139–180.

*Klosinski, P., Penczek, S.:* Teichoic Acids and Their Models: Membrane Biopolymers with Polyphosphate Backbones. Synthesis, Structure and Properties. Vol. 79, pp. 139–157.

*Kloosterboer, J. G.:* Network Formation by Chain Crosslinking Photopolymerization and its Applications in Electronics. Vol. 84, pp. 1–62.

*Knappe, W.:* Wärmeleitung in Polymeren. Vol. 7, pp. 477–535.

*Koenik, J. L.* see Mertzel, E. Vol. 75, pp. 73–112.

*Koenig, J. L.:* Fourier Transforms Infrared Spectroscopy of Polymers, Vol. 54, pp. 87–154.

*Kolařík, J.:* Secondary Relaxations in Glassy Polymers: Hydrophilic Polymethacrylates and Polyacrylates: Vol. 46, pp. 119–161.

*Kong, E. S.-W.:* Physical Aging in Epoxy Matrices and Composites. Vol. 80, pp. 125–171.

*Koningsveld, R.:* Preparative and Analytical Aspects of Polymer Fractionation. Vol. 7.

*Kovacs, A. J.:* Transition vitreuse dans les polymers amorphes. Etude phénoménologique. Vol. 3, pp. 394–507.

*Krässig, H. A.:* Graft Co-Polymerization of Cellulose and Its Derivatives. Vol. 4, pp. 111–156.

*Kramer, E. J.:* Microscopic and Molecular Fundamentals of Crazing. Vol. 52/53, pp. 1–56.

*Kraus, G.:* Reinforcement of Elastomers by Carbon Black. Vol. 8, pp. 155–237.

*Kratochvila, J.* see Mejzlik, J.: Vol. 81, pp. 83–120.

*Kreutz, W.* and *Welte, W.:* A General Theory for the Evaluation of X-Ray Diagrams of Biomembranes and Other Lamellar Systems. Vol. 30, pp. 161–225.

*Krimm, S.:* Infrared Spectra of High Polymers. Vol. 2, pp. 51–72.

*Kuhn, W., Ramel, A., Walters, D. H. Ebner, G. and Kuhn, H. J.:* The Production of Mechanical Energy from Different Forms of Chemical Energy with Homogeneous and Cross-Striated High Polymer Systems. Vol. 1, pp. 540–592.

*Kunitake, T.* and *Okahata, Y.:* Catalytic Hydrolysis by Synthetic Polymers. Vol. 20, pp. 159–221.

*Kurata, M.* and *Stockmayer, W. H.:* Intrinsic Viscosities and Unperturbed Dimensions of Long Chain Molecules. Vol. 3, pp. 196–312.

*Ledwith, A.* and *Sherrington, D. C.*: Stable Organic Cation Salts: Ion Pair Equilibria and Use in Cationic Polymerization. Vol. 19, pp. 1–56.

*Ledwith, A.* see Chiellini, E. Vol. 62, pp. 143–170.

*Lee, C.-D. S.* and *Daly, W. H.*: Mercaptan-Containing Polymers. Vol. 15, pp. 61–90.

*Legeay, G.* see Brosse, J.-C.: Vol. 81, pp. 167–224.

*LeMay, J. D., Kelley, F. N.*: Structure and Ultimate Properties of Epoxy Resins. Vol. 78, pp. 113–148.

*Lesná, M.* see Mejzlik, J.: Vol. 81, pp. 83–120.

*Lindberg, J. J.* and *Hortling, B.*: Cross Polarization — Magic Angle Spinning NMR Studies of Carbohydrates and Aromatic Polymers. Vol. 66, pp. 1–22.

*Lipatov, Y. S.*: Relaxation and Viscoelastic Properties of Heterogeneous Polymeric Compositions. Vol. 22, pp. 1–59.

*Lipatov, Y. S.*: The Iso-Free-Volume State and Glass Transitions in Amorphous Polymers: New Development of the Theory. Vol. 26, pp. 63–104.

*Lipatova, T. E.*: Medical Polymer Adhesives. Vol. 79, pp. 65–93.

*Lohse, F., Zweifel, H.*: Photocrosslinking of Epoxy Resins. Vol. 78, pp. 59–80.

*Lustoň, J.* and *Vašš, F.*: Anionic Copolymerization of Cyclic Ethers with Cyclic Anhydrides. Vol. 56, pp. 91–133.


*Madec, J.-P.* and *Maréchal, E.*: Kinetics and Mechanisms of Polyesterifications. II. Reactions of Diacids with Diepoxides. Vol. 71, pp. 153–228.

*Mano, E. B.* and *Coutinho, F. M. B.*: Grafting on Polyamides. Vol. 19, pp. 97–116.

*Maréchal, E.* see Madec, J.-P. Vol. 71, pp. 153–228.

*Mark, H. B., Jr.* see Diaz, A. F.: Vol. 84, pp. 113–140.

*Mark, J. E.*: The Use of Model Polymer Networks to Elucidate Molecular Aspects of Rubberlike Elasticity. Vol. 44, pp. 1–26.

*Mark, J. E.* see Queslel, J. P. Vol. 71, pp. 229–248.

*Maser, F., Bode, K., Pillai, V. N. R.* and *Mutter, M.*: Conformational Studies on Model Peptides. Their Contribution to Synthetic, Structural and Functional Innovations on Proteins. Vol. 65, pp. 177–214.

*Masuda, T.* and *Higashimura, T.*: Polyacetylenes with Substituents: Their Synthesis and Properties. Vol. 81, pp. 121–166.

*McGrath, J. E.* see Yorkgitis, E. M. Vol. 72, pp. 79–110.

*McIntyre, J. E.* see Dobb, M. G. Vol. 60/61, pp. 61–98.

*Meerwall v., E. D.*: Self-Diffusion in Polymer Systems. Measured with Field-Gradient Spin Echo NMR Methods, Vol. 54, pp. 1–29.

*Mejzlik, J., Lesná, M.* and *Kratochvila, J.*: Determination of the Number of Active Centers in Ziegler-Natta Polymerizations of Olefins. Vol. 81, pp. 83–120.

*Mengoli, G.*: Feasibility of Polymer Film Coating Through Electroinitiated Polymerization in Aqueous Medium. Vol. 33, pp. 1–31.

*Mertzel, E., Koenik, J. L.*: Application of FT-IR and NMR to Epoxy Resins. Vol. 75, pp. 73–112.

*Meyerhoff, G.*: Die viscosimetrische Molekulargewichtsbestimmung von Polymeren. Vol. 3, pp. 59–105.

*Millich, F.*: Rigid Rods and the Characterization of Polyisocyanides. Vol. 19, pp. 117–141.

*Möller, M.*: Cross Polarization — Magic Angle Sample Spinning NMR Studies. With Respect to the Rotational Isomeric States of Saturated Chain Molecules. Vol. 66, pp. 59–80.

*Morawetz, H.*: Specific Ion Binding by Polyelectrolytes. Vol. 1, pp. 1–34.

*Morgan, R. J.*: Structure-Property Relations of Epoxies Used as Composite Matrices. Vol. 72, pp. 1–44.

*Morin, B. P., Breusova, I. P.* and *Rogovin, Z. A.*: Structural and Chemical Modifications of Cellulose by Graft Copolymerization. Vol. 42, pp. 139–166.

*Mulvaney, J. E., Oversberger, C. C.* and *Schiller, A. M.*: Anionic Polymerization. Vol. 3, pp. 106–138.


*Nakase, Y., Kurijama, I.* and *Odajima, A.*: Analysis of the Fine Structure of Poly(Oxymethylene) Prepared by Radiation-Induced Polymerization in the Solid State. Vol. 65, pp. 79–134.

*Neuse, E.*: Aromatic Polybenzimidazoles. Syntheses, Properties, and Applications. Vol. 47, pp. 1–42.

*Nicolais, L.* see *Apicella, A.* Vol. 72, pp. 69–78.

*Noristi, L.* see *Barbé, P. C.:* Vol. 81, pp. 1–83.

*Nuyken, O., Weidner, R.:* Graft and Block Copolymers via Polymeric Azo Initiators. Vol. 73/74, pp. 145–200.

*Ober, Ch. K., Jin, J.-I.* and *Lenz, R. W.:* Liquid Crystal Polymers with Flexible Spacers in the Main Chain. Vol. 59, pp. 103–146.

*Okubo, T.* and *Ise, N.:* Synthetic Polyelectrolytes as Models of Nucleic Acids and Esterases. Vol. 25, pp. 135–181.

*Oleinik, E. F.:* Epoxy-Aromatic Amine Networks in the Glassy State Structure and Properties. Vol. 80, pp. 49–99.

*Osaki, K.:* Viscoelastic Properties of Dilute Polymer Solutions. Vol. 12, pp. 1–64.

*Osada, Y.:* Conversion of Chemical Into Mechanical Energy by Synthetic Polymers (Chemomechanical Systems). Vol. 82, pp. 1–47.

*Oster, G.* and *Nishijima, Y.:* Fluorescence Methods in Polymer Science. Vol. 3, pp. 313–331.

*Otsu, T.* see *Sato, T.* Vol. 71, pp. 41–78.

*Overberger, C. G.* and *Moore, J. A.:* Ladder Polymers. Vol. 7, pp. 113–150.

*Packirisamy, S.* see *Biswas, M.* Vol. 70, pp. 71–118.

*Papkov, S. P.:* Liquid Crystalline Order in Solutions of Rigid-Chain Polymers. Vol. 59, pp. 75–102.

*Patat, F., Killmann, E.* und *Schiebener, C.:* Die Absorption von Makromolekülen aus Lösung. Vol. 3, pp. 332–393.

*Patterson, G. D.:* Photon Correlation Spectroscopy of Bulk Polymers. Vol. 48, pp. 125–159.

*Penczek, S., Kubisa, P.* and *Matyjaszewski, K.:* Cationic Ring-Opening Polymerization of Heterocyclic Monomers. Vol. 37, pp. 1–149.

*Penczek, S., Kubisa, P.* and *Matyjaszewski, K.:* Cationic Ring-Opening Polymerization; 2. Synthetic Applications. Vol. 68/69, pp. 1–298.

*Penczek, S.* see *Klosinski, P.:* Vol. 79, pp. 139–157.

*Peticolas, W. L.:* Inelastic Laser Light Scattering from Biological and Synthetic Polymers. Vol. 9, pp. 285–333.

*Petropoulos, J. H.:* Membranes with Non-Homogeneous Sorption Properties. Vol. 64, pp. 85–134.

*Pino, P.:* Optically Active Addition Polymers. Vol. 4, pp. 393–456.

*Pitha, J.:* Physiological Activities of Synthetic Analogs of Polynucleotides. Vol. 50, pp. 1–16.

*Plate, N. A.* and *Noak, O. V.:* A Theoretical Consideration of the Kinetics and Statistics of Reactions of Functional Groups of Macromolecules. Vol. 31, pp. 133–173.

*Plate, N. A., Valuev, L. I.:* Heparin-Containing Polymeric Materials. Vol. 79, pp. 95–138.

*Plate, N. A.* see *Shibaev, V. P.* Vol. 60/61, pp. 173–252.

*Plesch, P. H.:* The Propagation Rate-Constants in Cationic Polymerisations. Vol. 8, pp. 137–154.

*Porod, G.:* Anwendung und Ergebnisse der Röntgenkleinwinkelstreuung in festen Hochpolymeren. Vol. 2, pp. 363–400.

*Pospišil, J.:* Transformations of Phenolic Antioxidants and the Role of Their Products in the Long-Term Properties of Polyolefins. Vol. 36, pp. 69–133.

*Postelnek, W., Coleman, L. E.,* and *Lovelace, A. M.:* Fluorine-Containing Polymers. I. Fluorinated Vinyl Polymers with Functional Groups, Condensation Polymers, and Styrene Polymers. Vol. 1 pp. 75–113.

*Queslel, J. P.* and *Mark, J. E.:* Molecular Interpretation of the Moduli of Elastomeric Polymer Networks of Know Structure. Vol. 65, pp. 135–176.

*Queslel, J. P.* and *Mark, J. E.:* Swelling Equilibrium Studies of Elastomeric Network Structures. Vol. 71, pp. 229–248.

*Rehage, G.* see *Finkelmann, H.* Vol. 60/61, pp. 99–172.

*Rempp, P. F.* and *Franta, E.:* Macromonomers: Synthesis, Characterization and Applications. Vol. 58, pp. 1–54.

*Rempp, P., Herz, J.* and *Borchard, W.*: Model Networks. Vol. 26, pp. 107–137.

*Richards, R. W.*: Small Angle Neutron Scattering from Block Copolymers. Vol. 71, pp. 1–40.

*Rigbi, Ζ.*: Reinforcement of Rubber by Carbon Black. Vol. 36, pp. 21–68.

*Rigby, D.* see Roe, R.-J.: Vol. 82, pp. 103–141.

*Roe, R.-J.* and *Rigby, D.*: Phase Relations and Miscibility in Polymer Blends Containing Copolymers. Vol. 82, pp. 103–141.

*Rogovin, Z. A.* and *Gabrielyan, G. A.*: Chemical Modifications of Fibre Forming Polymers and Copolymers of Acrylonitrile. Vol. 25, pp. 97–134.

*Roha, M.*: Ionic Factors in Steric Control. Vol. 4, pp. 353–392.

*Roha, M.*: The Chemistry of Coordinate Polymerization of Dienes. Vol. 1, pp. 512–539.

*Ross-Murphy, S. B.* see Clark, A. H.: Vol. 83, pp. 57–193.

*Rostami, S.* see Walsh, D. J. Vol. 70, pp. 119–170.

*Rozengerk, v. A.*: Kinetics, Thermodynamics and Mechanism of Reactions of Epoxy Oligomers with Amines. Vol. 75, pp. 113–166.

*Rubinson, J. F.* see Diaz, A. F.: Vol. 84, pp. 113–140.


*Safford, G. J.* and *Naumann, A. W.*: Low Frequency Motions in Polymers as Measured by Neutron Inelastic Scattering. Vol. 5, pp. 1–27.

*Sakaguchi, M.* see Kashiwabara, H.: Vol. 82, pp. 141–207.

*Saito, M.* see Kamide, K.: Vol. 83, pp. 1–57.

*Sato, T.* and *Otsu, T.*: Formation of Living Propagating Radicals in Microspheres and Their Use in the Synthesis of Block Copolymers. Vol. 71, pp. 41–78.

*Sauer, J. A.* and *Chen, C. C.*: Crazing and Fatigue Behavior in One and Two Phase Glassy Polymers. Vol. 52/53, pp. 169–224.

*Sawamoto, M.* see Higashimura, T. Vol. 62, pp. 49–94.

*Schmidt, R. G., Bell, J. P.*: Epoxy Adhesion to Metals. Vol. 75, pp. 33–72.

*Schuerch, C.*: The Chemical Synthesis and Properties of Polysaccharides of Biomedical Interest. Vol. 10, pp. 173–194.

*Schulz, R. C.* und *Kaiser, E.*: Synthese und Eigenschaften von optisch aktiven Polymeren. Vol. 4, pp. 236–315.

*Seanor, D. A.*: Charge Transfer in Polymers. Vol. 4, pp. 317–352.

*Semerak, S. N.* and *Frank, C. W.*: Photophysics of Excimer Formation in Aryl Vinyl Polymers, Vol. 54, pp. 31–85.

*Seidl, J., Malinský, J., Dušek, K.* und *Heitz, W.*: Makroporöse Styrol-Divinylbenzol-Copolymere und ihre Verwendung in der Chromatographie und zur Darstellung von Ionenaustauschern. Vol. 5, pp. 113–213.

*Semjonow, V.*: Schmelzviskositäten hochpolymerer Stoffe. Vol. 5, pp. 387–450.

*Semlÿen, J. A.*: Ring-Chain Equilibria and the Conformations of Polymer Chains. Vol. 21, pp. 41–75.

*Sen, A.*: The Copolymerization of Carbon Monoxide with Olefins. Vol. 73/74, pp. 125–144.

*Senturia, S. D., Sheppard, N. F. Jr.*: Dielectric Analysis of Thermoset Cure. Vol. 80, pp. 1–47.

*Sharkey, W. H.*: Polymerizations Through the Carbon-Sulphur Double Bond. Vol. 17, pp. 73–103.

*Sheppard, N. F. Jr.* see Senturia, S. D.: Vol. 80, pp. 1–47.

*Shibaev, V. P.* and *Platé, N. A.*: Thermotropic Liquid-Crystalline Polymers with Mesogenic Side Groups. Vol. 60/61, pp. 173–252.

*Shimada, S.* see Kashiwabara, H.: Vol. 82, pp. 141–207.

*Shimidzu, T.*: Cooperative Actions in the Nucleophile-Containing Polymers. Vol. 23, pp. 55–102.

*Shutov, F. A.*: Foamed Polymers Based on Reactive Oligomers, Vol. 39, pp. 1–64.

*Shutov, F. A.*: Foamed Polymers. Cellular Structure and Properties. Vol. 51, pp. 155–218.

*Shutov, F. A.*: Syntactic Polymer Foams. Vol. 73/74, pp. 63–124.

*Siesler, H. W.*: Rheo-Optical Fourier-Transform Infrared Spectroscopy: Vibrational Spectra and Mechanical Properties of Polymers. Vol. 65, pp. 1–78.

*Silvestri, G., Gambino, S.,* and *Filardo, G.*: Electrochemical Production of Initiators for Polymerization Processes. Vol. 38, pp. 27–54.

*Sixl, H.*: Spectroscopy of the Intermediate States of the Solid State Polymerization Reaction in Diacetylene Crystals. Vol. 63, pp. 49–90.

*Slichter, W. P.*: The Study of High Polymers by Nuclear Magnetic Resonance. Vol. 1, pp. 35–74.

*Small, P. A.:* Long-Chain Branching in Polymers. Vol. 18.

*Smets, G.:* Block and Graft Copolymers. Vol. 2, pp. 173–220.

*Smets, G.:* Photochromic Phenomena in the Solid Phase. Vol. 50, pp. 17–44.

*Sohma, J.* and *Sakaguchi, M.:* ESR Studies on Polymer Radicals Produced by Mechanical Destruction and Their Reactivity. Vol. 20, pp. 109–158.

*Solaro, R.* see Chiellini, E. Vol. 62, pp. 143–170.

*Sotobayashi, H.* und *Springer, J.:* Oligomere in verdünnten Lösungen. Vol. 6, pp. 473–548.

*Soutif, J.-C.* see Brosse, J.-C.: Vol. 81, pp. 167–224.

*Sperati, C. A.* and *Starkweather, Jr., H. W.:* Fluorine-Containing Polymers. II. Polytetrafluoroethylene. Vol. 2, pp. 465–495.

*Spiertz, E. J.* see Vollenbroek, F. A.: Vol. 84, pp. 85–112.

*Spiess, H. W.:* Deuteron NMR — A new Toolfor Studying Chain Mobility and Orientation in Polymers. Vol. 66, pp. 23–58.

*Sprung, M. M.:* Recent Progress in Silicone Chemistry. I. Hydrolysis of Reactive Silane Intermediates, Vol. 2, pp. 442–464.

*Stahl, E.* and *Brüderle, V.:* Polymer Analysis by Thermofractography. Vol. 30, pp. 1–88.

*Stannett, V. T., Koros, W. J., Paul, D. R., Lonsdale, H. K.,* and *Baker, R. W.:* Recent Advances in Membrane Science and Technology. Vol. 32, pp. 69–121.

*Staverman, A. J.:* Properties of Phantom Networks and Real Networks. Vol. 44, pp. 73–102.

*Stauffer, D., Coniglio, A.* and *Adam, M.:* Gelation and Critical Phenomena. Vol. 44, pp. 103–158.

*Stille, J. K.:* Diels-Alder Polymerization. Vol. 3, pp. 48–58.

*Stolka, M.* and *Pai, D.:* Polymers with Photoconductive Properties. Vol. 29, pp. 1–45.

*Stuhrmann, H.:* Resonance Scattering in Macromolecular Structure Research. Vol. 67, pp. 123–164.

*Subramanian, R. V.:* Electroinitiated Polymerization on Electrodes. Vol. 33, pp. 35–58.

*Sumitomo, H.* and *Hashimoto, K.:* Polyamides as Barrier Materials. Vol. 64, pp. 55–84.

*Sumitomo, H.* and *Okada, M.:* Ring-Opening Polymerization of Bicyclic Acetals, Oxalactone, and Oxalactam. Vol. 28, pp. 47–82.

*Szegö, L.:* Modified Polyethylene Terephthalate Fibers. Vol. 31, pp. 89–131.

*Szwarc, M.:* Termination of Anionic Polymerization. Vol. 2, pp. 275–306.

*Szwarc, M.:* The Kinetics and Mechanism of N-carboxy-α-amino-acid Anhydride (NCA) Polymerization to Poly-amino Acids. Vol. 4, pp. 1–65.

*Szwarc, M.:* Thermodynamics of Polymerization with Special Emphasis on Living Polymers. Vol. 4, pp. 457–495.

*Szwarc, M.:* Living Polymers and Mechanisms of Anionic Polymerization. Vol. 49, pp. 1–175.


*Takahashi, A.* and *Kawaguchi, M.:* The Structure of Macromolecules Adsorbed on Interfaces. Vol. 46, pp. 1–65.

*Takemoto, K.* and *Inaki, Y.:* Synthetic Nucleic Acid Analogs. Preparation and Interactions. Vol. 41, pp. 1–51.

*Tani, H.:* Stereospecific Polymerization of Aldehydes and Epoxides. Vol. 11, pp. 57–110.

*Tate, B. E.:* Polymerization of Itaconic Acid and Derivatives. Vol. 5, pp. 214–232.

*Tazuke, S.:* Photosensitized Charge Transfer Polymerization. Vol. 6, pp. 321–346.

*Teramoto, A.* and *Fujita, H.:* Conformation-dependent Properties of Synthetic Polypeptides in the Helix-Coil Transition Region. Vol. 18, pp. 65–149.

*Theocaris, P. S.:* The Mesophase and its Influence on the Mechanical Behvior of Composites. Vol. 66, pp. 149–188.

*Thomas, W. M.:* Mechanismus of Acrylonitrile Polymerization. Vol. 2, pp. 401–441.

*Tieke, B.:* Polymerization of Butadiene and Butadiyne (Diacetylene) Derivatives in Layer Structures. Vol. 71, pp. 79–152.

*Tobolsky, A. V.* and *DuPré, D. B.:* Macromolecular Relaxation in the Damped Torsional Oscillator and Statistical Segment Models. Vol. 6, pp. 103–127.

*Tosi, C.* and *Ciampelli, F.:* Applications of Infrared Spectroscopy to Ethylene-Propylene Copolymers. Vol. 12, pp. 87–130.

*Tosi, C.:* Sequence Distribution in Copolymers: Numerical Tables. Vol. 5, pp. 451–462.

*Tran, C.* see Yorkgitis, E. M. Vol. 72, pp. 79–110.

*Tsuchida, E.* and *Nishide, H.:* Polymer-Metal Complexes and Their Catalytic Activity. Vol. 24, pp. 1–87.

*Tsuji, K.:* ESR Study of Photodegradation of Polymers. Vol. 12, pp. 131–190.

*Tsvetkov, V.* and *Andreevà, L.:* Flow and Electric Birefringence in Rigid-Chain Polymer Solutions. Vol. 39, pp. 95–207.

*Tuzar, Z., Kratochvil, P.,* and *Bohdanecký, M.:* Dilute Solution Properties of Aliphatic Polyamides. Vol. 30, pp. 117–159.


*Uematsu, I.* and *Uematsu, Y.:* Polypeptide Liquid Crystals. Vol. 59, pp. 37–74.


*Valuev, L. I.* see Platé, N. A.: Vol. 79, pp. 95–138.

*Valvassori, A.* and *Sartori, G.:* Present Status of the Multicomponent Copolymerization Theory. Vol. 5, pp. 28–58.

*Vidal, A.* see Donnet, J. B. Vol. 76, pp. 103–128.

*Viovy, J. L.* and *Monnerie, L.:* Fluorescence Anisotropy Technique Using Synchrotron Radiation as a Powerful Means for Studying the Orientation Correlation Functions of Polymer Chains. Vol. 67, pp. 99–122.

*Voigt-Martin, I.:* Use of Transmission Electron Microscopy to Obtain Quantitative Information About Polymers. Vol. 67, pp. 195–218.

*Vollenbroek, F. A.* and *Spiertz, E. J.:* Photoresist Systems for Microlithography. Vol. 84, pp. 85–112.

*Voorn, M. J.:* Phase Separation in Polymer Solutions. Vol. 1, pp. 192–233.


*Walsh, D. J., Rostami, S.:* The Miscibility of High Polymers: The Role of Specific Interactions. Vol. 70, pp. 119–170.

*Ward, I. M.:* Determination of Molecular Orientation by Spectroscopic Techniques. Vol. 66, pp. 81–116.

*Ward, I. M.:* The Preparation, Structure and Properties of Ultra-High Modulus Flexible Polymers. Vol. 70, pp. 1–70.

*Weidner, R.* see *Nuyken, O.:* Vol. 73/74, pp. 145–200.

*Werber, F. X.:* Polymerization of Olefins on Supported Catalysts. Vol. 1, pp. 180–191.

*Wichterle, O., Šebenda, J.,* and *Králiček, J.:* The Anionic Polymerization of Caprolactam. Vol. 2, pp. 578–595.

*Wilkes, G. L.:* The Measurement of Molecular Orientation in Polymeric Solids. Vol. 8, pp. 91–136.

*Wilkes, G. L.* see Yorkgitis, E. M. Vol. 72, pp. 79–110.

*Williams, G.:* Molecular Aspects of Multiple Dielectric Relaxation Processes in Solid Polymers. Vol. 33, pp. 59–92.

*Williams, J. G.:* Applications of Linear Fracture Mechanics. Vol. 27, pp. 67–120.

*Wöhrle, D.:* Polymere aus Nitrilen. Vol. 10, pp. 35–107.

*Wöhrle, D.:* Polymer Square Planar Metal Chelates for Science and Industry. Synthesis, Properties and Applications. Vol. 50, pp. 45–134.

*Wöhrle, D.* see Kaneko, M.: Vol. 84, pp. 141–228.

*Wolf, B. A.:* Zur Thermodynamik der enthalpisch und der entropisch bedingten Entmischung von Polymerlösungen. Vol. 10, pp. 109–171.

*Wong, C. P.:* Application of Polymer in Encapsulation of Electronic Parts. Vol. 84, pp. 63–84.

*Woodward, A. E.* and *Sauer, J. A.:* The Dynamic Mechanical Properties of High Polymers at Low Temperatures. Vol. 1, pp. 114–158.

*Wunderlich, B.:* Crystallization During Polymerization. Vol. 5, pp. 568–619.

*Wunderlich, B.* and *Baur, H.:* Heat Capacities of Linear High Polymers. Vol. 7, pp. 151–368.

*Wunderlich, B.* and *Grebowicz, J.:* Thermotropic Mesophases and Mesophase Transitions of Linear, Flexible Macromolecules. Vol. 60/61, pp. 1–60.

*Wrasidlo, W.:* Thermal Analysis of Polymers. Vol. 13, pp. 1–99.


*Yamashita, Y.:* Random and Black Copolymers by Ring-Opening Polymerization. Vol. 28, pp. 1–46.

*Yamazaki, N.:* Electrolytically Initiated Polymerization. Vol. 6, pp. 377–400.

*Yamazaki, N.* and *Higashi, F.*: New Condensation Polymerizations by Means of Phosphorus Compounds. Vol. 38, pp. 1–25.

*Yokoyama, Y.* and *Hall, H. K.*: Ring-Opening Polymerization of Atom-Bridged and Bond-Bridged Bicyclic Ethers, Acetals and Orthoesters. Vol. 42, pp. 107–138.

*Yorkgitis, E. M., Eiss, N. S. Jr., Tran, C., Wilkes, G. L.* and *McGrath, J. E.*: Siloxane-Modified Epoxy Resins. Vol. 72, pp. 79–110.

*Yoshida, H.* and *Hayashi, K.*: Initiation Process of Radiation-induced Ionic Polymerization as Studied by Electron Spin Resonance. Vol. 6, pp. 401–420.

*Young, R. N., Quirk, R. P.* and *Fetters, L. J.*: Anionic Polymerizations of Non-Polar Monomers Involving Lithium. Vol. 56, pp. 1–90.

*Yuki, H.* and *Hatada, K.*: Stereospecific Polymerization of Alpha-Substituted Acrylic Acid Esters. Vol. 31, pp. 1–45.


*Zachmann, H. G.*: Das Kristallisations- und Schmelzverhalten hochpolymerer Stoffe. Vol. 3, pp. 581–687.

*Zaikov, G. E.* see Aseeva, R. M. Vol. 70, pp. 171–230.

*Zakharov, V. A., Bukatov, G. D.,* and *Yermakov, Y. I.*: On the Mechanism of Olifin Polymerization by Ziegler-Natta Catalysts. Vol. 51, pp. 61–100.

*Zambelli, A.* and *Tosi, C.*: Stereochemistry of Propylene Polymerization. Vol. 15, pp. 31–60.

*Zucchini, U.* and *Cecchin, G.*: Control of Molecular-Weight Distribution in Polyolefins Synthesized with Ziegler-Natta Catalytic Systems. Vol. 51, pp. 101–154.

*Zweifel, H.* see Lohse, F.: Vol. 78, pp. 59–80.

# Subject Index

Action spectrum 217
Adhesion 17, 25, 29, 50
Adhesives 5, 35–36, 55
Aging, chemical 38
–, physical 5, 38, 40, 49
Alkoxide cure 77
Amines 13–14, 77
–, cure 77
Anhydrides 80
Annealing 32
Aspherical lenses see Lenses
Asymmetric oxidation, electrochemical 128
Autoacceleration 7, 8, 10, 17
Autodeceleration 8

Batteries, polymer 181–185
–, rechargeable 180, 181
–, storage 113, 131
BDDA 26
Benzil ketals 12
Benzoin ethers 12
Benzophenone 13
1-Benzoyl-1-cyclohexanol 13, 30
Bisazides 101, 108
Bis-(hydroxyethyl)bisphenol-A dimethacrylate 37–39, 44–45
Bisphenol A resins 80
Branching 7,10
Breakdown voltage strength 82
Buffer coating 19, 21
Built On Mask (BOM) 1, 93, 94
1,4-Butanediol diacrylate 26

Cadmium sulfide 206
Carboxylate cure 77
Catalysis 173
Cathode Ray Tubes (CRT) 85–88, 92, 104, 105, 108
Cation exchange 189
Cationic epoxy systems 36
– photoinitiators 55
– polymerization 5, 7, 14, 17, 36, 45
Cavity-filling processes 72
Chain crosslinking (co)polymerization 5, 7, 37, 51, 52
– oxidation 43–44, 46
– polymerization, -process, -reaction 4, 11, 53
– transfer 5, 7, 10–11, 17, 26, 46–47, 50

Charge density 114
– storage 180
– –, polyacetylene 180
– –, polyaniline 180
– –, polypyrrole 180
– –, polythienylene 180
– transport 169
Chemical aging 38
– selectivity 120
– vapor deposition (CVD) 68
Chlorophyll 200, 203
Cladding 18, 19
Coating compositions 18, 37
– of glass fibers 20, 21
– processes 73
Coatings, dual cure 35–36, 55
–, protective 119
Color filters 86, 105
Compact discs 85–88, 102
Complementary metal-oxide semiconductor (CMOS) devices 66
Conductive polymers, electroactive 113
Conductivity 166, 167
–, electrical 120
Continuous atmospheric pressure reactor 69
Contrast Enhancement Layer (CEL) 85, 92–94, 97
Cottrell equation 169
Crosslink density 5, 8–10, 21, 25–26, 41, 50, 53
Crosslinked network systems, densely 7, 9, 11, 24, 37, 49, 52, 55
Crosslinking 87–108
– (co)polymerization 4–5, 7–8, 10–11, 37–38, 46, 51–52
–, intramolecular 51
– linear chains 7, 51
CVD (Chemical vapor deposition) 68
–, hot-wall plasma assisted 71
–, parallel-plate plasma assisted 70
Cyclic voltammogram 149, 154, 179, 184, 186, 188
Cyclization 7, 51–52

DEGDA 8
Deposition process 69, 71
Deuteration 42–43, 47
Deuterium isotope effect 46–47
Diacrylates 5, 15, 16, 26, 30, 37, 45
Diaryliodonium salts 14

Diazo-oxide 89–108
Dicyandiamides 80
α,α'-Dideutero 1,6-hexanediol diacrylate 47
Dielectric change 82
– coatings 16, 35–36, 55
– constant 82
Dielectrics, deposition 69
Di-epoxides 7, 14, 36, 45, 50–51
Di-ethyleneglycol diacrylate 8
Diffusion 8, 10, 48
– coefficient 169
Diffusional control of polymerization 8, 46
Dimensional stability 25, 29
Dimethacrylates 16, 30, 35, 41, 51
α,α'-Dimethoxy-α-phenylacetophenone 12–13,
    25, 30, 31, 38–40, 46, 50
Diode 194
Discs see Optical discs
Display applications 124, 131
– devices 113
Display devices,
– –, electrochromic 116
– –, polyaniline 187, 188
– –, polymeric viologen 187
– –, Prussian Blue 187
Divinyl compounds, monomers 7, 38, 44, 52, 54
DMPA 12–13, 25, 30–31, 38–40, 46, 50
Doping 154, 180, 196
Double bonds, conversion of 5, 8–10, 25, 30,
    38–40, 44, 49
Dry films 87, 89, 103
Dual cure coatings, systems 35–36, 55
– functionality coatings 36
Dye coating 208

EHA 26
Electrical conductivity 120
Electroactive conductive polymers 113
Electrocatalysis 173
Electrochemical asymmetric oxidation 128
–, polymerization 153, 157
– reaction 171
Electrochromic display devices 116
Electrochromism 187
–, multicolor 187
Electrode application 113
Electrodes, bilayer-coated 172, 216, 217
–, chemically selective 113
–, film 117
–, multilayer-coated 164
–, polymer-coated 143–163
Electron cycles 143
– exchange 168
– pumping 144, 146, 196, 212
– transfer 165, 168, 171
Electronic conduction 165
Electronics, polymer-based 192–195

Encapsulants 63 ff., 76, 82
–, electronic devices 63 ff.
–, glob-top 80
–, IC 64
–, inorganic 76
–, organic 76
–, silicone 76
Encapsulation 5, 36
–, chip-on-board 80
– techniques 67
– –, chip packaging 72
– –, economic impact 63 ff.
– –, on-chip 68
Energy cycles in nature 143
– storage 117, 126
Epoxides 8, 16–17, 33, 50, 55
Epoxies 79
Epoxy acrylates 16–17, 20–21, 30, 35, 38, 46,
    49–50
2-Ethoxy-ethoxy ethyl acrylate 50
Ethyleneglycol dimethacrylate 52
2-Ethylhexyl acrylate 26
Extractability 26–27

Film electrodes 117
Formaldehydes 80
Free volume 9, 31, 38–40, 49

Gelation 7, 10, 17, 38, 44, 46, 52
Glass transition 18, 21

Hardening 99, 100, 107
–, chemical 85,100
–, plasma 85, 100
–, UV 99
Hardness 25, 29–30, 32, 50
HCPK 13, 30
HDDA 8–10, 25–27, 37–42, 44–47, 54
HEBDM 37–39, 44–45
Heterojunction cell 212
1,6-Hexanediol diacrylate 8–10, 25–27, 37–42,
    44–47, 54–55
Hillocks 70
HMPP 13, 30, 42, 44, 47
Holograms, recording of 32–33
Hot-wall reduced-pressure reactor 68
Hybrid systems 18, 36, 55
Hydrogen abstraction 10, 12–14, 42–43
– evolution 178
– transfer 11, 13, 44–45
Hydroperoxides 11, 44
Hydroxycyclohexyl phenyl ketone 13, 30
2-Hydroxy-2-methyl-1-phenylpropan-1-one 13,
    30, 42, 44, 47

IC (Integrated Circuits) 4, 36, 63, 64, 85—108
— package 67
Image Reversal 85, 95, 96, 101, 108
— sensors 86, 105
Images 87, 91, 94—97, 104, 108
—, aerial 90—92, 95
—, enhanced 92
—, latent 90—108
—, negative 93—99
—, positive 93—96
—, relief 85, 87, 96, 108
Indene carboxylic acid 89—91, 95, 96, 99
Inhibition 7, 14, 46
Inhibitor 11
Inhomogeneity 7, 44—45, 48—49, 51—54
Initiation 5—6, 26, 45, 54
Initiator 5, 50
Integrated circuits (IC) 4, 36, 63, 64, 85—108
Interferogram 32
Ion exchange polymer 190
IPN 17, 51
Iron-arene salts 15
Isotope effect 46—47

Ketene 89
3-Ketocoumarins 13
Kinetics 5, 7, 43, 45—49

Laser, low temperature 71
Lattice 52
Lenses 4, 5, 27—29, 31, 55
—, photopolymerizable coatings 29
—, photoreplication 28
Lift-off 85, 101
Light pen 4, 27, 28
Liquid crystalline monomers 55
Liquid crystals 105
— —, displays 85, 86, 104, 105
— —, television 85, 86, 105
Liquid-junction device 201
—, semiconductors 204—206

Melamine 80
Merocyanine 199, 201, 204, 210
Metal phthalocyanine 148, 177, 185, 202, 203, 210
— porphyrin 175, 177, 203, 210
Methyl acrylate 34
— methacrylate 7, 38
Methylviologen 147, 170, 215
Microbending 19
Microelectrodes 122, 192, 195
Microelectronics 192
Microlithography 85, 104
MIS-type cell 200

MMA 7, 38
Mold 4
—, disc replication 23—27
—, lense replication 28—30
Multi-electron reaction 174
Multifunctional monomers 9

Nafion 148
Naphthoquinone diazide 89
Network formation 4—5, 37, 45—46, 51
Networks, interpenetrating 17, 51
Nitrenes 88, 98, 103
Novolak 80, 87—90, 95—102, 106—108
NVP 25—27, 50

Oligomers 15—17, 34
On-chip interconnections 82
Open-circuit voltage 182, 198
Optical density 114
— discs 5, 22—27, 29, 32, 55
— —, photopolymerizable coatings for 25
— —, photoreplication of 23—25, 29, 37
— —, read-out 23, 27
— —, systems 22
— fibers 5, 18—22, 35, 55
— —, coating of 20
— —, photopolymerizable coatings for 21
— storage 124
— waveguides 4—5, 34, 55
Oriented monomers 55
Oxime cure 77
Oxygen, post copolymerization of 11
Oxygen-assisted hydrogen transfer 45

Passivation layer 82
Patterns, making of 4, 33
Percolation model 51—54
Permeability control, electrochemical 191
Peroxides 5
Peroxy radicals 11, 17
Perylene-iodine complexes 136
Photoanode 119
Photocatalysis 204, 207
Photochemical diode 206
— process 212
Photocorrosion 126
Photocrosslinking 4
Photoelectroactivity 126
Photoelectrochemical catalysis, see Photo-catalysis
— cell 201, 211
Photoelectrodes 118
Photofragmentation 12
Photogalvanic cells 128, 212, 213
Photoinitiators 11—12, 14, 15, 33—34, 36

Photolysis 207
Photophysical process 196
Photopolymerization 4—51
Photoredox system 213
Photoregenerative cell 205
Photoreplication process, discs 24
— —, lenses 28
Photoresist systems 85
— —, negative 87, 88
— —, wet 86, 87, 102, 103
Photoresists 4—5, 32, 85—88, 90—100, 102—108
—, dry-film 86, 103
—, near UV 87
—, negative 85—88, 94, 102, 106, 108
—, positive 85—91, 94, 97, 101, 102, 106—108
Photosensors 118
Photosynthesis 144
Photovoltaic cells 127, 199
— —, energy diagram 200
— conversion systems 196—199
Phthalocyanine 148, 199
Physical aging 5, 38, 40, 49
Plasma deposition 70
— etching 85, 91, 97, 98, 100, 105
Plastic molding 73
PMMA 24, 28
P-n junction 198
Polyacetylene 8, 46, 113, 165, 180, 182, 198, 201
Polyamic acids 81
Polyaniline 113, 154, 160, 180, 183, 193
Polyaromatic hydrocarbons 161
Polyazulene 130
Polycarbonate 34—35
Poly(dimethyl-co-phenyl)siloxane
    acrylates 21—22
Polydimethylsiloxane 19
Polyester acrylates 16—17
— methacrylates 43
Polyether acrylates 16
— urethane acrylates 21—22
Polyimides 68
—, photodefinable 80, 81
—, ultra-low thermal coefficient expansion
    (TCE) 81
Polyisoprene 87, 88, 102, 106
Polymethylmethacrylate 24, 28, 96, 97, 100, 101
Poly(3-methylthiophene) 129, 193, 194
Polymeric (metal)-phthalocyanine 152, 162, 175,
    178, 179, 219
Polymers, electroactive conductive 113
—, electrochemical applications 113
Polynuclear complex see also Prussian Blue 161
Polyphenol 160
Polyphenylene 130, 184, 198
— sulfide 130, 165, 202
Polypyrrole 113, 123, 155—201
—, merocyanine 204
—, metal-phthalocyanine 202, 203

—, metal-porphyrin 203
—, polythienylene 202
Polysilicon, deposition 69
Polysiloxane acrylates 17
Polythiazyls 113, 134
— paste electrodes 135
Polythienothiofene 129
Polythiol/acrylate systems 35
Polythiophene 113, 129, 159, 180, 183, 198, 202
Polyurethane acrylates 17
Polyvinyl alcohol 87, 88, 93, 104, 106
— cinnamate 88, 102, 106
— ferrocene 152, 195
— phenol 87, 88, 107
Poly(4-vinylpyridine) 147, 149, 150, 170, 214
Polyviologen 148, 195
Porphyrin 176, 199
Portable Conformable Mask (PCM) 85, 96, 97,
    100
Printed circuits 85—88, 102, 103, 108
— — boards 4, 35, 49
Profile modification technique (Promote) 85,
    94—96, 101, 108
Propagation 6, 8—10, 14, 47
n-Propyl acrylate 8, 46
Protective coatings 119
Prussian Blue 184, 217

Radiation stimulated deposition 71
Random walk 52
Reactivity ratio 9, 52
Read-out systems 23, 27
Recording, holographic 33
Refractive index image recording 32
Replication 5, 23, 55
—, discs 23—25
—, lenses 28—31
Resistors 4, 35
Resist systems 85, 87, 88, 103, 106, 108
—, dry-film 103
—, wet 98
Resists 85, 87—92, 94—108
—, dry-developable 85, 98, 108
—, dry-film 103
—, negative 87, 88, 103
—, positive 87
Resole 80
Resolution 85, 87, 88, 90—92, 102, 105, 107, 108
Retardation of polymerization 7, 26
— of shrinkage 30
Rhodamine B 210
Ring formation 11, 51
Rochow process 76
$Ru(bpy)_3^{2+}$ 147, 148, 171, 205, 209, 215

Schottky junction 198
Screen printing 85, 103

Selectivity, chemical 120
Self-decelerating reaction 5, 39–40, 47–48
Semiconducting polymer 196
Semiconductors, inorganic 204–206
–, liquid-junction 206
Sensitization 208
– Ru(bpy)$_3^{2+}$ 209, 210
– metal-porphyrin 210
– metal-phthalocyanine 210
– merocyanine 210
Sensors, chemically selective 113
Separation energy 26–27
Sheet resistance 82
Short circuit current 182, 198
Shrinkage 5, 8–9, 16, 18, 29–31, 38–41, 49, 55
Silicon dioxide 68
– –, doped 68
– nitride 68
Silicone encapsulants 76
– gel 78
–, heat-curable 77
–, room temperature vulcanized (RTV) 76
Silicon-oxy-nitrite 68
Siloxane acrylates 16, 22
Simulation models 52–53
Small-scale integration (SSI) 63, 64
Solar energy conversion 134
Solder masks 4, 35
Solid-state device 198
Stability 86, 91, 99, 100, 107
Stabilization 85, 99
Step reaction, process 4, 11, 51, 53
Storage batteries 113, 131
Styrene 52
– unsaturated polyester 15
Surface coatings 73
Switching/conductive properties 114


TEGDA 37–39, 41, 45
Termination 5–11, 45–47, 52
Tetraazaannulene 175
Tetracyanoquinodimethane-
  -tetrathiafulvalinium 136
Tetra-ethyleneglycol diacrylate 37–39, 41, 45
Tetraphenylporphyrin 175, 203

Thickness, influence of 6
Thiol/ene systems 15
Thionine 213, 214
Thioxantones 13
TMPTA 13, 25
Toxicological properties 15
TPGDA 25
Transfer of hydrogen 11, 13, 44–45
– –, oxygen assisted 45
Transistors 192, 193
Trapped radicals, structure 42
Trapping of radicals 11, 46, 52–53
Triarylsulfonium salts 14, 41–44, 46, 52–54
N-(3-Trimethoxysilyl)propylpyrrole 127
Trimethylolpropane triacrylate 13, 25
Triphenyl silanol 15
Tripropyleneglycol diacrylate 25
Tris/(2,2'-bipyridine)ruthenium(II)
  (Ru(bpy)$_3^{2+}$) 147, 148, 205, 209
Trommsdorff effect 7
T(tan$\delta_{max}$) 40, 50–51


Ultraviolet (UV) 87–102
–, deep 87, 91, 94–100, 108, 109
–, near 87–98
Undoping 154, 180, 181
Urethane acrylates 16, 21, 33, 35–36, 38, 50


Vapor microsensor, chemical 34
Very large-scale integration (VLSI) 63, 64
Vinyl ethers 17, 55
N-Vinyl pyrrolidone 25–27, 50
Vitrification 5, 8–10, 45–47, 49
Voltammogram, cyclic 149, 154, 179, 184, 186,
  188
Volume relaxation 31, 32


Waveguides, optical 4–5, 34, 55
Wheathering 13, 50
Williams-Watts function, equation 48–49
Wires, coating of 5, 35


Yellowing 13–14, 17, 30